1989

Current Topics
in Membranes and Transport

VOLUME 31
**Molecular Neurobiology:
Endocrine Approaches**

Current Topics in Membranes and Transport

Edited by

Arnost Kleinzeller
Department of Physiology
University of Pennsylvania School of Medicine
Philadelphia, Pennsylvania

VOLUME 31
Molecular Neurobiology: Endocrine Approaches

Guest Editors

Jerome F. Strauss III
Department of Obstetrics and Gynecology
University of Pennsylvania School of Medicine
Philadelphia, Pennsylvania

Donald W. Pfaff
Department of Neurobiology and Behavior
The Rockefeller University
New York, New York

ACADEMIC PRESS, INC.
Harcourt Brace Jovanovich, Publishers
San Diego   New York   Berkeley   Boston
London   Sydney   Tokyo   Toronto

ACADEMIC PRESS, INC.
1250 Sixth Avenue
San Diego, California 92101

United Kingdom Edition published by
ACADEMIC PRESS INC. (LONDON) LTD.
24-28 Oval Road, London NW1 7DX

LIBRARY OF CONGRESS CATALOG CARD NUMBER: 70-117091

ISBN 0-12-153331-X (alk. paper)

PRINTED IN THE UNITED STATES OF AMERICA
87 88 89 90   9 8 7 6 5 4 3 2 1

Contents

Thyrotropin-Releasing Hormone: Role of Polyphosphoinositides in Stimulation of Prolactin Secretion

RICHARD N. KOLESNICK AND MARVIN C. GERSHENGORN

PART II.   STEROID HORMONE ACTION

Steroid Effects on Excitable Membranes

S. D. ERULKAR AND D. M. WETZEL

Estradiol-Regulated Neuronal Plasticity

CHARLES V. MOBBS AND DONALD W. PFAFF

CONTENTS

Steroid Hormone Influences on Cyclic AMP-Generating Systems

ALLAN HARRELSON AND BRUCE MCEWEN

PART III. MAGNOCELLULAR NEURONS

Expression of the Oxytocin and Vasopressin Genes

DIETMAR RICHTER AND HARTWIG SCHMALE

The Secretory Vesicle in Processing and Secretion of Neuropeptides

JAMES T. RUSSELL

Mammalian Neurosecretory Cells: Electrical Properties *in Vivo* and *in Vitro*

D. A. POULAIN AND J. D. VINCENT

Preface

In 1984, the National Institutes of Mental Health issued a report entitled "The Neuroscience of Mental Health" [DHHS Publication No. (ADM) 84-1363], which summarized the status of neuroscience research as viewed by an interdisciplinary panel. Among the panel's recommendations for emphasis in future research were the elucidation of the role of neurite growth factors; clarification of the molecular basis of the function of neurotransmitters/neuromodulators and the mechanisms by which receptor-mediated signals are translated into biological events; and establishing the structure of genes and molecular mechanisms of processing, storage, and release of neuropeptides. Studies of steroid and peptide interaction in the development and regulation of neural circuits and studies of gene expression mediated by hormones with a focus on the molecular mechanisms producing behavior were also targeted by the panel.

We might expect extraordinary difficulty in applying molecular techniques to brain tissue to address these research topics because of the cellular heterogeneity, the complexity of regulating influences, and the subtlety of its primary functions. Yet, at this time, two approaches are yielding significant data on molecular aspects of brain function. One is the cloning and characterization of cDNAs for mRNAs unique to the brain; the other is the use of experimental paradigms and reagents that are well established with other cell types for strategic advantage in dealing with neurons. This volume reports progress from a version of the second approach—the use of endocrine manipulations and reagents to probe functions of the nervous system.

The first part of this book describes current work on the actions of hypothalamic releasing factors and nerve growth factor from the characterization of their cell surface receptors to the mechanisms of signal transduction leading to cellular response. The second part considers the actions of gonadal steroids and glucocorticoids on neurons. Success in the cellular localization of steroid receptors and specific gene products that are regulated by steroids has significantly increased our knowledge of the organization of the brain and the integration of responses to endocrine signals that lead to behavior. The third part considers current research on magnocellular neurons. Studies of magnocellular neurons have, over the years, provided the framework for concepts of synthesis, storage, and secretion of neuropeptides. The three articles in this section treat recent advances in the understanding of the molecular biology of the oxytocin gene, posttranslational events, and the electrophysiology of magnocellular neurons. The information emerging from neuroendocrine studies using the experimental approaches characterized in this volume is paving the way for more refined experimentation which will lead us toward an understanding of brain function at the molecular level.

JEROME F. STRAUSS III
DONALD W. PFAFF

Yale Membrane Transport Processes Volumes

Joseph F. Hoffman (ed.). (1978). "Membrane Transport Processes," Vol. 1. Raven, New York.

Daniel C. Tosteson, Yu. A. Ovchinnikov, and Ramon Latorre (eds.). (1978). "Membrane Transport Processes," Vol. 2. Raven, New York.

Charles F. Stevens and Richard W. Tsien (eds.). (1979). "Membrane Transport Processes," Vol. 3: Ion Permeation through Membrane Channels. Raven, New York.

Emile L. Boulpaep (ed.). (1980). "Cellular Mechanisms of Renal Tubular Ion Transport": Volume 13 of *Current Topics in Membranes and Transport* (F. Bronner and A. Kleinzeller, eds.). Academic Press, New York.

William H. Miller (ed.). (1981). "Molecular Mechanisms of Photoreceptor Transduction": Volume 15 of *Current Topics in Membranes and Transport* (F. Bronner and A. Kleinzeller, eds.). Academic Press, New York.

Clifford L. Slayman (ed.). (1982). "Electrogenic Ion Pumps": Volume 16 of *Current Topics in Membranes and Transport* (A. Kleinzeller and F. Bronner, eds.). Academic Press, New York.

Joseph F. Hoffman and Bliss Forbush III (eds.). (1983). "Structure, Mechanism, and Function of the Na/K Pump": Volume 19 of *Current Topics in Membranes and Transport* (F. Bronner and A. Kleinzeller, eds.). Academic Press, New York.

James B. Wade and Simon A. Lewis (eds.). (1984). "Molecular Approaches to Epithelial Transport": Volume 20 of *Current Topics in Membranes and Transport* (A. Kleinzeller and F. Bronner, eds.). Academic Press, New York.

Edward A. Adelberg and Carolyn W. Slayman (eds.). (1985). "Genes and Membranes: Transport Proteins and Receptors": Volume 23 of *Current Topics in Membranes and Transport* (F. Bronner and A. Kleinzeller, eds.). Academic Press, Orlando.

Peter S. Aronson and Walter F. Boron (eds.). (1986). "$Na^+ - H^+$ Exchange, Intracellular pH, and Cell Function": Volume 26 of *Current Topics in Membranes and Transport* (A. Kleinzeller and F. Bronner, eds.). Academic Press, Orlando.

Gerhard Giebisch (ed.). (1987). "Potassium Transport: Physiology and Pathophysiology": Volume 28 of *Current Topics in Membranes and Transport* (F. Bronner and A. Kleinzeller, eds.). Academic Press, Orlando.

Part I
Peptide Action

Peptides and Slow Synaptic Potentials

STEPHEN W. JONES

Department of Physiology and Biophysics
Case Western Reserve University
Cleveland, Ohio 44106

AND

PAUL R. ADAMS

Department of Neurobiology and Behavior
State University of New York at Stony Brook
Stony Brook, New York 14853

I. INTRODUCTION

Chemical messengers are used in many different ways to convey information among cells. Much is known about two rather extreme examples: classical neurotransmitters, such as acetylcholine (ACh) at the neuromuscular junction, which act over a distance of less than 1 μm in less than 1 msec; and classical hormones, which act throughout the body for much longer times. Slow synaptic responses, often mediated by neuropeptides, have intermediate properties. The mechanisms of such responses, and their potential physiological roles, are matters of intense investigation.

The goal of this article is to compare and contrast the actions of neuropeptides with "classical" neurotransmitters, to see how cells are organized to respond to different messages. The bullfrog sympathetic ganglion will be emphasized as a model system containing synaptic responses on rapid (millisecond) and slow (seconds to minutes) time scales.

II. BULLFROG SYMPATHETIC GANGLIA

The paravertebral sympathetic ganglia of the bullfrog, particularly the most caudal ones (ninth and tenth), were among the first neuronal preparations to be used for intracellular recording (Nishi and Koketsu, 1960). The cell bodies are larger than those of most vertebrate neurons (some are over 50 μm in diameter), which allows impalement by microelectrodes with relatively little damage. The cells are electrically compact, as they have no dendrites, and the synapses are directly on the cell body or the proximal axon hillock. Thus the cells are nearly ideal for voltage clamp, with the notable exception that the voltage remains uncontrolled in much of the axon.

The ganglia are innervated by sympathetic preganglionic neurons, which have their cell bodies in the spinal cord. There are two classes of preganglionic inputs, distinguished on the basis of conduction velocity: C fibers, coming from the seventh and eighth spinal nerves, and the more rapidly conducting B fibers, which enter the ganglion chain at higher levels (Nishi *et al.*, 1965; Dodd and Horn, 1983a). It is possible to stimulate B and C fiber inputs separately. C fibers innervate C cells in the ganglia, and B fibers innervate B cells (Fig. 1). B cells are generally larger, and can have either rapidly or slowly conducting axons (*fast* or *slow* B cells; Dodd and Horn, 1983a). C cells have slowly conducting axons, possibly with rare exceptions (Tokimasa, 1984b). Both B and C fibers use acetylcholine as a neurotransmitter. In addition a peptide that immunologically resembles mammalian luteinizing hormone-releasing hormone (M-LHRH) is present in C fibers, but not in B fibers (Jan *et al.*, 1980b; but see Lascar *et al.*, 1982).

A. Fast EPSP

The most conspicuous synaptic potential in these cells is a rapid excitatory postsynaptic potential (EPSP), mediated by nicotinic acetylcholine receptors (AChRs) (Nishi and Koketsu, 1960). It results from direct opening of ion channels, upon activation of AChRs. The ion channels are roughly equally permeable to sodium and potassium, so that the cell is depolarized. In B cells, the fast EPSP is due largely to the action of a single B fiber, which has dozens of synaptic boutons on the postsynaptic cell (Weitsen and

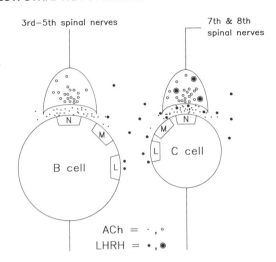

3rd—5th spinal nerves 7th & 8th spinal nerves

N

M

M

N

L C cell

B cell

L

ACh = ·, ○
LHRH = •, ◉

FIG. 1. Schematic diagram of synaptic transmission in bullfrog sympathetic ganglia. ACh is shown as being packaged in small synaptic vesicles, and the LHRH-like peptide in large, dense core vesicles (see Lascar *et al.*, 1982; Bowers *et al.*, 1986). During transmission, ACh is present mainly in the synaptic cleft, and does not diffuse to neighboring cells, in contrast to LHRH. N, Nicotinic AChR; M, muscarinic AChR; L, LHRH receptor. Muscarinic and LHRH receptors are shown in the extrasynaptic membrane, but this is speculative. Not drawn to scale.

Weight, 1977). Under normal conditions, that EPSP is suprathreshold for generation of a postsynaptic action potential. This neuron–neuron synapse is therefore fundamentally a relay synapse, like the neuromuscular junction. Smaller subthreshold fast EPSPs can sometimes be recorded, suggesting some degree of multiple innervation.

C cells have a similar fast EPSP, but it results from the summation of inputs from several preganglionic fibers rather than one, and its time course is somewhat longer, due to the properties of the AChR channel (Marshall, 1986). This difference between B and C cells is under developmental control: denervated B cells have slow, C cell-like AChRs (Marshall, 1985).

Several aspects of the molecular organization of synapses in the sympathetic ganglion closely resemble those of the neuromuscular junction. Specifically, there are discrete release sites where vesicles containing ACh are thought to fuse with the presynaptic membrane and a directly opposed postsynaptic membrane that contains a high density of AChRs (Marshall, 1981). It is likely that the ACh acts at near millimolar concentrations to locally saturate the AChRs, activating nearly all of them before being removed from the synapse by diffusion and by hydrolysis by cholinesterases. This is similar to the "saturated disk" model that has been proposed at the neuromuscular junction (Land *et al.*, 1980).

However, the postsynaptic AChR-containing area is relatively small, and the AChR density may be lower than that at the neuromuscular junction (Loring *et al.*, 1986). This could lead to a significant overflow of ACh beyond the synaptic cleft (L. Marshall, unpublished calculations). But given the relatively low affinity of nicotinic AChRs for ACh (tens of micromolar), any ACh that escapes would be unlikely to activate nicotinic AChRs at other synaptic regions. That is, individual quanta, each resulting from the release of a single packet of ACh at one active zone, should act independently, as at the neuromuscular junction. Much of this obviously is speculation; a quantitative study of the molecular machinery underlying the fast nicotinic EPSP in the ganglion is needed.

B. Slow IPSP

C cells, but not B cells, have a muscarinic slow inhibitory postsynaptic potential (IPSP) (Tosaka *et al.*, 1968; Dodd and Horn, 1983b). The IPSP may result from the ACh that escapes binding to the nicotinic AChRs, but that remains uncertain, as the location of the muscarinic receptors on the cell is not known. Similar muscarinic IPSPs have been studied in detail in parasympathetic neurons (Hartzell *et al.*, 1977) and in the heart.

The slow IPSP, like the fast EPSP, is due to opening of ion channels. However, the slow IPSP channels are selectively permeable to potassium, so that the response is hyperpolarizing and inhibitory. A more fundamental distinction is that the coupling between receptor activation and channel opening appears to be indirect. There is an absolute latency of 100 msec or longer for the slow IPSP. The latency is highly temperature sensitive, which rules out diffusion of ACh to distant receptors as the explanation of the delay, and strongly suggests involvement of a metabolic process (Hartzell *et al.*, 1977). The fast EPSP is over in 100 msec, reflecting rapid removal of ACh from the synapse. It is likely that the ACh that causes the slow IPSP has bound to the muscarinic AChR, dissociated, and been hydrolyzed before the IPSP even starts (see Nargeot *et al.*, 1982).

There is more direct evidence for a second messenger process for the slow IPSP in atrial cells, where the action of pertussis toxin and of GTP analogs strongly suggest mediation by a GTP-binding protein similar to that involved in muscarinic inhibition of adenylate cyclase (Pfaffinger *et al.*, 1985; Breitwieser and Szabo, 1985). However, the slow IPSP is unaffected by cyclic nucleotides, and muscarinic activation of potassium channels can be demonstrated in cell-free membrane patches (Soejima and Noma, 1984), which rules out the involvement of *cytoplasmic* second messengers. The simplest hypothesis is that receptor–channel coupling is mediated by direct binding of the activated GTP-binding protein to the channels, with the muscarinic receptor, GTP-binding protein, and potassium channel each being a distinct protein molecule (Pfaffinger *et al.*, 1985). Direct activation of the channel by application of a purified GTP-binding protein to inside-out patches has now been demonstrated (Codina *et al.*, 1987).

It is already clear that responses to acetylcholine that are mediated through different receptor types can have fundamentally different mechanisms, and time courses differing by several orders of magnitude.

C. Slow EPSP

B cells, but not C cells, have a muscarinic slow EPSP (Tosaka *et al.*, 1968). The slow EPSP rises to a peak in approximately 2 seconds, and lasts for nearly 1 minute. The ionic mechanisms of the slow EPSP are complex, but in general there is an increase in the resistance of the cell, suggesting a closing of ion channels (Kuba and Koketsu, 1976; Weight and Votava, 1970). Once again, a second messenger appears to be involved.

The depolarization during the slow EPSP is rather small, and repetitive stimulation of the preganglionic input is necessary for a maximal response (which is approximately 10 mV). This is in contrast to the much larger fast EPSP, and raises the question of the function of the slow EPSP in a cell that is basically a relay synapse to begin with.

The most likely role of the slow EPSP is to directly increase the excitability of the cell (Adams *et al.*, 1982b). Bullfrog sympathetic neurons generally fire only one or a few action potentials bunched at the beginning of a maintained depolarizing pulse. (Many other neurons can fire repetitively under normal conditions.) During the slow EPSP, however, sympathetic neurons are able to fire a train of action potentials (Fig. 2). This increase in excitability is not due simply to the depolarization during the slow EPSP; the change in firing pattern is not mimicked by an equivalent depolarization, and prevention of the depolarization during the slow EPSP (by manually clamping the cell to the original resting potential by passing hyperpolarizing current through the electrode) does not prevent the increase in excitability (Jones, 1985). That is, the slow EPSP is excitatory not because it depolarizes the cell, but because it closes channels in the membrane. In particular, it closes the potassium channels that give rise to the M-current (Brown and Adams, 1980).

1. THE M-CURRENT

During the early 1950s, Hodgkin and Huxley identified two distinct voltage-dependent currents in the squid giant axon: the sodium current and the potassium current. Those two currents were sufficient to explain generation and propagation of the action potential (Hodgkin and Huxley, 1952). Similar currents are present in axons of vertebrates (Frankenhaeuser and

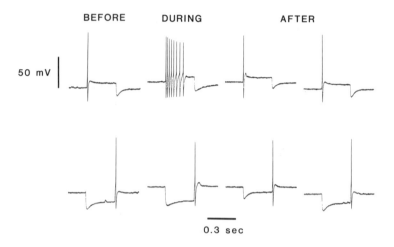

Fig. 2.   Increase in excitability during a slow EPSP. Records are responses to 0.3-second depolarizing currents (above) and hyperpolarizing currents (below). Two records are shown after the slow EPSP: after return to the original resting potential (extreme right), and manually clamped to the peak potential reached during the slow EPSP (second from right). Slow changes in voltage are due mainly to changes in the amount of active M-current. From Jones and Adams (1987).

Huxley, 1964). Since then, it has become painfully obvious that the electrical behavior of neuronal cell bodies cannot be fully explained by those two currents. For example, bullfrog sympathetic neurons have at least eight distinct voltage-dependent currents: two sodium, five potassium, and one calcium (Adams *et al.*, 1986; Jones, 1987b). These currents are involved not only in generation of the action potential, but also in regulation of excitability—that is, setting the threshold, and control of repetitive activity. The M-current plays a particularly important role in this.

The M-current is a potassium current that is activated by depolarization in the critical region between the resting potential and threshold for generation of an action potential (Adams *et al.*, 1982a). It does not inactivate. Once it is activated at a particular voltage, it remains turned on until the voltage is changed. Much of the resting potassium conductance of the cell may be M-current (but see Tosaka *et al.*, 1983). It turns on and off slowly at physiological voltages; at the point where it is half maximally activated, -35 mV, the time constant is approximately 150 msec. This means that the M-current does not respond much to large, rapid changes in membrane potential (e. g., the action potential), but it is highly sensitive to small, slow changes. For example, a maintained depolarizing current will not immediately activate M-current, but over tens of milliseconds the M-current will turn on, which (since it is a potassium current) will tend to hyperpolarize the cell back toward the resting potential. If the depolarizing current is strong enough to reach threshold, the M-current will not have an immediate effect. But a maintained depolarization can turn on enough M-current to bring the cell below threshold within approximately 100 msec, terminating firing (Fig. 2). This is the main mechanism underlying spike frequency adaptation in bullfrog sympathetic neurons, although other currents undoubtedly contribute (see below).

Under voltage clamp, the M-current appears as a slow change (or "relaxation") in the current upon voltage steps in the region between -60 and -30 mV. Hyperpolarized to that voltage, the M-current is turned off in the steady-state; at more depolarized voltages, other currents dominate the picture. The M-current was originally described using a standard voltage clamp method using two intracellular electrodes, one for recording voltage and the other to pass current. However, the small amplitude and slow time course of the M-current make it an ideal subject for a single-electrode voltage clamp, where one electrode is switched at a high rate (3–10 kHz) between passing current and recording voltage. Although the voltage clamp control is less complete with the single electrode, that procedure typically causes less damage to the cell, and allows stable recording for longer times.

The most striking feature of the M-current is its sensitivity to neurotransmitters (Adams et al., 1982b). In particular, muscarinic agonists block nearly all of the M-current, without affecting its kinetics. This effect is largely responsible for the slow EPSP in the bullfrog sympathetic ganglion (Fig. 3). At a voltage where considerable M-current is activated (e. g., -40 mV), inhibition of the M-current causes a net inward current. (Reduction of a potassium current, which is an outward current at that voltage, results in an inward current.) At a voltage where the M-current is turned off (e. g., -60 mV), inhibition of the M-current has no effect. In addition, the M-current relaxations seen upon the voltage steps to and from -40 mV are correspondingly reduced. Other currents in the cell are insensitive, or much less sensitive, to muscarinic agonists (Adams et al., 1982b).

This mechanism explains several features of the slow EPSP that were initially quite confusing. First, the size of the slow EPSP reflects not only the driving force on potassium (as expected for a decrease in a voltage-insensitive potassium current) but also the amount of M-current that is active. For that reason, the slow EPSP does not reverse to an outward current below the equilibrium potential for potassium. However, the M-current itself clearly does reverse at or near the potassium equilibrium potential (judged from the "tail currents," measured during the few milliseconds while the M-current is turning off following a hyperpolarizing step); and if extracellular potassium is elevated so that its equilibrium potential is in the range where the M-current is turned on, the slow EPSP can be reversed.

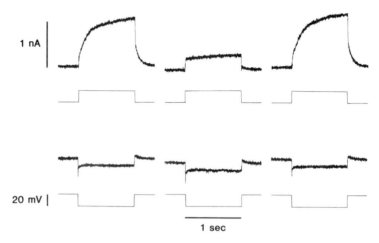

Fig. 3.   Inhibition of the M-current during a slow EPSP. (above) Currents in response to a 20-mV depolarizing command from the holding potential of −60 mV. The M-current slowly turns on during the depolarization, producing an outward current (outward currents are shown upward by convention). At the end of the step, the M-current turns back off over tens of milliseconds. Records in the left column are before, records in the center during, and records at right after the EPSP. (below) Response to hyperpolarizing pulses in the region where the M-current is turned off. The small inward (downward) shift in current during the slow EPSP is the "additional inward current" (see also Fig. 5).

2.   OTHER EFFECTS

The M-current is inhibited during the slow EPSP, and inhibition of the M-current qualitatively explains the increased excitability that sympathetic neurons display during the slow EPSP. However, there are other effects that can be demonstrated, particularly in response to exogenous muscarinic agonists, in many cells (Kuba and Koketsu, 1976; Kuffler and Sejnowski, 1983; Jones 1985).

In about half of the cells, muscarinic agonists cause an additional inward current, due to an increase in conductance. The extrapolated reversal potential is close to 0 mV, so that the effect gets larger with hyperpolarization. This current may be due to an increase in conductance to sodium (Kuba and Koketsu, 1976), but that is not firmly established. The effect of this on the cell is also not clear. A tonic depolarization, particularly one due to an increase in conductance, might even have an inhibitory effect due to inactivation of the voltage-dependent sodium current. If the depolarization is prevented by manually clamping the cell to its resting potential, the additional inward current reduces the excitability of the cell, presumably because the increased conductance acts as a shunt (Jones, 1985).

Another effect that can be demonstrated in many cells is a reduction in the after-hyperpolarization (AHP) (Tokimasa, 1984a; Pennefather *et al.*, 1985a). Most neurons, even squid axons, have an AHP following an action potential due to prolonged activation of potassium currents. The AHP in B cells is particularly dramatic, as it can be tens of millivolts in amplitude and can last for half a second. This slow AHP is due to a calcium-dependent potassium current; influx of calcium during the action potential opens potassium channels. The kinetics and pharmacology of the AHP current indicate that it is different from the M-current and from the potassium currents that repolarize the action potential. The slow time course of the AHP may reflect the removal of calcium from the cytoplasm (Smith *et al.*, 1983). The AHP current is reduced by muscarinic agonists, but maximal doses block only about 30% of the current (Pennefather *et al.*, 1985a). It now appears that the AHP current assists the M-current in preventing repetitive firing, and that full removal of adaptation requires inhibition of both M-current and AHP current (Pennefather *et al.*, 1985b).

There are some indications that muscarinic agonists affect some or all of the other voltage-dependent currents of B cells (Koketsu, 1984). However, evidence for most of those effects is rather indirect, and other explanations are possible. In particular, the reductions in the rate of rise and fall of the action potential may be due to shunting by the additional inward current. Regardless of the resolution of these issues, it is clear that activation of muscarinic receptors has multiple effects on B cells.

3. SECOND MESSENGERS

There is suggestive evidence that second messengers interpose between muscarinic receptor activation and inhibition of the M-current in B cells. The time course of M-current inhibition is slow, even though the ACh responsible for the slow EPSP is removed rapidly. More than one conductance is affected, in contrast to the fast EPSP where a single channel type is opened. And (as will be discussed below) activation of each of several distinct receptor types can inhibit nearly all of the M-current. But all this is only suggestive evidence, and it does not tell us which second-messenger systems might be involved.

Initial studies argued strongly against involvement of cyclic nucleotides or calcium (Adams *et al.*, 1982b). Intracellular injection of either cAMP or cGMP, or external application of membrane-permeable analogs, was without effect on the M-current. Injection of calcium into the cell, or removal of extracellular calcium, was without effect on the M-current.

Another possibility is activation of C kinase, a protein kinase that is activated by phospholipid and diacylglycerol (DAG), in the presence of normal levels of intracellular calcium (Nishizuka, 1986). The physiological stimulus that activates C kinase appears to be hydrolysis of a minor membrane lipid, phosphatidylinositol 4,5-bisphosphate (PIP$_2$), by phospholipase C. This effect occurs in response to activation of a variety of receptors, including muscarinic receptors in at least some cell types (Hokin, 1985).

The involvement of C kinase in M-current inhibition has been tested by the use of phorbol esters, tumor promoters that replace diacylglycerol as an activator of C kinase (Adams and Brown, 1986). Phorbol ester had clear effects, but they were not what would be naively expected if C kinase activation mediated muscarinic inhibition of the M-current. First, phorbol ester alone did inhibit the M-current, but incompletely. Second, phorbol ester lowered the sensitivity of the *remaining* M-current to muscarine, by shifting the dose–response relation to higher concentrations. Phorbol ester did not prevent the additional inward current (that is, the muscarinic conductance increase). So, phorbol ester partially mimics the muscarinic response, and partially prevents it! One possible explanation is that C kinase does not mediate muscarinic inhibition of the M-current, but it indirectly modulates whatever second-messenger system is in fact involved.

Or perhaps C kinase activation is the mediator, but there is a rapid negative feedback loop that uncouples muscarinic receptors from M-channels (Fig. 4). This is not entirely an ad hoc hypothesis. Muscarinic

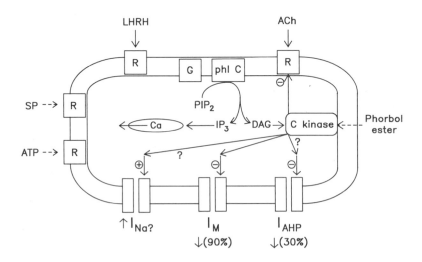

Fig. 4. Highly speculative diagram of second-messenger mechanisms mediating slow excitatory actions in bullfrog sympathetic B neurons. Each receptor (R), when activated, activates a GTP-binding protein (G), which in turn couples to phospholipase C (phl C). Dashed lines indicate actions of exogenous drugs. SP, substance P.

receptors cause PIP_2 hydrolysis in the PC12 cell line (which resembles sympathetic neurons in some respects). Preincubation for just 1–3 minutes with phorbol ester prevents that effect (Vicentini *et al.*, 1985). The loss of sensitivity may be due to internalization of muscarinic receptors, which occurs in response to phorbol ester in a different cell line (Liles *et al.*, 1986), although on a somewhat slower time scale (10–30 minutes). If such effects are taking place in bullfrog sympathetic neurons, it remains unclear how muscarinic agonists can cause an inhibition of up to 85% of the M-current that can persist for several minutes. Perhaps activation of C kinase by phorbol esters differs subtly in timing or compartmentalization from physiological activation via muscarinic receptors.

Receptor-mediated phosphatidylinositol turnover produces an additional second messenger, inositol trisphosphate (IP_3). IP_3 acts to release calcium from internal stores, probably endoplasmic reticulum. It is possible that an increase in intracellular calcium acts synergistically with activation of C kinase to mediate muscarinic responses. Since muscarinic receptor activation does not lead to opening of calcium-dependent potassium channels in bullfrog sympathetic neurons, any IP_3-induced increase in intracellular calcium must be rather small. In adrenal medulla, which is developmentally similar to sympathetic ganglia, the muscarinic response causes only a slight increase in intracellular calcium, not enough to trigger secretion (Cheek and Burgoyne, 1985).

One approach to unraveling second-messenger mechanisms in the bullfrog ganglion cells is to use patch-clamp recording techniques. This can be done with either explant cultures from the ganglion (Galvan *et al.*, 1984) or completely dissociated cells. Preliminary results suggest that internal perfusion of cells with patch electrodes in the whole-cell mode reduces the amplitude of the M-current, and reduces its sensitivity to muscarinic agonists (Galvan *et al.*, 1984). This is similar to the effects of phorbol esters, and suggests that some crucial component of the internal environment is being dialyzed out of the cell (or is being inadvertently added). The missing ingredient is yet to be established.

Patch-clamp recording can also be used to find the M-current channel itself, and to study its sensitivity to neurotransmitters and second messengers. This work is in progress.

D. Late, Slow EPSP

Both B and C cells have an EPSP that lasts for several minutes, and is not due to the action of ACh (Nishi and Koketsu, 1968). This *late, slow EPSP* is due to the action of an LHRH-like peptide (Jan *et al.*, 1979, 1980a,b), for which most of the classical criteria for identifying a substance as a neurotransmitter have been satisfied (Jan and Jan, 1982). In particular, the peptide is present in presynaptic terminals (as deduced by im-munocytochemistry); it is released upon nerve stimulation or by depolariza-tion in high extracellular potassium; the late, slow EPSP is closely mim-icked by exogenous LHRH agonists; and LHRH antagonists block both the late, slow EPSP and the action of LHRH. Two pieces of information are still missing. First, no inactivation mechanism has been identified for the LHRH in the ganglion. Second, the exact chemical identity of the LHRH-like peptide has not been proved.

1. MECHANISMS OF LHRH ACTION

As is the case for muscarinic agonists, the action of LHRH agonists on B cells is often complex. In fact, on any given cell, the actions of muscarinic and LHRH agonists are essentially identical (e. g., Kuffler and Sejnowski, 1983). In short, LHRH agonists depolarize B cells, and increase their ex-citability. Under voltage clamp (Fig. 5), there is a consistent inhibition of the M-current, a less consistent increase in conductance that is particularly visible at hyperpolarized potentials, and a slight inhibition of the AHP cur-rent (Adams and Brown, 1980; Katayama and Nishi, 1982; Jones, 1985; Pennefather et al., 1985a).

Second-messenger mechanisms underlying the action of LHRH have not been specifically investigated in the bullfrog ganglion. Obviously, they could well be similar or identical to the second-messenger mechanisms for muscarinic actions. It is worth noting that the hormonal action of M-LHRH in the pituitary now seems to involve both a small rise in in-tracellular calcium, partly from intracellular stores, and activation of C kinase (Chang et al., 1986).

2. PHARMACOLOGY OF LHRH ACTION

The remarkable identity between the biophysical mechanisms underlying the actions of muscarine and LHRH might suggest that they act upon a common receptor. But that is clearly not the case. The effects of muscarinic agonists are blocked by nanomolar concentrations of classic muscarinic an-tagonists such as atropine and scopolamine, without any effect on the response to LHRH. Conversely, LHRH analogs that antagonize the actions of M-LHRH in the pituitary are selective antagonists of LHRH actions on B cells. Similarly, the slow EPSP is selectively blocked by muscarinic agonists, and the late, slow EPSP is selectively blocked by LHRH an-tagonists.

The lack of good antagonists has plagued work on most neuropeptides. But there are several specific LHRH antagonists, resulting from extensive characterization of LHRH receptors in mammalian pituitary. The phar-

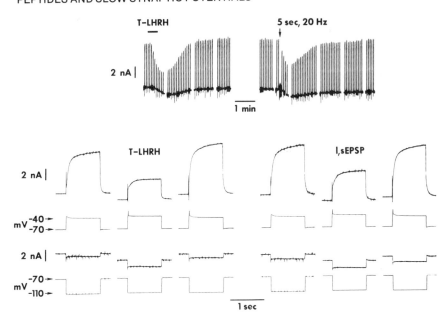

FIG. 5. T-LHRH and the late, slow EPSP. (above) Time course of the responses at a slow time scale; each vertical line is the current flowing in response to a clamp step from −70 to −40 mV. (below) Selected fast records, arranged as in Fig. 3. During each response, the amplitude of the current upon a hyperpolarizing step increased, reflecting the increased conductance during the additional inward current. From Jones (1985).

macology of LHRH effects on B cells is qualitatively similar in many respects to that of pituitary, for example in the actions of many agonist and antagonist analogs of M-LHRH (Jan and Jan, 1982). But there are also clear differences. In particular, the LHRH of salmon (T-LHRH) is approximately 20 times more potent than is M-LHRH (Jan et al., 1983; Jones et al., 1984; Jones, 1985). The opposite is the case for pituitary (Sherwood et al., 1983). "Superagonist" analogs of T-LHRH, designed by analogy with such analogs of M-LHRH, are more potent than T-LHRH (Jan et al., 1983; Jones, 1985).

T-LHRH is not the only nonmammalian LHRH to be discovered in recent years. Two LHRHs from chickens have been sequenced (King and Millar, 1982; Miyamoto et al., 1984). Two lamprey LHRHs are known, one of which has been sequenced (Sherwood et al., 1986). There is some evidence for as yet unsequenced LHRH-like peptides in teleost fish (Sherwood et al., 1984; King and Millar, 1985). Recently, the known peptides have been tested on the bullfrog ganglion. One of them (chicken II LHRH) is extremely potent, with 50% inhibition of the M-current at about 1nM (S. Jones, unpublished). Chicken II LHRH is at least 100 times more active than T-LHRH, the second best analog. To conclusively identify the neurotransmitter for the late, slow EPSP it will be necessary to determine its primary structure (by amino acid sequencing, or genetic analysis). Rough calculations suggest that amino acid sequencing would require ganglia from several hundred bullfrogs.

3. TIMING OF LHRH ACTION

The most striking feature of the late, slow EPSP in a B cell is its time course. A record of the late, slow EPSP looks like a perfectly traditional synaptic potential, except for the time scale, which is approximately 10^5 times longer than for a fast EPSP. The time course is even more puzzling, given that the slow and the late, slow EPSPs involve the same conductance changes. That is, both events have the same consequences, but the late, slow EPSP lasts perhaps 10 times longer.

The simplest explanation, proposed by Jan et al. (1980b), is that the time course of the late, slow EPSP is determined by the presence of LHRH in the extracellular space. In contrast, the time couse of the slow EPSP is presumably determined by the second-messenger processes, not by receptor kinetics or by diffusion of ACh.

There is considerable evidence that the LHRH-like peptide is not rapidly removed following release. First, the peptide is present only in nerve terminals upon the C cells, but perfectly good late, slow EPSPs can be recorded from B cells. This implies that LHRH must be able to diffuse several micrometers from the point of release to the site of action. [This mechanism has led Jan et al. (1983) to call this action "nonsynaptic." That may be literally true on an anatomist's definition of synapse as the specialized close apposition between the pre- and postsynaptic membranes, but it is confusing to talk of a nonsynaptic synaptic potential.] Further evidence for the continued presence of LHRH is that rapid local application of an LHRH antagonist can terminate an ongoing late, slow EPSP (Jan and Jan, 1982). Other explanations are possible, such as maintained release of LHRH over several minutes following each action potential, but these seem less likely. However, assuming that ACh and LHRH are contained in different classes of synaptic vesicle, it is possible that their release is regulated differentially.

But then why does the late, slow EPSP ever end? The simplest possibility is that the LHRH simply diffuses away until its concentration goes below that which activates significant numbers of receptors. If this is the case, one of the traditional criteria for identification of a neurotransmitter, presence of a specific mechanism for inactivation of the transmitter, would not be met. That would not cast doubt on the role of LHRH in the late, slow EPSP; it would simply raise the issue of whether we would prefer to call its action something other than neurotransmission, or whether we would prefer to modify the definition of neurotransmitter (which has been done before).

However, it is also possible that there is a specific mechanism for removal of LHRH from the synapse, possibly an extracellularly located peptidase, but that the rate of inactivation is exceedingly slow. Involvement of peptidases in the actions of other neuropeptides has been proposed, but definitive proof is lacking.

The timing of the late, slow EPSP could also be influenced by the affinity of the LHRH receptor for the endogenous transmitter. Chicken II LHRH inhibits the M-current by 50% at approximately 1 nM; such a high potency is considered typical for a peptide acting as a hormone. But is it appropriate for a neurotransmitter? Since the dissociation constant (K_d) for binding is the ratio of dissociation rate to association rate, high affinity binding (low K_d) requires a slow dissociation rate. For comparison, the binding of ACh to nicotinic receptors is very fast, near the diffusion limit, but to allow rapid termination of its action (with time constant close to 1 msec) the K_d is greater than 10 μM. A K_d of 1 nM would predict a minimum dissociation time constant of approximately 30 seconds. That is roughly consistent with the time course of termination of the late, slow EPSP upon rapid application of LHRH antagonists (Jan and Jan, 1982).

Lower affinity LHRH analogs such as T-LHRH act with a time course very similar to that of the late, slow EPSP itself (Fig. 5). This suggests that the action of exogenous T-LHRH is terminated in the same way as is the natural transmitter for the late, slow EPSP (i.e., diffusion or proteolysis). However, a "superagonist" that is roughly 10 times more potent than T-LHRH (Jan *et al.*, 1983; Jones, 1985) acts considerably more slowly than T-LHRH (Fig. 6). Perhaps its higher potency is due to a slow dissociation rate, slower than that of the natural transmitter for the late, slow EPSP.

4. LHRH ACTIONS ON C CELLS

It may seem rather odd to concentrate on the late, slow EPSP in B cells, when the peptide responsible is present only in terminals upon C cells, and C cells do have a late, slow EPSP. The primary reason for this is technical; C cells are smaller and are more difficult to record from.

Basically, LHRH actions on C cells are similar to those on B cells. LHRH and the late, slow EPSP depolarize C cells with an increase in excitability (Jan and Jan, 1982; Dodd and Horn, 1983b). Interestingly, the late, slow EPSP is more rapid (both in time to peak and in time to decay) in C cells (Jan and Jan, 1982). The more rapid rising phase can be easily explained by the closer apposition between release sites and binding sites, but the more rapid time to decay is puzzling. If the action of LHRH is terminated simply by diffusion, its concentration would still be the highest at all times near the

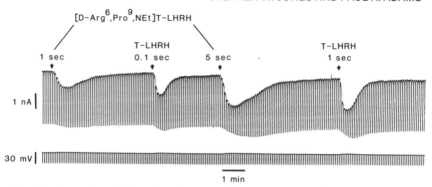

FIG. 6.   Comparison of the action of T-LHRH to a "superagonist." Each peptide was applied by pressure at 40 psi from a different barrel of a three-barrel pipette positioned near the cell. The applications were too brief to produce a steady-state effect, and the concentration at the cell was probably much less than that in the pipette (100 μM T-LHRH, 1 μM superagonist). Regardless of duration of application, or of the size of the response, the response to the superagonist was significantly slower. The M-current was inhibited by 55% in the larger responses to each agent (not shown). Note also that the time course of the effect at the hyperpolarized potential, reflecting the additional inward current, was slower in all responses (Katayama and Nishi, 1982; Jones, 1985). The holding potential was approximately -30 mV.

site of release. Perhaps the affinity or density of LHRH receptors on C cells differs from that of B cells; or, if there is a mechanism for inactivating LHRH, it may be preferentially located near C cells.

There is also a difference in the electrophysiological effects of LHRH on B and C cells. LHRH agonists do inhibit the M-current in C cells (Jones, 1987a), but they have an additional effect, a *decrease* in conductance at hyperpolarized potentials, that is rare or absent in B cells. This response resembles in some respects a muscarinic inhibition of a chloride current in rat sympathetic neurons (Brown and Selyanko, 1985).

C Cells have a M-current that can be inhibited by LHRH, and a muscarinic receptor. However, muscarinic agonists hyperpolarize C cells (Dodd and Horn, 1983b), and do not inhibit the M-current (Jones, 1987a). This indicates that the coupling between receptors and ion channels can vary even among cells in a single ganglion.

E. Substance P and Nucleotides

Two other classes of drugs, presumably acting through independent receptors, inhibit the M-current (and sometimes have other effects as well): substance P and related peptides called *tachykinins* (Adams *et al.*, 1983; Akasu *et al.*, 1983a), and ATP and some related nucleotides (Gruol *et al.*, 1981; Adams *et al.*, 1982b; Akasu *et al.*, 1983b).

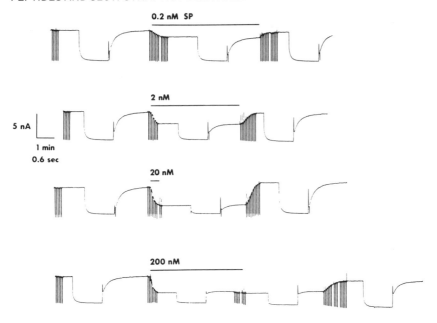

FIG. 7,   Dose–response relations for inhibition of the M-current by substance P(SP) in one cell. Records on slow and fast time scales are intermingled. Substance P was applied by bath perfusion during the times marked by the vertical bars. This was one of the more sensitive cells encountered. High sensitivity to substance P appeared to be correlated with rapid responses such as these, suggesting a short diffusion path between the bath perfusion and the cell.

The action of substance P is strikingly variable (Jones, 1985). On any given cell its effect is clearly dose-dependent (Fig. 7), but the dose that inhibits 50% of the M-current can vary from 2 nM to 2 μM, with some cells not responding even at 2–10 μM. A 2 nM I$_{50}$ would correspond roughly to published K_d values from binding studies. The reason for the variability is unclear. It seems unlikely that receptors on different cells in the same ganglion would have intrinisic binding affinities differing by three orders of magnitude. If spare receptors are the explanation, that would imply that the true K_d is   2 μM or higher. Another possibility is that a variable amount of substance P is hydrolyzed in the ganglion before reaching its receptors, as has been reported elsewhere (Watson, 1983). If that is the case, it might provide clues to the role of peptidases in general.

Several analogs of substance P, called tachykinins, also inhibit the M-current. The pharmacology of this site differs from certain other substance-P receptors in that C-terminal fragments, and some proposed substance-P antagonists, are inactive (Jones, 1985).

In the absence of selective antagonists, it is difficult to prove that substance P is acting on a distinct receptor. One clue is that the action of substance P tends to desensitize over a few minutes, which is not observed for muscarinic or LHRH agonists (Nishi *et al.*, 1980; Jones, 1985). Furthermore, there is no cross-desensitization between substance P and the other classes of agonist (Jan and Jan, 1982), suggesting that the desensitization is at the receptor level.

Jan and Jan (1982) found a few nerve fibers in the bullfrog sympathetic ganglion that were immunoreactive for substance P. However, there was no evidence that those processes made synaptic contacts. In general, there is no evidence that substance P is a neurotransmitter in this ganglion, so that the presence of receptors appears to be purely accidental.

ATP, UTP and certain other nucleotides also inhibit the M-current. ATP is present in synaptic vesicles along with ACh, and appears to be coreleased, but blockade of the slow EPSP by atropine fails to reveal a purinergic slow EPSP. The role of this receptor also remains mysterious.

III.   OTHER SLOW SYNAPTIC ACTIONS

At this point, an obvious question is how generally applicable the mechanisms of slow synaptic transmission in bullfrog sympathetic ganglia will be to other systems, particularly the mammalian central nervous system. This section is a brief, and rather arbitrary, look at the diversity of slow synaptic mechanisms. The emphasis will be on what ion channels are affected, and by what second messengers.

There are several difficulties with constructing even a superficial overview of this field. First, few systems have been studied in detail, largely due to the technical difficulties in dealing with most vertebrate neurons. For example, the complex and anatomically beautiful dendritic arbors of many neurons are electrically too large to be effectively voltage clamped, so that uncontrolled currents arising in remote regions complicate interpretation. Second, preparations that electrophysiologists prefer are generally shunned by biochemists, and vice versa, so that information on the metabolic and electrophysiological consequences of receptor activation are rarely directly comparable.

For a more encyclopedic account of actions of neurotransmitters in vertebrate nervous systems, see the book edited by Rogawski and Barker (1985).

A.   Excitatory Actions

Many cell types show slow EPSPs tht resemble those of the bullfrog ganglion. In many cases, those EPSPs are associated with an increase in excitability, and a decrease in conductance. To what extent can these effects be explained by an inhibition of M-current?

The M-current exists, and is inhibited by neurotransmitters, in several preparations. Muscarine inhibits the M-current in mammalian sympathetic neurons (Constanti and Brown, 1981; Hashiguchi et al., 1982; Freschi, 1983; Brown and Selyanko, 1985), hippocampal pyramidal cells (Halliwell and Adams, 1982), olfactory cortex neurons (Constanti and Galvan, 1983), cultured spinal cord neurons (Nowak and Macdonald, 1983) and smooth muscle (Sims et al., 1985). Substance P may inhibit the M-current in spinal neurons (Nowak and Macdonald, 1981) and in nerve terminals of the chick ciliary ganglion (Dryer and Chiappinelli, 1985). Bradykinin inhibits the M-current of neuroblastoma–glioma hybrid cells, apparently via activation of C kinase (Higashida and Brown, 1986).

However, there are clear examples of slow EPSPs that do not involve M-current inhibition. In myenteric plexus, ACh, substance P, serotonin, and slow EPSPs depolarize cells and inhibit a slow AHP by blocking a calcium-dependent potassium current (Grafe et al., 1980; North and Tokimasa, 1983). This current differs from the AHP current of bullfrog sympathetic neurons in being partially active at rest. Since this calcium-dependent potassium current is not very voltage sensitive, the slow EPSP in myenteric neurons has a well-defined reversal potential near the equilibrium potential for potassium.

Norepinephrine, acting through a β-adrenergic receptor, inhibits an AHP-like current in hippocampus (Madison and Nicoll, 1986a), and this effect is mediated by cAMP (Madison and Nicoll, 1986b). Athough high concentrations of muscarinic agonists inhibit the M-current in hippocampal neurons, a slow muscarinic EPSP in those cells involves inhibition of AHP current and a less well understood blockade of resting potassium current (Madison et al., 1987). There are ambiguous effects of phorbol esters on hippocampal neurons that might suggest involvement of C kinase in muscarinic actions there (Baraban et al., 1985; Malenka et al., 1986).

There is convincing evidence that substance P is the transmitter for a slow EPSP in mammalian inferior mesenteric sympathetic ganglia (Tsunoo et al., 1982). Those cells have a M-current that is inhibited by muscarine, but the effect of substance P appears to be quite different, and variable from cell to cell (Brown et al., 1985).

Voltage-dependent potassium currents other than the M-current may be inhibited by neurotransmitters in some cells. A transient outward current (A-current) appears to be blocked by norepinephrine, acting through α_1 receptors, in the raphe nucleus (Aghajanian, 1985); by somatostatin in pituitary (Dufy et al., 1983); and by muscarinic agonists in cultured hippocampal neurons (Nakajima et al., 1986). The inward ("anomalous") rectifier, which contributes to the resting potential in some cells, is inhibited by glutamate in retinal horizontal cells (Kaneko and Tachibana, 1985) and by substance P in globus pallidus (Stanfield et al., 1985).

Clearly, slow EPSPs can involve blockade of a variety of ion channels, and can be mediated through more than one second-messenger system.

B. Inhibitory Actions

Muscarinic agonists hyperpolarize some central neurons by an increase in potassium conductance (McCormick and Prince, 1986; Egan and North 1986). Enkephalins, acting thru μ and δ receptors; norepinephrine, through α_2 receptors; γ-aminobutyric acid (GABA), through $GABA_B$ receptors; serotonin, through 5-HT1a receptors; somatostatin; and adenosine all have similar effects (North, 1986; Andrade et al., 1986; Trussell and Jackson, 1986). These effects resemble the slow IPSP in the heart, and may also involve receptor–channel coupling via direct binding of GTP-binding proteins (Andrade et al., 1986; Trussell and Jackson, 1986).

Several transmitters inhibit voltage-dependent calcium currents. This has been clearly established in dorsal root ganglion cells for enkephalins, acting through κ receptors; GABA, through $GABA_B$ receptors; and norepinephrine (Dunlap and Fischbach, 1981; Werz and MacDonald, 1984). These effects are mimicked by activation of C kinase (Rane and Dunlap, 1986; see also DiVirgilio et al., 1986) and involve a pertussis toxin-sensitive GTP-binding protein (Holz et al., 1986; Hescheler et al., 1987).

Many neurotransmitters are known to inhibit transmitter release by an action on nerve terminals. Since it is rarely possible to record intracellularly from nerve terminals, conclusive evidence on the ionic mechanisms involved is lacking. There is a lively controversy over whether the effects are due to increases in potassium current or to decreases in calcium current. Probably both mechanisms will be found to exist in different cases.

IV. CONCLUSIONS

Knowledge has progressed considerably since the days when the only well-characterized examples of neurotransmitter actions were rapid, direct opening of ion channels, such as at the neuromuscular junction. The number of known (or suspected) transmitters, and of known (or suspected) cellular actions of transmitters, is each increasing without apparent limit. It is important to see whether any unifying principles are emerging from the chaos.

One point is that the factors determining the time course of a synaptic potential can be qualitatively different. In the bullfrog ganglion, the fast EPSP is terminated by dissociation of ACh from the receptor (with a time constant of a few milliseconds); the slow EPSP and IPSP reflect the time course of the second messengers involved; and the late, slow EPSP is terminated only upon removal of the peptide transmitter from the extracellular space. That is, for the cholinergic postsynaptic potentials, the transmitter is removed so rapidly that the time course of the response depends largely on the receptor and effector mechanisms involved. Even though the slow and the late, slow EPSPs involve the same ion channels, their time courses are dramatically different because of differences in metabolism of the transmitters involved.

It is likely that the concentration of transmitter, and the affinity of the receptor, also differ qualitatively for different synaptic potentials. ACh locally saturates its receptors during the fast EPSP; whether this is the case at other synapses is unknown.

The energetic cost of producing a peptide, synthesized as a protein precursor, must be much greater than that for a small classical transmitter such as ACh. This undoubtably has led to economical use of peptides. However, the larger size and complexity of peptides, and the possibility of rapid coevolution of peptides and receptors, may allow a greater diversity of messages to be conveyed. This idea would fit with the use of classical transmitters to convey a rapid, simple "yes or no" message, with peptides being reserved for "modulatory" actions. However, it is clear that there is not always a fundamental distinction between the messages conveyed by peptides and other transmitters, as is shown by the identical action (except in time course) of the slow and the late, slow EPSPs in B cells.

The same neurotransmitter can have a multitude of actions, and different neurotransmitters can converge upon a single ion channel. This is to be expected when second messengers are involved. It is commonplace for several different transmitters (or hormones) to be coupled to, for example, cAMP in one cell, and cases are known where different second-messenger systems lead to phosphorylation of a single protein (even on the same amino acid residue). Also, one second-messenger system can have different effects in different cells; for example, C kinase may inhibit calcium currents in one cell and potassium currents in another.

The involvement of second-messenger systems, and the existence of synaptic potentials lasting minutes, indicate that there need be no fundamental distinction between the cellular actions of neurotransmitters and of hormones. This is a reminder that some of the actions of hormones on cells are mediated through actions on ion channels. That is, stimulation of the release of one hormone by another may involve a depolarization and subsequent calcium entry, just as can occur during neurotransmission. Conversely, the most important action of some neurotransmitters may be on the metabolic activity of the target cell, and might even be electrically silent (see Bowers et al., 1986).

ACKNOWLEDGMENTS

Work in the authors' laboratories is supported by NIH grants NS 24471 and 18579.

REFERENCES

Adams, P. R., and Brown, D. A. (1980). Luteinizing hormone-releasing factor and muscarinic agonists act on the same voltage-sensitive K⁺-current in bullfrog sympathetic neurones. *Br. J. Pharmacol.* **68**, 353–355.

Adams, P. R., and Brown, D. A. (1986). Effects of phorbol ester on slow potassium currents of bullfrog ganglion cells. *Biophys. J.* **49**, 215a.

Adams, P. R., Brown, D. A., and Constanti, A. (1982a). M-currents and other potassium currents in bullfrog sympathetic neurons. *J. Physiol (London)* **330**, 537–572.

Adams, P. R., Brown, D. A., and Constanti, A. (1982b). Pharmacological inhibition of the M-current. *J. Physiol. (London)* **332**, 223–262.

Adams, P. R., Brown, D. A., and Jones, S. W. (1983). Substance P inhibits the M-current in bullfrog sympathetic neurones. *Br. J. Pharmacol.* **79**, 330–333.

Adams, P. R., Jones, S. W., Pennefather, P., Brown, D. A., Koch, C., and Lancaster, B. (1986). Slow synaptic transmission in frog sympathetic ganglia. *J. Exp. Biol.* **124**, 259–285.

Aghajanian, G. K. (1985). Modulation of a transient outward current in serotonergic neurons by α_1-adrenoceptors. *Nature (London)* **315**, 501–503.

Akasu, T., Nishimura, T, and Koketsu, K. (1983a). Substance P inhibits the action potentials in bullfrog sympathetic ganglion cells. *Neurosci. Lett.* **41**, 161–166.

Akasu, T., Hirai, K., and Koketsu, K. (1983b). Modulatory actions of ATP on membrane potentials of bullfrog sympathetic ganglion cells. *Brain Res.* **258**, 313–317.

Andrade, R., Malenka, R. C., and Nicoll, R. A. (1986). A GTP binding protein may directly couple 5-HT1a and GABA$_B$ receptors to potassium (K) channels in rat hippocampal pyramidal cells. *Soc. Neurosci. Abstr.* **12**, 15.

Baraban, J. M., Snyder, S. H., and Alger, B. E. (1985). Protein Kinase C regulates ionic conductance in hippocampal pyramidal neurons: Electrophysiological effects of phorbol esters. *Proc. Natl. Acad. Sci. U.S.A.* **82**, 2538–2542.

Bowers, C. W., Jan, L. Y., and Jan, Y. N. (1986). A substance P-like peptide in bullfrog autonomic nerve terminals: Anatomy biochemistry and physiology. *Neuroscience* **19**, 343–356.

Breitwieser, G. E., and Szabo, G. (1985). Uncoupling of cardiac muscarinic and β-adrenergic receptors from ion channels by a guanine nucleotide analogue. *Nature (London)* **317**, 538–540.

Brown, D. A., and Adams, P. R. (1980). Muscarinic suppression of a novel voltage-sensitive K⁺ current in a vertebrate neurone. *Nature (London)* **283**, 673–676.

Brown, D. A., and Selyanko, A. A. (1985). Two components of muscarine-sensitive membrane current in rat sympathetic neurones. *J. Physiol. (London)* **358**, 335–363.

Brown, D. A., Adams, P. R., Jones, S. W., Griffith, W. H., Hills, J., and Constanti, A. (1985). Some ionic mechanisms of slow peptidergic neuronal excitation. *Regul. Peptides* **4** (Suppl.), 157–161.

Chang, J. P., McCoy, E. E., Graeter, J., Tasaka, K., and Catt, K. J., (1986). Participation of voltage-dependent calcium channels in the action of gonadotropin-releasing hormone. *J. Biol. Chem.* **261**, 9105–9108.

Cheek, T. R., and Burgoyne, R. D. (1985). Effect of activation of muscarinic receptors on intracellular free calcium and secretion in bovine adrenal chromaffin cells. *Biochim. Biophys. Acta* **846**, 167–173.

Codina, J., Yatani, A., Grenet, D., Brown, A. M., and Birnbaumer, L. (1987). The α subunit of the GTP binding protein Gk opens atrial potassium channels. *Science* **236**, 442–445.

Constanti, A., and Brown, D. A. (1981). M-currents in voltage-clamped mammalian sympathetic neurones. *Neurosci. Lett.* **24**, 289–294.

Constanti, A., and Galvan, M. (1983). M-current in voltage-clamped olfactory cortex neurones. *Neurosci. Lett.* **39**, 65–70.

DiVirgilio, F., Pozzan, T., Wollheim, C. B., Vincentini, L. M., and Meldolesi, J. (1986). Tumor promoter phorbol myristate acetate inhibits Ca^{2+} influx through voltage-gated Ca^{2+} channels in two secretory cell lines, PC12 and RINm5F. *J. Biol. Chem.* **261**, 32–35.

Dodd, J., and Horn, J. P. (1983a). A reclassification of B and C neurones in the ninth and tenth paravertebral sympathetic ganglia of the bullfrog. *J. Physiol. (London)* **334**, 255–269.

Dodd, J., and Horn, J. P. (1983b). Muscarinic inhibition of sympathetic C neurones in the bullfrog. *J. Physiol. (London)* **334**, 271–291.

Dryer, S. E., and Chiappinelli, V. A. (1985). Substance P depolarizes nerve terminals in an autonomic ganglion. *Brain Res.* **336**, 190–194.

Dufy, B., Dufy-Barbe, L., and Barker, J. L. (1983). Electrophysiological assays of mammalian cells involved in neurohormonal communication, *In "Methods in Enzymology."* (P. M. Conn, ed.), Vol. 103, pp. 93–111. Academic Press, New York.

Dunlap, K., and Fischbach, G. D. (1981). Neurotransmitters decrease the calcium conductance activated by depolarization of embryonic chick sensory neurones. *J. Physiol. (London)* **317**, 519–535.

Egan, T. M., and North, R. A. (1986). Acetylcholine hyperpolarizes central neurones by acting on an M_2 muscarinic receptor. *Nature (London)* **319**, 405–407.

Frankenhaeuser, B., and Huxley, A. F. (1964). The action potential in the myelinated nerve fibre of *Xenopus laevis* as computed on the basis of voltage clamp data. *J. Physiol. (London)* **171**, 302–315.

Freschi, J. E. (1983). Membrane currents of cultured rat sympathetic neurons under voltage clamp. *J. Neurophysiol.* **50**, 1460–1478.

Galvan, M., Satin, L. S., and Adams, P. R. (1984). Comparison of conventional microelectrode and whole-cell patch clamp recordings from cultured bullfrog ganglion cells. *Soc. Neurosci. Abstr.* **10**, 146.

Grafe, P., Mayer, C. J., and Wood, J. D. (1980). Synaptic modulation of calcium-dependent potassium conductance in myenteric neurones in the guinea-pig. *J. Physiol. (London)* **305**, 235–248.

Gruol, D. L., Siggins, G. R., Padjen, A., and Forman, D. S. (1981). Explant cultures of adult amphibian sympathetic ganglia: Electrophysiological and pharmacological investigation of neurotransmitter and nucleotide action. *Brain Res.* **223**, 81–106.

Halliwell, J. V., and Adams, P. R. (1982). Voltage-clamp analysis of muscarinic excitation in hippocampal neurons. *Brain Res.* **250**, 71–92.

Hartzell, H. C., Kuffler, S. W., Stickgold, R., and Yoshikami, D. (1977). Synaptic excitation and inhibition resulting from direct action of acetylcholine on two types of chemoreceptors on individual amphibian parasympathetic neurones. *J. Physiol. (London)* **271**, 817–846.

Hashiguchi, T., Kobayashi, H., Tosaka, T., and Libet, B. (1982). Two muscarinic depolarizing mechanisms in mammalian sympathetic neurons. *Brain Res.* **242**, 378–382.

Higashida, H., and Brown, D. A. (1986). Two polyphosphatidylinositide metabolites control two K^+ currents in a neuronal cell. *Nature (London)* **323**, 333–335.

Hescheler, J., Rosenthal, W., Trautwein, W., and Schultz, G. (1987). The GTP-binding protein, G_o, regulates neuronal calcium channels. *Nature (London)* **325**, 445–447.

Hodgkin, A. L., and Huxley, A. F. (1952). A quantitative description of membrane current and its application to conduction and excitation in nerve. *J. Physiol. (London)* **117**, 500–544.

Holz IV, G. G., Rane, S. G., and Dunlap, K. (1986). GTP-binding proteins mediate transmitter inhibition of voltage-dependent calcium channels. *Nature (London)* **319**, 670–672.

Hokin, L. E. (1985). Receptors and phosphoinositide-generated second messengers. *Annu. Rev. Biochem.* **54**, 205–235.

Jan, L. Y., and Jan, Y. N. (1982). Peptidergic transmission in sympathetic ganglia of the frog. *J. Physiol. (London)* **327**, 219–246.

Jan, Y. N., Jan, L. Y., and Kuffler, S. W. (1979). A peptide as possible transmitter in sympathetic ganglia of the frog. *Proc. Natl. Acad. Sci. U.S.A.* **76**, 1501–1505.

Jan, Y. N., Jan, L. Y., and Kuffler, S. W. (1980a). Further evidence for peptidergic transmission in sympathetic ganglia. *Proc. Natl. Acad. Sci. U.S.A.* **77**, 5008–5012.

Jan, Y. N., Jan, L. Y., and Brownfield, M. S. (1980b). Peptidergic transmitters in synaptic boutons of sympathetic ganglia. *Nature (London)* **288**, 380–382.

Jan, Y. N., Bowers, C. W., Branton, D., Evans, L., and Jan, L. Y. (1983). Peptides in neuronal function: Studies using frog autonomic ganglia. *Cold Spring Harbor Symp. Quant. Biol.* **48**, 363–374.

Jones, S. W. (1985). Muscarinic and peptidergic excitation of bull-frog sympathetic neurones. *J. Physiol. (London)* **366**, 63–87.

Jones, S. W. (1987a). A muscarine-resistant M-current in C cells of bullfrog sympathetic ganglia. *Neurosci. Lett.* **74**, 309–314.

Jones, S. W. (1987b). Sodium currents in dissociated bull-frog sympathetic neurones. *J. Physiol. (London)* **389**, 605–627.

Jones, S. W., and Adams, P. R. (1987). The M-current and other potassium currents of vertebrate neurons. *In* "Neuromodulation" (L. K. Kaczmarek and I. B. Levitan, eds.), pp. 159–186. Oxford University Press, New York.

Jones, S. W., Adams, P. R., Brownstein, M. J., and Rivier, J. E. (1984). Teleost luteinizing hormone-releasing hormone: Action on bullfrog sympathetic ganglia is consistent with role as neurotransmitter. *J. Neurosci.* **4**, 420–429.

Kaneko, A., and Tachibana, M. (1985). Effects of L-glutamate on the anomalous rectifier potassium current in horizontal cells of *Carassius auratus* retina. *J. Physiol. (London)* **358**, 169–182.

Katayama, Y., and Nishi, S. (1982). Voltage-clamp analysis of peptidergic slow depolarizations in bullfrog sympathetic ganglion cells. *J. Physiol. (London)* **333**, 305–313.

King, J. A., and Millar, R. P. (1982). Structure of chicken hypothalamic luteinizing hormone-releasing hormone. I. Structural determination on partially purified material. *J. Biol. Chem.* **257**, 10722–10729.

King, J. A., and Millar, R. P. (1985). Multiple molecular forms of gonadotropin-releasing hormone in teleost fish brain. *Peptides* **6**, 689–694.

Koketsu, K. (1984). Modulation of receptor sensitivity and action potentials by transmitters in vertebrate neurones. *Jpn. J. Physiol.* **34**, 945–960.

Kuba, K., and Koketsu, K. (1976). Analysis of the slow excitatory postsynaptic potential in bullfrog sympathetic ganglion cells. *Jpn. J. Physiol.* **26**, 651–669.

Kuffler, S. W., and Sejnowski, T. J. (1983). Peptidergic and muscarinic excitation at amphibian sympathetic synapses. *J. Physiol. (London)* **341**, 257–278.

Land, B. R., Salpeter, E. E., and Salpeter, M. M. (1980). Acetylcholine receptor site density affects the rising phase of miniature endplate currents. *Proc. Natl. Acad. Sci. U.S.A.* **77**, 3736–3740.

Lascar, G., Taxi, J., and Kerdelhué, B. (1982). Localisation immunocytochimique d'une substance du type gonadolibérine (LH-RH) dans les ganglions sympathiques de Grenouille. *C. R. Acad. Sci., Ser. III* **294**, 175–179.

Liles, W. C., Hunter, D. D., Meier, K. E., and Nathanson, N. M. (1986). Activation of protein kinase C induces rapid internalization and subsequent degradation of muscarinic acetylcholine receptors in neuroblastoma cells. *J. Biol. Chem.* **261**, 5307–5313.

Loring, R. H., Sah, D. W. Y., Landis, S. C., and Zigmond, R. E. (1986). Toxin F selectively blocks nicotinic transmission in cultured sympathetic neurons and binds to sites near synapses. *Soc. Neurosci. Abstr.* **12**, 237.

McCormick, D. A., and Prince, D. A. (1986). Acetylcholine induces burst firing in thalamic reticular neurones by activating a potassium conductance. *Nature (London)* **319**, 402–405.

Madison, D. V., and Nicoll, R. A. (1986a). Actions of noradrenaline recorded intracellularly in rat hippocampal CA1 pyramidal neurones, in vitro. *J. Physiol. (London)* **372**, 221–224.

Madison, D. V., and Nicoll, R. A. (1986b). Cyclic adenosine 3′,5′-monophosphate mediates β-receptor actions of noradenaline in rat hippocampal pyramidal cells. *J. Physiol. (London)* **372**, 245–259.

Madison, D. V., Lancaster, B., and Nicoll, R. A. (1987). Voltage clamp analysis of cholinergic action in the hippocampus. *J. Neurosci.* **7**, 733–741.

Malenka, R. C., Madison, D. V., Andrade, R., and Nicoll, R. A. (1986). Phorbol esters mimic some cholinergic actions in hippocampal pyramidal neurons. *J. Neurosci.* **6**, 475–480.

Marshall, L. M. (1981). Synaptic localization of α-bungarotoxin binding which blocks nicotinic transmission at frog sympathetic neurons. *Proc. Natl. Acad. Sci. U.S.A.* **78**, 1948–1952.

Marshall, L. M. (1985). Presynaptic control of synaptic channel kinetics in sympathetic neurones. *Nature (London)* **317**, 621–623.

Marshall, L. M. (1986). Different synaptic channel kinetics in sympathetic B and C neurons of the bullfrog. *J. Neurosci.* **6**, 590–593.

Miyamoto, K., Hasegawa, Y., Nomura, M., Igarashi, M., Kanagawa, K., and Matsuo, H. (1984). Identification of the second gonadotropin-releasing hormone in chicken hypothalamus: Evidence that gonadotropin secretion is probably controlled by two distinct gonadotropin-releasing hormones in avian species. *Proc. Natl. Acad. Sci. U.S.A.* **257**, 10729–10732.

Nakajima, Y., Nakajima, S., Leonard, R. J., and Yamaguchi, K. (1986). Acetylcholine raises excitability by inhibiting the fast transient potassium current in cultured hippocampal neurons. *Proc. Natl. Acad. Sci. U.S.A.* **83**, 3022–3026.

Nargeot, J., Lester, H. A., Birdsall, N. J. M., Stockton, J., Wassermann, N. H., and Erlanger, B. F. (1982). A photoisomerizable muscarinic antagonist. Studies of binding and of conductance relaxations in frog heart. *J. Gen. Physiol.* **79**, 657–678.

Nishi, S., and Koketsu, K. (1960). Electrical properties and activities of single sympathetic neurons in frogs. *J. Cell. Comp. Physiol.* **55**, 15–30.

Nishi, S., and Koketsu, K. (1968). Early and late afterdischarges of amphibian sympathetic ganglion cells. *J. Neurophysiol.* **31**, 109–121.

Nishi, S., Soeda, H., and Koketsu, K. (1965). Studies on sympathetic B and C neurons and patterns of preganglionic innervation. *J. Cell. Comp. Physiol.* **66**, 19–32.

Nishi, S., Katayama, Y., Nakamura, J., and Ushijima, H. (1980). A candidate substance for the chemical transmitter mediating the late slow EPSP in bullfrog sympathetic ganglia. *Biomed. Res.* **1** (Suppl.), 144–148.

Nishizuka, Y. (1986). Studies and perspectives of protein kinase C. *Science* **233**, 305–312.

North, R. A. (1986). Receptors on individual neurones. *Neuroscience* **17**, 899–907.

North, R. A., and Tokimasa, T. (1983). Depression of calcium-dependent potassium conductance of guinea-pig myenteric neurones by muscarinic agonists. *J. Physiol. (London)* **342**, 253–266.

Nowak, L. M., and Macdonald, R. L. (1981). Substance P decreases a potassium conductance of spinal cord neurons in cell culture. *Brain Res.* **214**, 416–423.

Nowak, L. M., and Macdonald, R. L. (1983). Muscarine-sensitive voltage-dependent potassium current in cultured murine spinal cord neurons. *Neurosci. Lett.* **35**, 85–91.

Pennefather, P., Lancaster, B., Adams, P. R., and Nicoll, R. A. (1985a). Two distinct Ca-dependent K currents in bullfrog sympathetic ganglion cells. *Proc. Natl. Acad. Sci. U.S.A.* **82**, 3040–3044.

Pennefather, P., Jones, S. W., and Adams, P. R. (1985b). Modulation of repetitive firing in bullfrog sympathetic ganglion cells by two distinct K currents, I_{AHP} and I_M. *Soc. Neurosci. Abstr.* **11**, 148.

Pfaffinger, P. J., Martin, J. M., Hunter, D. D., Nathanson, N. M., and Hille, B. (1985). GTP-binding proteins couple cardiac muscarinic receptors to a K channel. *Nature (London)* **317**, 536–538.

Rane, S. G., and Dunlap, K. (1986). Kinase C activator 1,2-oleoylacetylglycerol attenuates voltage-dependent calcium current in sensory neurons. *Proc. Natl. Acad. Sci. U.S.A.* **83**, 184–188.

Rogawski, M. A., and Barker, J. L. (1985). "Neurotransmitter Actions in the Vertebrate Nervous Systems." Plenum, New York.

Sherwood, N., Eiden, L., Brownstein, M., Spiess, J., Rivier, J., and Vale, W. (1983). Characterization of a teleost gonadotropin-releasing hormone. *Proc. Natl. Acad. Sci. U.S.A.* **80**, 2794–2798.

Sherwood, N. M., Harvey, B., Brownstein, M. J., and Eiden, L. E. (1984). Gonadotropin-releasing hormone (Gn-RH) in striped mullet (*Mugil cephalus*), milkfish (*Chanos chanos*), and rainbow trout (*Salmo gairdneri*): Comparison with salmon Gn-RH. *Gen. Comp. Endocrinol.* **55**, 174–181.

Sherwood, N. M., Sower, S. A., Marshak, D. R., Fraser, B. A., and Brownstein, M. J. (1986). Primary structure of gonadotropin-releasing hormone from lamprey brain. *J. Biol. Chem.* **261**, 4812–4819.

Sims, S. M., Singer, J. J., and Walsh Jr., J. V. (1985). Cholinergic agonists suppress a potassium current in freshly dissociated smooth muscle cells of the toad. *J. Physiol. (London)* **367**, 503–529.

Smith, S. J., MacDermott, A. B., and Weight, F. F. (1983). Detection of intracellular Ca^{2+} transients in sympathetic neurones using arsenazo III. *Nature (London)* **304**, 350–352.

Soejima, M., and Noma, A. (1984). Mode of regulation of the ACh-sensitive K-channel by the muscarinic receptor in rabbit atrial cells. *Pflugers Arch.* **400**, 424–431.

Stanfield, P. R., Nakajima, Y., and Yamaguchi, K. (1985). Substance P raises neuronal membrane excitability by reducing inward rectification. *Nature (London)* **315**, 498–501.

Tokisama, T., (1984a). Muscarinic agonists depress calcium dependent g_K in bullfrog sympathetic neurons. *J. Auton. Nerv. Syst.* **10**, 107–116.

Tokimasa, T. (1984b). Calcium-dependent hyperpolarizations in bullfrog sympathetic neurons. *Neuroscience* **12**, 929–937.

Tosaka, T., Chichibu, S., and Libet, B. (1968). Intracellular analysis of slow inhibitory and excitatory postsynaptic potentials in sympathetic ganglia of the frog. *J. Neurophysiol.* **31**, 396–409.

Tosaka, T., Takasa, J., Miyazaki, T., and Libet, B. (1983). Hyperpolarization following activation of K+ channels by excitatory postsynaptic potentials. *Nature (London)* **305**, 148–150.

Trussell, L. O., and Jackson, M. B. (1986). A GTP-binding protein mediates an adenosine-activated K+ current independently of cAMP. *Soc. Neurosci. Abstr.* **12**, 15.

Tsunoo, A., Konishi, S., and Otsuka, M. (1982). Substance P as an excitatory transmitter of primary afferent neurons in guinea-pig sympathetic ganglia. *Neuroscience* **7**, 2025–2037.

Vicentini, L. M., De Virgilio, F., Ambrosini, A., Pozzan, T., and Meldolesi, J. (1985). Tumor promoter phorbol 12-myristate, 13-acetate inhibits phosphoinositide hydrolysis and cytosolic Ca^{2+} rise induced by the activation of muscarinic receptors in PC12 cells. *Biochem. Biophys. Res. Commun.* **127**, 310–317.

Watson, S. P. (1983). Rapid degradation of [³H] substance P in guinea-pig ileum and rat vas deferens *in vitro*. *Br. J. Pharmacol.* **79**, 543–552.

Weight, F. F., and Votava, J. (1970). Slow synaptic excitation in sympathetic ganglion cells: Evidence for synaptic inactivation of potassium conductance. *Science* **170**, 755–758.

Weitsen, H. A., and Weight, F. F. (1977). Synaptic innervation of sympathetic ganglion cells in the bullfrog. *Brain Res.* **128**, 197–211.

Werz, M. A., and MacDonald, R. L. (1984). Dynorphin reduces calcium-dependent action potential duration by decreasing voltage-dependent calcium conductance. *Neurosci. Lett.* **46**, 185–190.

Molecular Mechanisms of Neurite Formation Stimulated by Insulin-like Factors and Nerve Growth Factor

DOUGLAS N. ISHII

Department of Physiology and Biophysics
Colorado State University
Fort Collins, Colorado 80523

AND

JOHN F. MILL

Laboratory of Molecular Biology
National Institute of Neurological and Communicative Disorders and Stroke
National Institutes of Health
Bethesda, Maryland 20892

I. INTRODUCTION

Orderly development of the nervous system requires that specific connections be formed between neurons and their often distant target cells. The importance of a neuroendocrine relationship was recognized early in studies with the classic neurotropic protein, nerve growth factor (NGF). The elegant pioneering studies of Rita Levi-Montalcini, Vicktor Hamburger,

Stan Cohen, and their colleagues, together with the contributions of other investigators, have shown clearly that NGF can stimulate axon elongation in sensory and sympathetic neurons and guide the growth cone toward a remote target. The molecular mechanism by which NFG stimulates neurite formation is a partially solved puzzle that continues to fascinate the neurobiologist.

Evidence has begun to accumulate supporting the suggestion that insulin and its homologs may be part of a broad family of neuritogenic polypeptides (Recio-Pinto and Ishii, 1984). There is evidence that neurite assembly is likely to be under multiple regulatory control by polypeptides. This article reviews the data supporting these suggestions, and the molecular data concerned with the initiation of neurite elongation stimulated by NGF and the insulin-like factors. Thus, the discussion will center on the earliest detectable interaction of neuritogenic agents with the cell, namely binding to specific sites, and other early biochemical events clearly in the pathway leading to neurite formation. In this regard, of substantial interest is the effect on tubulin mRNA levels, because of the important contribution that microtubules make to the construction of the axonal cytoskeleton. Phosphorylation appears to be an important transmembrane event shared by neuritogenic agents, but its precise role in neurite formation remains to be determined. Helpful publications may be consulted on other topics relating to neurite growth, such as axonal transport (Grafstein and McQuarrie, 1978; Schwartz, 1979), insertion of membranes into the plasmalemma surface of neurites (Pfenninger and Maylie-Pfenninger, 1981), growth cone function (Tennyson, 1970; Yamada et al., 1971; Bunge, 1973; Bray, 1973), arborization (Lopresti et al., 1973; Shankland and Goodman, 1982; Roederer and Cohen, 1983), adhesion to substrate (Letourneau, 1975a,b; Carbonetto et al., 1982), and synaptogenesis (Purves and Lichtman, 1978; Sonderegger et al., 1983).

II. NEURITOGENIC FACTORS AND DEVELOPMENT OF THE NERVOUS SYSTEM

A. Nerve Growth Factor

NGF's sequence of 118 amino acids has been elucidated; there are three intrachain disulfide bonds (Angeletti and Bradshaw, 1971). NGF is produced in various tissues, and is extracted under neutral conditions from mouse submaxillary glands as a 7 S complex composed of α, β, and γ, subunits (Varon et al., 1967). All of the NGF activity is present in the β-subunit (Stach and Shooter, 1974), which is ordinarily present as a non-covalently linked dimer of 26,518 Da. The other subunits appear to be involved in the processing and storage of β-NGF. The NGF cDNA has been cloned (Scott et al., 1983; Ullrich et al., 1983). The human NGF gene maps to chromosome 1 (Francke et al., 1983).

Administration of NGF into developing vertebrates causes hypertrophy and hyperplasia of sensory and sympathetic ganglia (Levi-Montalcini and Hamburger, 1951; Levi-Montalcini and Cohen, 1960). The hypertrophy and hyperplasia can probably be attributed to accelerated differentiation, generalized effects on cell growth, and enhancement of survival. There is no evidence that NGF is a mitogen. Extensive destruction and atrophy of these tissues can occur on administration of the anti-NGF antiserum (Levi-Montalcini and Booker, 1960; Gorin and Johnson, 1980; Johnson *et al.*, 1980). Although the literature on NGF is extensive and the data highly informative, we shall limit the discussion to its effects on neurite formation and survival, because that literature will provide the conceptual framework for interpretation of much of the new findings concerning insulin and its homologs. The interested reader can consult several excellent reviews for other aspects of the NGF literature (Levi-Montalcini and Angeletti, 1968; Varon, 1975; Mobley *et al.*, 1977; Bradshaw, 1978; Vinores and Guroff, 1980; Harper and Thoenen, 1980).

1. Neurite Outgrowth

Administration of excessive amounts of NGF can cause hyperinnervation of normal and nontarget tissues of sympathetic neurons (Levi-Montalcini and Angeletti, 1968). Even in adult animals, there is an increase of the axonal terminal network (Bjerre *et al.*, 1975). The chemotactic properties of NGF can be demonstrated both *in vivo* and *in vitro*. For example, administration of excessive NGF into the brain results in the abnormal entry of paravertebral sympathetic fibers into and up the spinal cord to the brain stem (Menesini-Chen *et al.*, 1978). NGF can cause neurite outgrowth from cultured sensory and sympathetic neurons; the fibers will grow in the direction of the highest NGF concentration (Campenot, 1977; Gunderson and Barrett, 1980). The importance of these observations becomes evident when it is recognized that target cells can produce (Shelton and Reichardt, 1984) and release NGF. Thus, neurites grow toward explants of target tissues (Chamley *et al.*, 1973). It is believed that NGF produced by target tissues can attract the growing neurite. Once synapses have formed, the cells are ensured of a source of NGF, which undergoes retrograde transport (Hendry *et al.*, 1974), travels back toward the neuronal soma, to support and modify neuronal function. It is suggested that NGF transported in a retrograde fashion is biologically active, because hypertrophy and increased tyrosine hydroxylase activity occur in sympathetic neurons (Hendry, 1977). These profound effects support the assessment that NGF is a neurotropic agent.

2. Survival

NGF's support of sensory and sympathetic cell survival (Levi-Montalcini and Angeletti, 1968) may contribute toward adjustment to the proper number of neurons necessary for function in later life. There is extensive destruction of sensory neurons when anti-NGF antiserum is adminstered to the developing fetus (Johnson *et al.,* 1980). A similar prenatal dependence on NGF can be shown for adrenal chromaffin cells (Aloe and Levi-Montalcini, 1979). The sympathetic neurons develop a requirement for NGF during the late prenatal period (Coughlin *et al.,* 1977). The anti-NGF antiserum can cause loss of about 80% of these neurons when administered early in the postnatal period. Neuroblasts are characteristically overproduced during development, which probably ensures that all available target cells are inner-vated. Subsequently, superfluous neurons are eliminated by a process of pro-grammed cell death (Hamburger and Oppenheim, 1982). The hyperplasia observed when excess NGF is administered to developing vertebrates is the consequence of improved survival of neuroblasts (Charlwood *et al.,* 1974), and a more rapid rate of maturation. A general view that has emerged is that there is competition among certain growing neurites for the source of NGF. Synaptogenesis perhaps secures through retrograde axonal transport the NGF vital for survival, and unsuccessful neurons are among those eventually pruned during development. While this discourse is an oversimplification of the complex events involved in the final sculpting of the nervous system, it does present the critical role that survival factors are thought to play.

B. Insulin and Insulin-Like Growth Factors

Four decades have passed since the momentous discovery of nerve growth factor. It is the recent assessment of Hans Thoenen and colleagues (Barde *et al.,* 1983) that NGF is still the only neurotropic factor with an established physiological role. The discovery of new neuritogenic factors is expected to greatly extend the present understanding of the development of the nervous system. New evidence that indicates that insulin and its homologues belong to a wide family of neuritogenic polypeptides is presented below.

1. THE RAT IGF-II GENE

The amino acid sequence of human IGF-II, or somatomedin A, has been determined (Rinderknecht and Humbel, 1978a,b). Multiplication-stimulating activity is the homologous rat equivalent of IGF-II (Marquardt *et al.,* 1981). IGF-II shares amino acid sequence homology with proinsulin in the A and B domains, suggesting a common gene ancestry (Blundell and Humbel, 1980). The rat IGF-II cDNA has been independently cloned by several groups (Dull *et al.,* 1984; Whitfield *et al.,* 1984; Soares *et al.,* 1985). and there are some differences in the data. The rat pre-pro-IGF-II consists of a signal peptide of 23 residues, the IGF-II sequence, and a trailer polypeptide of 89 residues (Soares *et al.*, 1985). The base sequence (Dull *et al.*, 1984; Soares *et al.*, 1985) agrees with the IGF-II amino acid sequence (Marquardt *et al.*, 1981), except that the first residue in the C domain is a Ser rather than a Gly. The cDNA sequencing data of Dull *et al.* (1984) and Soares *et al.* (1985) for the coding region of the pre-pro-IGF-II are in complete agreement, but differ in six residues from that of Whitfield *et al.* (1984). There were, in addition, disparities in the data of Dull *et al.* (1984) and Soares *et al.* (1985) concerning the sequence of the 5'-noncoding region, which are now resolved (Soaris *et al.,* 1986).

cDNA sequence comparison of the insulin, IGF-I, IGF-II, and relaxin genes (Dull *et al.*, 1984) indicates that the derived proteins form a family of polypeptides that are related in structure, most likely due to a common evolutionary origin. Thus, it is not surprising that high concentrations of insulin and IGFs can cross-occupy one another's receptors, and cause many of the same actions in various cells (Zapf *et al.*, 1978). The human IGF-II and insulin genes map to chromosome 11, and the IGF-I gene maps to chromosome 12 (Brissenden *et al.*,1984; Tricoli *et al.*, 1984). There is some evidence suggesting that NGF may distantly belong to the insulin–IGF–relaxin grouping (Bradshaw, 1978). NGF's amino acid sequence is about 25% homologous to that of proinsulin.

2. EXPRESSION OF IGF-II TRANSCRIPTS IN RAT BRAIN AND OTHER TISSUES

Various rat tissues were examined by Northern blotting, and a 3.4-kb transcript corresponding to pre-pro-IGF-II was detected (Soares *et al.*, 1985). Brain, heart, liver, lung, and muscle are observed to express this transcript in neonatal rats. Interestingly, the highest expression is observed in heart and muscle. There is a developmental change in the expression of the 3.4-kb transcript. For example, the expression of this transcript is greatly diminished in adult muscle and liver. The developmental decrease in transcript levels is consistent with the observation that serum IGF-II levels are higher in neonatal than in adult rat serum (Moses *et al.*, 1980; Daughaday *et al.*, 1982).

Of major interest to this discussion is the expression of the pre-pro-IGF-II transcript in brain. Prior to this observation, there was no conclusive evidence to show that IGF-II is produced in brain, although its presence had been detected by immunochemical methods. In light of this new finding, the blood–brain barrier obviously can pose no impediment to the action of IGF-II in brain. This observation firmly supports the hypothesis that the IGFs have a role in the development of the nervous system, including the central nervous system (CNS).

3. EFFECTS ON NEURITE FORMATION IN SENSORY
   AND SYMPATHETIC NEURONS

Neurons have a high requirement for transferrin (Bottenstein, 1980). Other investigations have, by and large, overlooked the effects of low insulin concentrations, because neurons survive poorly in NGF-deficient cultures, unless supplemented with transferrin. Cultured 12- to 13-day-old embryonic chick sympathetic ganglion cells respond by neurite formation to physiological concentrations of insulin and IGF-II (Recio-Pinto et al., 1986). Concentrations of insulin and IGF-II as low as 0.1 nM can enhance neurite formation, and the half-maximally effective concentration is 0.4–0.6 nM for both polypeptides in sympathetic cell cultures. The sensitivity of response is the same whether neurons are cultured with or without serum; evidently other serum components are not required in the response. The morphology of cultures treated with insulin and NGF is shown in Fig. 1. After several weeks, the neurites form an extensive network that covers the entire culture surface. Cultures treated with the insulin homologs are not obviously different in morphology from NGF-treated cultures.

In sensory cell cultures, the half-maximally effective concentration is 0.1 nM for IGF-II. This sensitivity is about the same as in sympathetic cell cultures. In contrast, insulin is 100-fold less potent in dorsal root ganglion cell cultures. We speculate that this difference is due to either a tissue specificity, or age dependence, in the response to insulin. With respect to the latter possibility, the sensory ganglion cultures were from slightly younger (10-day-old) embryos.

4. EFFECT OF INSULIN AND IGF-II ON SURVIVAL OF NEURONS

NGF has an important effect on the survival of sensory and sympathetic neurons, as discussed. This important property is potentially shared with the insulin homologs. The effect of various concentrations of insulin on survival of sympathetic neurons was compared to the effects of NGF (Recio-Pinto et al., 1986). Insulin and IGF-II both support the survival of sympathetic neurons. Insulin supports the survival of about half as many neurons as does NGF. The combination of insulin and NGF appears to have no greater effect than NGF alone. In addition, insulin and IGF-II both enhance the survival of sensory neurons.

5. INSULIN ACTS DIRECTLY ON NEURONS

Since insulin acts on a variety of cell types, it was important to exclude the possibility that insulin might act directly on the ganglionic nonneuronal cells, which, in turn, might mediate an indirect response from the neurons.

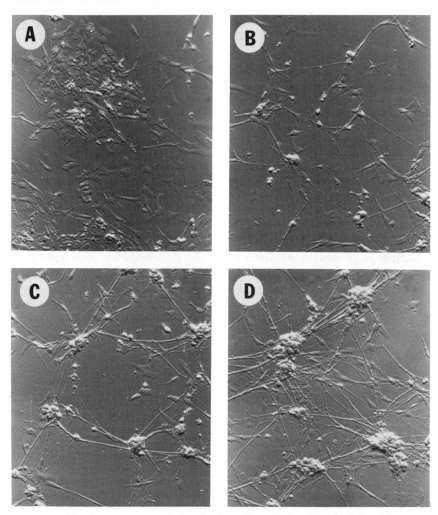

FIG. 1.   Morphology of sympathetic cell cultures from 12-day-old embryonic chicks treated with insulin and NGF. The cells were cultured for 2 days in the L-15 plating medium (supplemented with 0.1 mg/ml transferrin and 10% fetal calf serum) under conditions described previously (Recio-Pinto *et al.*, 1986). (A) Untreated; (B) plus 1 μM insulin; (C) plus 0.4 nM NGF; (D) plus 1 μM insulin and 0.4 nM NGF. Modulation contrast, × 164.

Nonneuronal cells can be virtually eliminated from cultures by treatment with Methocel (Mains and Patterson, 1973) or cytosine arabinofuranoside. The latter is toxic to proliferating nonneuronal cells, but not to the quiescent neurons. Under both conditions, insulin can stimulate neurite outgrowth (long interconnecting networks formed after several weeks), and enhance neuronal survival (Recio-Pinto *et al.*, 1986). Therefore, insulin acts directly on the neurons.

6. INSULIN HOMOLOGS AND NGF
CAN ACT ON THE SAME NEURONS

The combination of insulin and NGF does not increase the *number* of surviving neurons in culture over that in cultures treated with NGF alone. The lack of additive effect suggests either that insulin is less efficacious and can support the survival of fewer NGF-responsive neurons, or that insulin acts on a subpopulation of NGF-responsive neurons. Whichever may be correct, it is evident that insulin and NGF can both act on at least some of the same neurons. This inference is supported by the observation that insulin, IGF-II and NGF can all act on the same neuroblastoma cell (Recio-Pinto and Ishii, 1984).

7. INSULIN STIMULATES PRECOCIOUS NEUROTRANSMISSION
IN RETINAL CELL CULTURES

Retinal ganglion cells will form functional synapses when cocultured with striated muscle cells (Puro *et al.*, 1982). Treatment of cocultures with low concentrations of insulin causes a precocious transition of the cultures from a non-transmitting stage to one in which stimulation of the cholinergic retinal neurons leads to acetylcholine release (Puro and Agardh, 1984). This induction of evocable synaptic transmission can be blocked by cycloheximide, but not by actinomycin D, suggesting that protein synthesis, but not transcription, is required. These results provide further support for the hypothesis that the insulin homologs play a role in the development of the nervous system.

III. NEURITE FORMATION IN HUMAN NEUROBLASTOMA CELLS

An impediment to the study of the mechanism of NGF and IGF action in sensory and sympathetic neurons is the requirement of these factors for survival. Dying, untreated cultures are not adequate as experimental controls. Moreover, the heterogeneity in cell populations can complicate further the interpretation of results. A reasonable alternative experimental system is proffered by human neuroblastoma cells, which can survive under serum-free conditions for at least a week without a loss in numbers (Sonnenfeld and Ishii, 1982). The cells resume growth when serum is reintroduced. The various attractive properties of the human neuroblastoma SH-SY5Y cell for the study of neurite outgrowth under multiple regulatory control have been discussed (Ishii *et al.*, 1985).

A. Effects of Nerve Growth Factor

SH-SY5Y is a cloned cell line (Biedler *et al.*, 1978) responsive to NGF by enhanced neurite outgrowth (Perez-Polo *et al.*, 1979; Sonnenfeld and Ishii,

1982), veratridine-dependent Na^+ uptake (Perez-Polo et al., 1979), and protein content (Sonnenfeld and Ishii, 1982; Spinelli et al., 1982). The neurites have an ultrastructure similar to that of developing sympathetic ganglion cells, and end in structures typical of neuronal growth cones (Burmeister and Lyser, 1982). Enhanced ultrastructural maturation can be shown following treatment with NGF (Perez-Polo et al., 1979). Cell lines responsive and resistant to neuritogenic agents such as NGF (Sonnenfeld and Ishii, 1982), tumor promoters (Spinelli et al., 1982), and insulin (Recio-Pinto and Ishii, 1984) are described.

B.   Effects of Insulin and IGF-II

Physiological concentrations of insulin homologs can reversibly stimulate neurite outgrowth in SH-SY5Y cells, whether serum is present or not (Recio-Pinto and Ishii, 1984). The average length of neurites and the proportion of cells in culture with neurites is increased. The half-maximally effective insulin concentration is about 4 nM, but this value is reduced to 0.1 nM, when bacitracin, a protease inhibitor, is present in cultures. The value for IGF-II is 0.5 nM, and its dose–response curve is also shifted to the left when bacitracin is present. These results in the clonal SH-SY5Y cells show further that insulin homologs can directly enhance neurite outgrowth.

Insulin stimulates the increased incorporation of radioactive leucine and uridine into the acid-insoluble fraction. Cycloheximide inhibits the increased incorporation of radioactive leucine and neurite outgrowth, the latter reversibly. The inhibition of leucine incorporation is probably reversible also, but was not studied. The enhanced incorporation of radioactive uridine, and neurite outgrowth, are inhibited by actinomycin D. Other RNA synthesis inhibitors, such as 5-bromo-2'-deoxyuridine, also antagonize insulin-mediated neurite outgrowth. Thus, it appears that rapid protein and RNA synthesis is required for the neurite outgrowth response to insulin.

C.   Specificity of Insulin and IGF-II Effects

Insulin and IGF-II preparations possibly might be contaminated with NGF. Various tests clearly exclude this possibility.

First, synthetic human insulin is as potent as porcine and bovine insulin (Recio-Pinto and Ishii, 1984). It is improbable that synthetic insulin contains the same amount of an active contaminant as preparations obtained from tissue sources. The synthetic insulin was produced by bacterial synthesis of A and B chains, proper linkage of chains and purification. The product was free of proinsulin; bacterial proteins and small peptides were not detected.

Anti-NGF antiserum (Ishii and Shooter, 1975) inhibits the neurite outgrowth response to NGF in sympathetic and SH-SY5Y cells, but not to insulin or IGF-II (Recio-Pinto and Ishii, 1984; Recio-Pinto et al., 1986). In contrast, antiinsulin antiserum inhibits the response to insulin.

Calculations show that if NGF were an active contaminant, it would have to constitute 22% of the weight of insulin preparations to cause the observed effects. Nevertheless, insulin and IGF-II concentrations orders of magnitude greater than those necessary to enhance neurite formation fail to inhibit the binding of ^{125}I-labeled NGF to its specific sites. IGF-II (Recio-Pinto and Ishii, 1984) is more potent than NGF (Sonnenfeld and Ishii, 1982) on a weight basis. Therefore, the effects of IGF-II cannot be explained on the basis of contamination, even if all of it were NGF.

Finally, there is a qualitative difference in the effects of insulin, IGF-II, and NGF on neurite outgrowth. Under serum-free conditions, SH-SY5Y cells are responsive to insulin homologs. In contrast, NGF is inactive under these conditions. The reason for this shall be discussed below.

IV. MECHANISM OF NEURITE FORMATION

A. Receptors Modulating Neurite Formation

It is appropriate to discuss the properties of the specific receptors through which the initial encounter between the neuritogenic agent and the cell is made. What is the evidence that particular binding sites are indeed receptors? What are the properties of the receptor classes that mediate similar functions? Are similar cellular events triggered?

1. NGF BINDING SITES

a. Evidence for Two Types of NGF Binding Sites. Several groups have contributed to the initial detection of NGF binding sites (Herrup and Shooter, 1973; Banerjee et al., 1973; Frazier et al., 1973). Binding is heterogeneous, as revealed both by nonlinear Scatchard plots and complex dissociation rates, and was initially thought to be due to negative cooperativity (Frazier et al., 1973). The heterogeneous binding might alternatively be due to multiple classes of binding sites. This latter possibility is now greatly favored. Evidence for two types of binding sites has been obtained in sensory (Sutter et al., 1979), sympathetic (Olender and Stach, 1980), PC12 (Schechter and Bothwell, 1981), and human neuroblastoma (Sonnenfeld and Ishii, 1982, 1985) cells. The higher and lower affinity sites are referred to as Type I and Type II, respectively (Sutter et al., 1979). Sometimes, they are more descriptively referred to as *slow* and *fast* NGF sites, respectively (Schechter and Bothwell, 1981), based on their relative dissociation rates. Furthermore, the dissociation rates of the two sites differ in sensitivity to temperature, the slow sites being more sensitive. These points are illustrated in Fig. 2. When the ^{125}I-labeled NGF concentration is sufficiently high to permit binding to both sites, the dissociation rate from PC12 cells at 37 °C is biphasic. At lower ligand concentrations, the rate is monophasic. The dissociation rate from fast NGF sites is very rapid, estimated from other experiments to be 40–50 seconds at both 4 and 37 °C (Sutter et al., 1979; Sonnenfeld and Ishii, 1985). On the other hand, dissociation from the slow NGF sites has a half-time of about 25.6 minutes ($N = 6$ experiments) at 37 °C, but becomes extremely slow at 4 °C.

The stability of the binding to slow NGF sites at the lower temperature provides the basis for separate detection of the two types of sites by an assay (Sonnenfeld and Ishii, 1982, 1985), which is a modification of the method used by Sutter et al. (1979). Rapid equilibrium sedimentation is necessary for full retention of binding to fast NGF sites. The use of needle-nosed microcentrifuge tubes facilitates the detection of fewer than 1000 sites per cell.

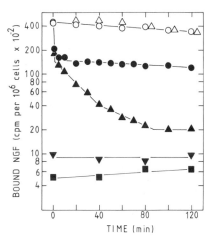

FIG. 2. Time course of loss of bound ^{125}I-labeled NGF from PC12 cells at 4 and 37 °C. The cells (2 × 10^6/ml) were preincubated for 1 hour in Roswell Park Memorial Institute (RPMI) medium 1640 that contained 5 mg/ml bovine serum albumin, 25 mM HEPES, buffered at pH 7.4, and 10 ng/ml ^{125}I-labeled NGF at 37 °C. The experiment was initiated by dividing and treating the cultures as follows: incubated at 37 °C (△); incubated at 4 °C (○); incubated at 37 °C with 10 μg/ml NGF (▲); and incubated at 4 °C with 10 μg/ml NGF (●). Cultures preincubated with 10 μg/ml NGF from the onset were simply continued without change: 37 °C (▼); 4 °C (■). Other methods were as described elsewhere (Sonnenfeld and Ishii, 1985). Note that NGF bound to the fast sites dissociates rapidly at both high and low temperatures, but NGF bound to the slow sites is quite stable at low temperature. This forms the basis for the separate assay of fast and slow NGF sites. Incubation for 10 minutes at 4 °C with 10 μg/ml NGF results in total displacement of fast NGF sites and leaves the "residual" binding, composed of slow NGF sites and nonspecific binding.

The use of this assay is shown in the time course (Fig. 3) and Scatchard plot (Fig. 4) experiments. In the time course experiment (Fig. 3), total binding may be resolved into fast and predominantly slow NGF components. Binding to the slow NGF component is temperature sensitive, and less binding is observed at the lower temperature. At 37 °C, there is a peak in total binding followed by a decline, which can be attributed to the fast NGF sites. Binding to the slow NGF sites plateaus and appears to be relatively stable after 30 minutes.

The complex total binding observed in Scatchard plots is also consistent with two types of sites (Fig. 4). Following binding of ^{125}I-labeled NGF at 37 °C, the plot of total binding is shown to be curvilinear. However, the separate contribution of the two types may be found by rapidly cooling the cells and adding excess NGF. All of the fast ^{125}I-labeled NGF binding is rapidly and quantitatively lost, and the residual radioactivity shown by the high-affinity Scatchard line represents binding to the slow NGF sites. Here, there were about 3600 slow NGF sites per cell with $K_d = 3 \times 10^{-11}M$. The fast sites are calculated as the difference between total and residual binding. There were about 20,000 fast sites per cell with $K_d = 1.4 \times 10^{-9}M$. Later it will be shown for the SH-SY5Y cell that the number of sites detected is dependent on the culture conditions.

Alternatively, the upwardly concave Scatchard plot of total binding (Fig. 4) may be an indication of negative cooperativity. Negative cooperativity has been reported in the binding of other polypeptides (De Meyts, 1976). In negative cooperativity, the decreased average affinity arising from increased fractional occupancy of sites is assumed to be at least partially due to an increased dissociation rate of bound ligand. The addition of excess nonradioactive ligand would, therefore, increase the dissociation rate over that measured by infinite dilution alone. In fact this type of behavior is observed (Fig. 5). However, the loss rate measured by infinite dilution alone is additionally dependent on the cell concentration, being slower with increasing cell density. This could occur only if there were release and rebinding of ^{125}I-labeled NGF; higher cell densities would release larger concentrations of ^{125}I-labeled NGF for rebinding, resulting in an apparent slower rate of dissociation. We, and others, have observed that NGF released from cells, in fact, can be rebound. These results show that in this type of experiment one must be concerned not only with the extent of dilution of the free ligand, but also with the final site density. The curvilinear Scatchard plot, then, does not seem to arise as a case of negative cooperativity, and is more likely due to multiple types of sites.

The association rates have been measured (Sonnenfeld and Ishii, 1985), and the ratio of rate constants gives values that closely agree with the K_d values obtained from Scatchard plots. The K_d value of slow NGF sites is 1–3 × $10^{-11}M$ in sensory (Sutter et al., 1979; Olender et al., 1981) and sympathetic

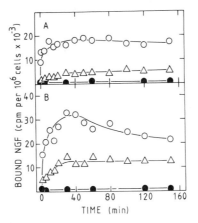

FIG. 3. Time course and temperature dependence of ^{125}I-labeled NGF binding to PC12 cells. The cells (2×10^6/ml were incubated at 4 (A) or 37 °C (B) in the medium described in Fig. 2 together with 10 ng/ml ^{125}I-labeled NGF, both in the presence and absence of 10 μg/ml NGF. At various times, aliquots were removed for assay of total (○) and residual (△) radioactivity (see Fig. 2). The fast sites are measured as the difference between total and residual binding. Slow sites are the differences between residual and nonspecific (●) binding. Note that binding to the slow NGF sites is temperature sensitive. There is a down modulation of binding which is confined to the fast NGF sites.

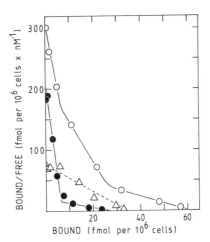

FIG. 4. Scatchard plot of ^{125}I-labeled NGF binding in PC12 cells. The cells were incubated for 1 hour at 37 °C with various concentrations of radiolabeled NGF. The cells were rapidly assayed for total (○), residual (●), and fast NGF site (△) binding, as described in Figs. 2 and 3. The cell-free supernatant was assayed to find the free ^{125}I-labeled concentration. The concentrations of bound NGF were computed using M_r of 26,500 for the NGF dimer. The curvilinear plot of total binding is due to multiple types of sites. In other cell lines in which only the slow NGF sites are present, only the high-affinity Scatchard line is detected (Sonnenfeld and Ishii, 1985).

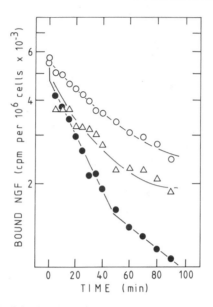

Fig. 5.   Effect of cell density, infinite dilution, and infinite dilution with added excess NGF on the dissociation rate of bound ^{125}I-labeled NGF in PC12 cells. The cells (2 × 10^6/ml) were incubated with 10 ng/ml ^{125}I-labeled NGF for 1 hour at 37 °C. The cells were washed three times by successive centrifugation at 600 g and resuspension in 5 ml cold RPMI 1640 medium containing 1 mg/ml serum albumin. It was estimated that the free radioactivity was diluted by a factor of one million. The cells were finally resuspended in RPMI 1640 medium with 5 mg/ml albumin, divided into aliquots, and incubated at 37 °C under the following conditions: 4 × 10^6 cells/ml (○); 1 × 10^6 cells/ml (△); 4 × 10^6 cells/ml plus 10 μg/ml NGF (●). Radioactivity remaining bound to the cells at various times was assayed. The dissociation rate is dependent on the cell concentration in "infinite dilution," and is incompatible with the concept of negative cooperativity.

(Olender and Stach, 1980) cells. That of fast NGF sites is 0.5–2 × 10^{-9}M. Because fast and slow NGF sites have the same association rates in sensory (Sutter *et al.*, 1979), human neuroblastoma, and PC12 (Sonnenfeld and Ishii, 1985) cells, they differ in affinity largely on the basis of distinctly separate dissociation rate constants. It has been suggested that the rate constants of association to fast and slow sites in PC12 cells are unequal (Schechter and Bothwell, 1981). However, the rates, rather than the rate constants, were measured, leading in turn to a miscalculation of equilibrium constants.

The fast and slow NGF sites in intact cells differ in other fundamental ways. The former has greater sensitivity to trypsin (Schechter and Bothwell, 1981). The kinetic parameters, and differential sensitivity to trypsin, of fast and slow NGF sites found in intact sensory, PC12, human neuroblastoma, and human melanoma cells are retained in their isolated membrane fractions (Riopelle *et al.*, 1980; Lyons *et al.*, 1983; Buxser *et al.*, 1983).

Therefore, the binding parameters of fast and slow NGF sites in sensory, sympathetic, human neuroblastoma, and PC12 cells are in many ways similar, and a unified view becomes evident. Of course, the precise association and dissociation rates may vary somewhat between different types of intact cells, partially the consequence of measuring the binding of mouse NGF to chick, rat, and human tissues. Another possibility that has not received careful study is that the binding is dependent on a rapid activation and inactivation. If this were the case, the differential kinetics in intact cells may reflect differences in metabolic rates. The binding parameters may also vary somewhat between sites measured on intact cells and in extracts, due to a change in the physical properties of the microenvironment.

b. *Size of NGF Binding Sites.* The NGF binding sites solubilized from rabbit sympathetic ganglia have a molecular weight of about 135,000, based on their apparent hydrodynamic radius (Costrini *et al.*, 1979). Cross-linking studies reveal two labeled species of M_r 112,000 and 143,000, and there is some indication that the smaller is derived from the larger species through limited proteolysis (Massague *et al.*, 1981). Cross-linking studies in chick sensory ganglion cells show the presence of two major labeled bands on sodium dodecylsulfate–polyacrylamide gel electrophoresis (SDS-PAGE) of M_r 105,000 and 145,000 (Hosang and Shooter, 1985). The lower molecular weight form is believed to represent NGF cross-linked to fast sites.

Only a single NGF–receptor complex of M_r 148,000–158,000 was initially detected in PC12 cells (Massague and Czech, 1982), but a subsequent preliminary report indicates the presence of M_r 98,000, 138,000, and 190,000 complexes (Puma *et al.*, 1983). Hosang and Shooter (1985) have detected bands of M_r 100,000, 158,000, and 225,000. Binding to the M_r 158,000 band is trypsin resistant, and is retained when cells are incubated at 0 °C with excess NGF prior to cross-linking, suggesting that this species is the slow NGF site. The M_r 100,000 band is believed to be the fast NGF site. The M_r 225,000 species is speculated to be a precursor. In considering molecular sizes estimated from cross-linking studies, it should be kept in mind that monomeric or dimeric NGF may be cross linked. Thus, the lower limit of sizes for the unbound binding sites is estimated as M_r 87,000 and 143,000. A fragment M_r 10,000 can be removed from both sites by treatment with neuraminidase.

Following a 1500-fold purification by affinity chromatography, two major NGF binding peptides of M_r 85,000 and 200,000 are observed in melanoma cell extracts (Puma *et al.*, 1983). Binding to melanoma cells has properties similar to that of the fast NGF sites in other cells, but the biological function of NGF sites on melanoma cells is unclear. Monoclonal antibodies to the A875 melanoma cell NGF-binding component have been raised (Ross *et al.*, 1984). Two NGF-binding species are found to be immunoprecipitated, and their apparent molecular weights are 75,000 and

150,000 when electrophoresed on nonreducing gels (Grob et al., 1985). The inclusion of 2-mercaptoethanol in the gel results in a single species of 75,000 Da. The authors suggest that the binding component can exist in either the monomeric or the covalently linked dimeric form.

Overall, it appears that the fast NGF site is M_r 85,000–105,000, the slow site is suggested to be 135,000–158,000, and there is another binding form of 190,000–225,000. Whether the three species arise from separate genes, truncated forms of the same transcript, from posttranslational processing, or a combination of these events has yet to be decided. Another possibility, because NGF is internalized, is that certain cross-linking agents that can readily enter cells may detect other forms of binding, perhaps to complexes involved in NGF degradation.

   c. *Relationship of Fast and Slow NGF Sites to Each Other.* Because there are two types of sites, there is the potential that the sites are related. Landreth and Shooter (1980) have found that binding to fast NGF sites in PC12 cells develops more rapidly, and they observed a lag in binding to slow sites. Following the initiation of binding, when cells are resuspended in fresh medium without radioactive NGF, there is a subsequent redistribution such that binding to slow NGF sites occurs at the expense of binding to fast sites. These observations have lead Landreth and Shooter (1980) to propose that NGF binding can convert fast sites to slow sites.

   More recently, it was found that while certain lines of human neuroblastoma cells, such as MC-IXC, have both fast and slow NGF sites, some lines, such as SH-SY5Y, have only slow NGF sites (Sonnenfeld and Ishii, 1982). Moreover, binding of NGF to slow sites is not prevented by the complete removal of fast NGF sites with trypsin in PC12 (Schechter and Bothwell, 1981) and MC-IXC (Sonnenfeld and Ishii, 1985) cells. These findings suggest that prior NGF binding to fast sites is not required for binding to slow sites. One might argue, however, that a few fast sites escape trypsin treatment. If such fast sites were rapidly and completely converted to slow sites, it might appear that fast sites were no longer present. This argument does not hold, as discussed below.

   Binding to PC12 cells was reexamined to determine whether there indeed is NGF-mediated conversion of low- to high-affinity sites (Sonnenfeld and Ishii, 1985). Why does binding to fast NGF sites develop more rapidly? There are many, many more fast than slow NGF sites. Thus, simply as a consequence of the law of mass action, binding would be expected to occur more rapidly to fast sites. One must keep in mind that this observation merely reflects the relative *rate* of binding, a parameter dependent on the ligand and binding-site concentrations. What is important is that the *rate constants* of association of the fast and slow sites are essentially identical (Sutter et al.,1979; Sonnenfeld and Ishii, 1985). Therefore, the observation (Landreth and Shooter, 1980), that the rate of NGF binding to fast sites is more rapid does not support the conversion hypothesis.

Is the NGF redistribution experiment of Landreth and Shooter (1980) susceptible to reasonable interpretation other than conversion? Models of efficient and inefficient conversion were considered, but support for the conversion theory is not found (Sonnenfeld and Ishii, 1985). For example, conversion could be very inefficient and dependent on a rapid turnover of binding to fast sites. If this were correct, the association rate constant of slow sites would be substantially smaller than that of fast sites, which is not the case. Alternatively, conversion could be efficient. Most fast sites would then be rapidly converted to slow sites. In a model of efficient NGF-mediated conversion, there would be no release of NGF from fast sites during rapid conversion to slow sites, which might explain why the association rate constants are the same. However, it was shown that the "conversion" is completely inhibited by the presence of excess unlabeled NGF (Sonnenfeld and Ishii, 1985). Therefore, prebound ^{125}I-labeled NGF cannot be directly converted from fast to slow sites. Prebound radioactive NGF has to be released prior to conversion.

The most satisfactory and straightforward explanation of these observations is that the ^{125}I-labeled NGF bound to the large numbers of fast NGF sites is released to the incubation medium, then is simply rebound by the more limited numbers of high-affinity slow sites. The very rapid dissociation from fast sites would tend to facilitate this process. The addition of excess NGF would inhibit the reuptake process. NGF released from fast sites in this type of experiment can readily be taken up by slow sites. Indeed, the rebinding of ^{125}I-labeled NGF is inherent in the concept of the "unstirred layer" (Shooter *et al.*, 1981).

Other indications that NGF-mediated conversion does not occur include the observation that the association rates of both fast and slow sites in PC12 and neuroblastoma cells are consistent with simple bimolecular processes (Sonnenfeld and Ishii, 1985). At $1.9 \times 10^{-11} M$ ^{125}I-labeled NGF, binding to the fast sites is evenly distributed to the soma and nerve fibers of sensory neurons (Carbonetto and Stach, 1982). In contrast, slow sites are five-fold more numerous on the nerve fibers. This assymmetrical distribution does not lend itself to a simple model of conversion. Finally, the fast sites in PC12 cells are suggested to be M_r 100,000 and the slow sites M_r 158,000. Assuming that this difference in size is correct, we are unaware of an example of ligand-mediated conversion in which so substantial an increase in covalent size can occur. These observations do not, however, rule out the possibility of other relationships between the fast and slow sites. For example, they might be related as two distinct binding forms of the same gene product.

d. Relationship of Fast and Slow Sites to Neurite Formation in Sensory, Sympathetic, and PC12 Cells. The presence of two types of binding sites has complicated the definitive assignment of one or both sites as being the receptors for NGF-mediated neurite formation in sensory, sympathetic, and PC12 cells. Tight binding, saturability, and inability of other polypeptides to compete for binding are properties consistent with receptors. It is, nevertheless, recognized that, in themselves, these properties do not prove sites are receptors. Indeed, these properties are even observed in the binding of polypeptides to noncellular materials (Cuatrecasas and Hollenberg, 1975). Thus, these sites alternatively may be transport, degradative, or nonspecific limited-capacity sites (Goldstein *et al.*, 1971). The binding and response can be inhibited when NGF is modified by exposure to bromosuccinimide (Banerjee *et al.*, 1973; Cohen *et al.*, 1980a) or through complexation to monoclonal antibodies (Zimmermann *et al.*, 1981). This does not necessarily prove that the sites to which binding is inhibited are receptors, because the modification of NGF might inhibit binding as well to other undetected sites which may be the true receptors. Certain tumor-promoting artificial sweeteners, such as saccharin, can reversibly inhibit NGF binding and associated neurite outgrowth in sensory ganglia (Ishii, 1982a,b). This pharmacological antagonism does indicate that at least one of the two sites is indeed the receptor for this function. However, none of the studies above reveals conclusively which of the two sites is the receptor in sensory, sympathetic, or PC12 cells.

It might be thought that the correlation of occupancy and response would reveal which of the two sites is the receptor. The saturation isotherm for total NGF binding does not fit well the dose–response relationship, shown for PC12 cells in Fig. 4. However, when the individual saturation isotherms of fast and slow NGF sites are considered separately, that of the slow NGF sites appears to correlate best with the dose-dependent neurite outgrowth. This is consistent with the observation that the K_d of slow NGF sites (Sutter *et al.*, 1979) is close to the concentration of NGF that provokes a half-maximal neurite outgrowth response in cultured sensory ganglia (Ishii, 1978). At NGF concentrations close to the K_d of slow sites, only a small fraction of the fast sites is estimated to be occupied (Sutter *et al.*, 1979). While this correlation supports the suggestion that the slow NGF sites are the receptors for neurite formation, support is provided only within the limited context of simple occupancy theory, but not of "spare receptor" theory (Stephenson, 1956). For example, if spare receptors were present, a full neurite outgrowth response could occur when only a fraction of available fast NGF sites was occupied. Even at low NGF concentrations, a significant number of fast sites may be occupied because they are present in numbers at least 10-fold greater than the slow sites (Sutter *et al.*, 1979; Claude *et al.*, 1982). Thus, the relative importance of fast and slow NGF sites to functions in sensory, sympathetic, and PC12 cells is not completely established.

e. The Slow Sites Are the NGF Receptors for Neurite Formation in Human Neuroblastoma Cells. The absence of fast NGF sites in SH-SY5Y cells has permitted the assignment of the slow NGF site as the putative receptor for NGF-enhanced neurite formation, protein synthesis, veratridine-dependent Na^+ uptake, and other responses (Sonnenfeld and Ishii, 1982, 1985). However, the absence of fast sites does not quite prove by simple default that slow NGF sites are the receptors, even though their K_d fits closely the half-maximally effective concentration of NGF for neurite outgrowth. The strongest support for the argument, in fact, is derived instead from studies which show concomitant, reversible inactivation of slow NGF binding and the neurite outgrowth response in SH-SY5Y cells (Recio-Pinto et al.,1984). These studies will be considered next.

f. Insulin and IGF-II Can Concomitantly Modulate Binding to Slow Sites and Response to NGF. When cultured under serum-free conditions, SH-SY5Y cells lose their capacity to respond to NGF by neurite outgrowth (Recio-Pinto et al., 1984). This is not due to a generalized loss in ability to form neurites in serum-free medium, because the response to insulinlike factors (Recio-Pinto and Ishii, 1984) and tumor promoters (Spinelli and Ishii, 1983) is retained. The loss can be reversed by the addition of a small amount of insulin. The addition of 1 nM insulin by itself causes but a small increase in neurite formation. In the presence of 1 nM insulin, the response to various concentrations of NGF was tested in serum-free medium. There was a synergistic potentiation of neurite formation. NGF was now found to provoke a dose-dependent increase in neurite outgrowth. IGF-II could also prevent the loss of responsiveness to NGF in serum-free medium. The insulinlike factors ordinarily endogenous to serum probably help to mediate the response to NGF. Several observations support this likelihood. For example, the spontaneous level of neurite formation in serum-containing cultures can be decreased by the addition of antiinsulin antiserum. Moreover, the normal response to NGF in serum-containing cultures is inhibited by the antiinsulin antiserum. The antiserum is specific, for it does not inhibit the response to tumor promoters. The synergistic potentiation of neurite outgrowth could also be shown in PC12 cells.

The mechanism by which insulin modulates the response to NGF was studied (Recio-Pinto et al., 1984). It was shown that SH-SY5Y cells cultured in serum-free medium concomitantly lose the capacity to bind ^{125}I-labeled NGF to the high-affinity slow sites, along with loss of the neurite formation response. The addition of serum, or insulin, causes the binding capacity to return. In cells cultured in serum-containing medium supplemented with insulin, there is an increase in the number of slow NGF sites. The affinity constant and dissociation rate of the slow NGF sites are unchanged. The increase in the number of sites is evident on both a per-cell and per-milligram-protein basis. IGF-II also can increase the binding capacity of slow NGF sites in SH-SY5Y cells. Insulin increases NGF binding capacity in other lines of human neuroblastoma cells as well. RNA and protein synthesis seem to be required, but it is not presently known whether insulin homologues reactivate preexisting sites, or increase the synthesis of new sites. Whatever the mechanism, it is evident that the corresponding activation and inactivation of slow sites, and the corresponding activation and inactivation of the neurite outgrowth response, provide potent support for the argument that the slow sites are, indeed, the NGF receptors in SH-SY5Y cells. These results further show that neurite formation may be under complex regulation by multiple soluble proteins.

g. A Speculative Role for Fast NGF Sites. The absence of fast sites on responsive lines of human neuroblastoma cells indicates they are not directly required for the neurite outgrowth response. But, given their large numbers and relatively high affinity, it would be illogical to suppose they played no role in the response to NGF. One possible function is that fast sites regulate other NGF responses, such as neuronal survival. However, one might include for consideration roles not unique to neurons, particularly when the distribution of fast NGF sites is recalled. Fast NGF sites are present on ganglionic satellite cells, as well as on cells of neuronal lineage (Sutter *et al.*, 1979; Carbonetto and Stach, 1982). Fast NGF sites may serve to concentrate NGF in the microenvironment of the responsive neuron. This could provide a biologically important mechanism to augment the binding of a polypeptide, present only in nanomolar concentrations, to the limited number of slow sites. This putative role in augmenting NGF binding might also explain why fast sites are present on satellite cells. For example, fast sites are prominent on Schwann (Zimmermann and Sutter, 1983) and Schwannoma (Zimmermann and Sutter, 1983; Sonnenfeld *et al.*, 1986) cells.

The theories of contact guidance (Letourneau, 1975a,b) and chemotactic guidance by a soluble factor (Gundersen and Barrett, 1980) are not necessarily mutually exclusive. A formulation incorporating both concepts may be proposed. In contact guidance, it is theorized that the growth cone has greater adhesion along certain cellular "tracks," which serve to guide the axon toward a distant target. It has long been observed that regrowth of severed nerves follows the path marked by the residual Schwann cells. In a related observation, brain neurons climb along glial processes during development (Rakic, 1981). It is possible that diffusible NGF is concentrated along "tracts" by numerous cell-surface sites with the affinity and rapid dissociation rates characteristic of fast NGF sites. A slow dissociation rate would not adequately serve to present the free NGF constantly required for rebinding by the higher affinity slow sites. The increased concentration of NGF in the microenvironment would then act to chemotactically guide the advancing growth cone in a manner which includes a form of contact guidance. In fact, neurites preferentially grow over cultured nonneurons which have the fast NGF sites (Zimmermann and Sutter, 1983). The dimeric form for NGF is biologically active (Rice and Stach, 1976). A bivalent molecule might be available for binding jointly to the cell surface and the advancing growth cone. Shared binding to a bivalent molecule may help promote the adhesion of a growth cone along the surface of an existing pioneer neurite, and promote cohesion between neuritic surfaces, leading to nerve fiber bundle formation. Of course, this mechanism does not require that a neuritogenic agent be dimeric. Among other factors, the cell-adhesion molecule (CAM) is recognized as playing a prominent role in contact guidance (Rutishauser et al., 1978).

2. Insulin and IGF Binding Sites

a. Insulin Binding Sites in Brain. Regions of the brain that lack tight capillary junctions, such as the infundibular hypothalamus, are accessible to circulating insulin. Systemic administration of radioactive insulin results in the limited labeling of the circumventricular organs of the brain, and the endothelial microvessels (van Houten et al., 1979). Other studies suggest insulin binding sites may be more widespread, and binding to crude brain homogenates (Havrankova et al., 1978a; Pacold and Blackard, 1979) and plasma membrane (Posner et al., 1974; Laundau et al., 1983; Kappy et al., 1984; Zahniser et al., 1984) fractions is observerd. The binding is heterogeneous, and curvilinear Scatchard plots are obtained. Binding of high (K_d of 3–11 nM) and low affinity is observed to be relatively uniform in membrane fractions obtained from many adult brain regions, such as cerebral cortex, hippocampus, striatum, and cerebellum. The external plexiform layer of the olfactory bulb, and the hypothalamus, in contrast, are especially rich in sites. Binding to brain tissues peaks in 15-day-old rats (Young et al., 1980; Havrankova et al., 1981; Kappy et al., 1984), and 16-to 18-day-old chicks (Hendricks et al., 1984). It is curious that these regions also contain the highest concentration of immunoreactive insulin (Havrankova et al., 1978b). The affinity and dissociation rates of insulin from sites in brain and liver tissue seem to be identical. Certain differences are reported, however. Down regulation in response to high insulin concentrations is not observed in brain tissues, in contrast to peripheral tissues (Boyd et al., 1982; Zahniser et al., 1984). Specific binding is detected on both neuronal (van Houten et al., 1979) and nonneuronal cells (Raizada et al., 1982), as well as on microvessels. Hence, some care in interpreting binding data is necessary in samples derived from a complex tissue.

b. Physical Properties of Insulin Binding Sites. The structure of the insulin receptor has been studied in peripheral tissues. Gel electrophoresis under nonreducing conditions shows that the insulin receptor is an M_r 300,000–350,000 complex (Jacobs et al., 1979). Under reducing conditions, however, 125,000 to 130,000-Da subunits are observed (Jacobs et al., 1977;

Yip *et al.*, 1978). Cross-linking studies show there are α- (125,000 daltons) and β- (90,000) glycoprotein subunits. A model for the subunit structure has been proposed (Czech and Massague, 1982). A cDNA encoding the insulin receptor has been cloned (Ebina *et al.*, 1985).

  c. IGF Binding Sites in Brain. Binding sites for IGF-I and IGF-II have been observed in particulate fractions from brain (Sara *et al.*, 1983; Goodyer *et al.*, 1984) and cultured pituitary cells (Rosenfeld *et al.*, 1984). Interestingly, in membrane preparations from anterior pituitary, hypothalamus, and brain, the specific binding of IGF-II is severalfold higher than that of IGF-I, and insulin binding is quite low (Goodyer *et al.*, 1984). IGF-I and IGF-II bind to sites which have a substantial preference for IGFs over insulin.

  d. Physical Properties of IGF Binding Sites. The physical properties of the IGF binding sites have also been studied almost exclusively in peripheral tissues. There are two subtypes of IGF binding sites (Kasuga *et al.*, 1981; Rechler *et al.*, 1983). The type I site has a higher affinity for IGF-I than IGF-II, and a low affinity for insulin. It has an M_r of 350,000 and dissociates in the presence of reducing agents into subunits of 90,000 and 125,000 Da (Massague and Czech, 1982). Because it is structurally similar to the insulin receptor, the monoclonal antibodies raised against insulin receptors can immunoprecipitate type I sites well. The type II site has a higher affinity for IGF-II than IGF-I, and insulin does not bind to this site. It appears to be composed of a single M_r 225,000 species that is not cross-linked through disulfide bonds to other macromolecules (Massague and Czech, 1982). A biological function has, as yet, not been attributed to the type II sites.

  e. Insulin and IGF Binding Sites: Relationship to Neurite Formation. Although insulin and IGF binding sites have been observed in brain tissues, a clear correlation with function has proved difficult to demonstrate. Binding parameters derived from a complex tissue are not readily interpretable, particularly when further complicated by the potential presence of multiple types of sites. Since insulin and IGF can enhance neurite formation in cloned human neuroblastoma cells (Recio-Pinto and Ishii, 1984), an opportunity to correlate binding with a neuronal function of understood importance is presented in a clonal, intact cell system. Our studies (Recio-Pinto and Ishii, 1987) using radioactive insulin, IGF-I, and IGF-II show that both insulin and IGF-type binding sites are simultaneously present on SH-SY5Y cells. Displacement studies reveal which sites mediate neurite outgrowth. Insulin at concentrations of 1 nM does not compete for binding to the IGF binding sites, but does enhance neurite outgrowth. Similarly, IGF-I and IGF-II at concentrations of 1 nM do not compete for binding to the insulin sites, but do enhance neurite formation. The most reasonable interpretation is that insulin, at low concentrations, is acting through the insulin binding sites. Likewise, the IGFs at low concentration are acting through the IGF binding sites. At high ligand concentrations, cross-occupancy of both insulin and IGF binding sites occurs. This is likely to be the basis for the observations that the dose–response curves for enhancement of neurite formation (Recio-Pinto and Ishii, 1984), and tubulin mRNA levels (Mill *et al.*, 1985), are very broad. These studies, when considered together with the reports on insulin and IGF binding sites in brain, suggest that brain sites may regulate morphological differentiation and, potentially, survival of neurons during development. Thus, there are distinct receptors for NGF, insulin, and IGFs. Insulin and IGFs do not bind to NGF receptors, and NGF does not bind to insulin sites.

B. Tubulin mRNA, Tubulin, and Microtubules

Construction of neurites involves, at the very least, transcription of genes coding for cytoskeletal proteins, processing of nuclear transcripts, synthesis of cytoskeletal proteins, anterograde axonal transport of subunits, and assembly of subunits into fibrillar elements. Which of these steps is critically influenced by the neuritogenic agents? The major fibrillar elements in neurites include microfilaments (actin-containing filaments of about 7-nm diameter), neurofilaments (neuron-specific filaments of about 10-nm diameter), and microtubules. The early studies of Levi-Montalcini and Angeletti (1968) have shown that NGF-induced neurite outgrowth and hypertrophy in sensory and sympathetic cells are attended by excess production of microtubules and neurofilaments. Drugs that disrupt neurotubules, such as colchicine and vinblastine, can completely inhibit neurite formation (Yamada *et al.*, 1971). This demonstrates clearly the important contribution microtubules make to neurite structure. Because insulin and NGF have profound effects on microtubules, their regulation of neurotubules will be the main subject of this section.

1. TUBULIN AND MICROTUBULES

Microtubules constitute between 5 and 10% of the total cytoplasmic protein. They are filaments that are about 25 nm in diameter, have a 10-nm central core, and are about 10–20 μm in length. One end, of at least some microtubules, appears to be attached to the plasma membrane (Chalfie and Thomson, 1979). Microtubules do not extend to the plasmalemma or the filopodia of growth cones, and probably serve to build up the neurite behind the advancing growth cone. They appear to have more than a simple structural role. The relationship of microtubules to axonal transport has been reviewed (Grafstein, 1977; Schwartz, 1979).

Heterodimer subunits, consisting of 57-kDa α- and 54-kDa β-tubulin, assemble to form the microtubules. The polymerization is an energy-requiring process, driven by GTP, and has been observed to begin around a limited number of organizing centers in some cells (Frankel, 1976). The amino acid sequences of the tubulins are known to be highly conserved. Recent findings reveal the presence of multiple isotypes of tubulins. For example, 7 out of 14 C-terminal residues are different in the products from two human β-tubulin genes. Moreover, in the chick, the α- and β-tubulins are each the products of only four functional genes, yet seven α- and 10 β-tubulins may be distinguished by two-dimensional gel electrophoresis, suggesting that important posttranslational modifications may occur (Sullivan and Wilson, 1984). There is circumstantial evidence, but as yet no definitive proof, that the isotypes may serve different cellular functions. For example, multiple isotypes of tubulin have been observed in single neurons (Gozes and Sweadner, 1981). Also, there is a preferential expression of one of the β-tubulin genes in chick brain, leading to the speculation that the gene serves a specialized brain requirement, such as in axon formation or in glial functions (Lopata et al., 1983). The distribution of α- and β-tubulin in brain is not the same (Cummings et al., 1982), suggesting that there may be other roles for tubulin than that provided for by the classic heterodimer model. Another possibility is that the multiple transcriptional units reflect the needs of several independent regulatory systems, rather than the need for production of functionally distinct proteins.

2.   TUBULIN GENES AND mRNAs

The mammalian genome contains about 15–20 gene sequences for each of the α- and β-tubulins, as detected by nucleic acid hybridization to genomic blots (Hall et al., 1983; Lemischka et al., 1981; Bond and Farmer, 1983). In the chicken, only four of the tubulin multigenes contribute useful transcripts (Cleveland et al., 1980). Likewise, only two to four of the mammalian genes appear, in general, to be functionally important. All of the functional mammalian β-tubulin genes consist of four exons and three introns of varying length. The α- and β-tubulin genes are unlinked and dispersed across the chromosomes, indicating the possibility for independent regulation of each gene (Havercroft and Cleveland, 1984). The other DNA sequences appear to be pseudogenes which contain a series of stop codons within coding sequences. The pseudogenes contains no introns, are flanked by short direct repeat sequences, and have representations of polyadenylation 3 ' to their AATAAA consensus sequence (Cleveland, 1983).

α-Tubulin mRNA is detected as a single size class of 1.8 kb in Northern blots of rat brain RNA (Bond and Farmer, 1983). On the other hand, β-tubulin mRNA appears as two size classes of 1.8 and 2.6 kb, representing transcripts from the different genes.

Free tubulin may modulate tubulin mRNA levels and, thereby, tubulin synthesis. For example, colchicine can depolymerize microtubules, increase the free tubulin pool size, and decrease synthesis of tubulin in mouse 3T6 cells (Ben Ze'ev *et al.*, 1979). Moreover, the amounts of specific tubulin mRNAs and protein are decreased in Chinese hamster ovary (CHO) cells by a variety of drugs that destabilize microtubules (Cleveland *et al.*, 1981). The reduced content of transcripts is not due to decreased transcription, as shown by nuclear run-off experiments (Cleveland, 1983). Possibly, the rate of nuclear RNA processing is diminished, or the stability of cytoplasmic mRNAs is lowered. At any rate, free tubulins may feed back to modulate tubulin synthesis. This conclusion is strongly supported by an experiment in which microinjection of tubulin resulted in a large decrease in tubulin synthesis in CHO cells (Cleveland *et al.*,1983).

3. Tubulin mRNA Levels during Brain Development

There is a sharp decline in the production of α- and 1.8-kb β-tubulin mRNAs in rat brain during the postnatal period (Bond and Farmer, 1983). This event coincides temporally with the period of cessation of cell multiplication, terminal differentiation, and neurite outgrowth. As the neurons mature, there is a decrease in transcription, and tubulin stockpiled during development may be mobilized into the growing neurites. Support for this suggestion is derived from studies with cultured cells (described below). A difficulty to a straightforward explanation, of course, is the presence of multiple neuronal and cell populations, whose ratios or metabolic rates may be altered during development, leading, in turn, to a change in overall brain tubulin content.

4. Tubulin Protein and mRNA Levels during Regeneration of the Optic Nerve

Following crush injury to the goldfish optic nerve, regeneration is associated with an increase in total tubulin content. β-Tubulin levels are reported to rise more than α-tubulin levels (Neumann *et al.*, 1983). Once again, there is the suggestion that the β-tubulins may have a pronounced role in neurite formation. Alternatively, the β-tubulin band detected by sodium dodecyl sulfate (SDS) gel electrophoresis may have included other proteins with the same relative mobility. The levels of total tubulin mRNAs are increased in regenerating optic nerves. However, since the α- and β-tubulin transcripts were not distinguished separately, it is not known whether the difference in levels of α- and β-tubulins results from unequal amounts of their mRNAs. The stimulus for regeneration of the goldfish optic nerve is not precisely identified.

5. NGF'S EFFECTS ON TUBULIN AND MICROTUBULE LEVELS

The accumulation of microtubules in NGF-treated neurons (Levi-Montalcini and Angeletti, 1966) may be due to several mechanisms. The microtubules become more resistant to depolymerization induced by colchicine following treatment of PC12 cells with NGF (Black and Greene, 1982). In addition to stabilization, NGF may stimulate microtubule assembly partly by effects on the microtubule-associated proteins (MAP). Two high-molecular-weight MAP (MAP 1 and MAP 2) have been purified (Borisy et al., 1975; Herzog and Weber, 1978). It is believed that the MAP may serve to increase the formation of microtubules and cross-link them with other cytoskeletal proteins (Kim et al., 1979). The in vitro assembly of microtubules is increased in the presence of MAP 1 and MAP 2 (Murphy et al., 1977). For these reasons, the increase in MAP 1 content caused by NGF in PC12 cells (Greene et al., 1983) is, potentially, propitious for microtubule assembly.

There is circumstantial evidence, but no definitive proof, for the suggestion that NGF may directly increase the assembly of microtubules from a preexisting pool of tubulin (Levi-Montalcini et al., 1974). For example, NGF binds to tubulin (Calissano and Cozzari, 1974), but NGF has a large electrostatic charge and adsorbs nonspecifically to a large number of substances. NGF-induced polymerization of tubulin has been shown in vitro (Levi et al., 1975), but only upon addition of supraphysiological NGF concentrations.

An increase in the concentration of free tubulin is likely to promote microtubule assembly. However, there is controversy as to whether NGF increases tubulin content. Hier et al. (1972) and Kolber et al. (1974) have reported increases. On the other hand, there are several reports of no effect in cultured sensory and sympathetic ganglia (Yamada and Wessells, 1971; Stoeckel et al., 1974). These early studies were based on a single-point colchicine-binding assay. It was then recognized that colchicine binding in samples is subject to variable decay rates. With a technically improved assay, the content of tubulin in the soluble fraction of sensory ganglia was found to be invariant during neurite outgrowth (Mizel and Bamberg, 1975). Part of the difficulty in interpreting these data may be the nature of ganglion cell cultures. Substantial amounts of tubulin are present in the associated nonneuronal cells, which proliferate during the experiment. Moreover, the tubulin specifically in the neurites can be only a fraction of the total tubulin content in the cultures. Therefore, an increase in tubulin content, expressed as a fraction of total soluble protein, could be missed due to the complicating dilutional effect of the proliferating nonneuronal cells, and the fact that only part of the tubulin pool is in the neurites. Finally, an increase in the content of soluble tubulin may only be transient, given the presence of a feedback regulatory system. Whether an increase is detected, or not, may depend on the time at which NGF-treated samples are assayed.

The prevailing interpretation today is that NGF may increase microtubule assembly from a preexisting pool of tubulin subunits. Recent data suggest that NGF may additionally have effects on tubulin mRNA levels (see below).

6. INSULIN'S EFFECTS ON TUBULIN AND ACTIN mRNA
   CONTENT IN HUMAN NEUROBLASTOMA CELLS

Insulin increases the content of α- and β-tubulin transcripts in the cloned SH-SY5Y cell (Mill et al., 1985). The response curves show the same dependence on dose for the increases in tubulin mRNA content and neurite formation. The broad dose–response curves for each of these functions additionally correlate closely with the saturation of insulin binding, keeping in mind the cross-occupancy of IGF binding sites at higher concentrations.

The increase in tubulin mRNA levels is detected within a few hours of addition of insulin (Fig. 6). Interestingly, although the neurite levels remain elevated after 1 day of treatment, the content of tubulin mRNAs subsequently declines. We interpret these data to suggest that elevated tubulin mRNA levels are required during the initiation of neurite extension. Thereafter, lower levels may be sufficient to sustain formed neurites. This might be the case, even though the microtubules are longer, if the concentration of microtubule ends was not substantially changed. The concentration of ends would influence the overall rate at which microtubules disassemble. This interpretation is in agreement with the observation that the concentration of tubulin transcripts declines sharply during rat brain development (Bond and Farmer, 1983).

The increase in the content of tubulin mRNA in SH-SY5Y cells is specific, because the content is increased relative to total RNA. In addition, insulin does not substantially increase the levels of actin mRNA. One should not infer, however, that actin is unimportant to neurite formation. Actin filaments are prominent in the growth cone behind the filopodia. Drugs, such as cytochalasin, that disrupt actin filaments also inhibit neurite extension (Yamada et al., 1971).

7. IGF-IIs EFFECTS ON TUBULIN mRNA LEVELS IN
   HUMAN NEUROBLASTOMA CELLS

The broadness of insulin's dose–response curve indicates that cross-occupancy of IGF binding sites at high concentrations should contribute further to increased levels of tubulin transcripts. It seems likely that insulin and IGF-II may act through similar mechanisms to stimulate neurite formation.

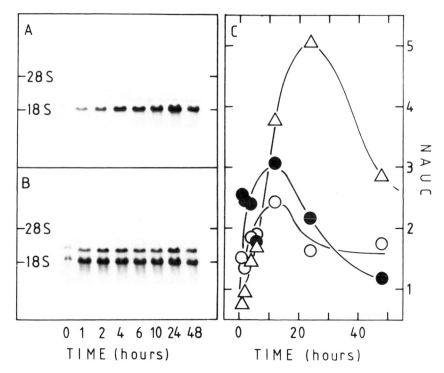

FIG. 6.   Time course of insulin's effect on α- and β-tubulin mRNA levels in SH-SY5Y cells. The cells were incubated in serum-free medium with 0.1 μM insulin for various lengths of time. RNA was extracted, purified, and equivalent amounts (25 μg) were electrophoresed through 0.8% agarose gels containing formaldehyde. Ethidium bromide staining was used to determine the position of 18 and 28 S ribosomal RNAs. The RNA was transferred to nitrocellulose, and the resultant blots were hybridized to nick-translated [³²P]cDNA probes. The human cDNA probes pKα1 and pDβ1 contained the entire coding sequences of α- and β-tubulin, respectively, and varying amounts of 3′- and 5′-noncoding regions. (A) Autoradiogram of a filter hybridized to pKα1. (B) The same filter boiled, washed, and rehybridized to pDβ1. (C) In a separate experiment, the autoradiograms, prepared as in (A) and (B), were scanned on a densitometer, and the area under each curve was measured. The plots show the normalized area under the curve (NAUC) for each experimental point relative to untreated cultures. (△) α-tubulin mRNA; (○) 1.8-kb β-tubulin mRNA; (●) 2.6-kb β-tubulin mRNA. From Mill *et al.* (1985).

It is, therefore, not surprising to find that physiological concentrations of IGF-II also elevate tubulin mRNA content (Mill *et al.*, 1985).

8.   NGFs EFFECT ON TUBULIN mRNA LEVELS IN SENSORY NEURONS

Do other neuritogenic polypeptides, such as NGF, share the capacity to increase tubulin transcript levels? Embryonic chick sensory neurons were cultured for 3 days with 20 ng/ml NGF, which was necessary to sustain neuronal survival during treatment with cytosine arabinoside used to eliminate the nonneuronal cells (Recio-Pinto et al., 1986). Cytosine arabinoside spares the neurons, which are at a postmitotic stage in development. Thereafter, the cytosine arabinoside was washed out and the incubations were continued for 2 more days, some cultures receiving a high concentration (20 ng/ml) of NGF, while others received a lower concentration (1 ng/ml). The high concentration was chosen to be very close to the amount of NGF which is just sufficient to produce a maximal neurite formation response, based on dose–response curves (Recio-Pinto et al., 1986). This dose is, therefore, within the physiologically relevant range. The RNA from the cultures treated with the two doses of NGF was analyzed as described by Mill et al. (1985). Briefly, total RNA was extracted, and equivalent amounts of RNA were electrophoresed on denaturing formaldehyde gels. Following transfer to nitrocellulose, the blots were successively hybridized to the nick-translated cDNA probes pKα1 and pDβ1. The resulting autoradiograms are shown in Fig. 7. It is evident that the higher concentration of NGF caused an increase in the amounts of α- and β-tubulin mRNAs, relative to total RNA. NGF is also observed to elevate tubulin mRNA levels in PC12 cells (Fernyhough and Ishii, 1987).

Thus, we conclude that the two separate classes of neuritogenic polypeptides, represented by the insulin homologs and NGF, share the capacity to elevate the content of tubulin transcripts. These results, furthermore, support the reasonable prediction that insulin homologs and NGF can increase the synthesis of tubulins. If this were correct, then the content of soluble tubulin may rise, perhaps only transiently, during the period of increased microtubule assembly and neurite extension. On the other hand, the content of polymerized tubulin may be expected to remain increased, together with the extended neurite.

9. RELATIONSHIP BETWEEN THE EFFECTS OF NEURITOGENIC AGENTS ON TUBULIN TRANSCRIPT LEVELS AND CELLULAR GROWTH RATES

Insulin can increase the population growth rate of SH-SY5Y cells when serum is present (Recio-Pinto and Ishii, 1984). The cells, however, become essentially quiescent in serum-free medium. Because insulin can increase neurite outgrowth and tubulin mRNA levels under both conditions of culture, there is no correlation of these effects with the cell population growth rate. The neurites retract during cell division, then reextend, in NGF-treated SH-SY5Y cell cultures. This behavior is observed in dividing CNS neurons also. In the case of the sensory neurons, cell division is not a consideration because the experiment described above with NGF was done

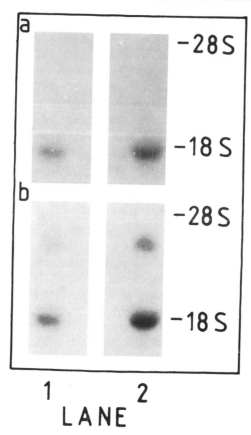

FIG 7.   Effect of NGF on levels of α- and β-tubulin mRNAs in cultured embryonic chick sensory neurons. A single-cell culture (1.5×10^7 cells per 100-mm dish) was prepared from 10-day-old sensory ganglia, as described (Recio-Pinto et al., 1986). The cells were cultured for 2 days with 20 ng/ml NGF, then treated with 2.7 μg/ml cytosine arabinoside for 3 days to eliminate the dividing nonneuronal cells. The cultures were washed three times with 5 ml culture medium, which was RPMI 1640 containing 12% serum. Thereafter, cells were cultured for 2 days with either 1.0 (lane 1) or 20 (lane 2) ng/ml NGF. The procedure for extraction and analysis of the RNA by Northern blot was the same as described in Fig. 6. Each gel lane contained 16.7 μg RNA. (a) Hybridization with ^{32}P-labeled pK∝1; (b) hybridization with pDβ1. NGF is shown to increase the levels of α- and β-tubulin mRNAs in sensory neurons.

on neurons at an early postmitotic stage of development. And, too, there is no evidence that NGF is a mitogen (see Mobley et al., 1977). Clearly, the effect of neuritogenic agents on tubulin transcript levels is independent of any effect on the cell population growth rate.

10. COOPERATION BETWEEN INSULIN HOMOLOGS AND NGF IN NEURITE FORMATION

Insulin acts on the same, or a subpopulation, of NGF-responsive neurons (Recio-Pinto et al., 1986). This suggests the possibility for cooperation between insulin and NGF in the construction of neurites. In fact, evidence in support of this possibility has been obtained. Under serum-free conditions NGF loses its capacity to enhance neurite formation in SH-SY5Y cells (Recio-Pinto et al., 1984). This is not due to a generalized inability to form neurites, for neurites can still form in response to insulin or tumor promoters (Spinelli and Ishii, 1983). Rather, there is a concomitant loss of the capacity to bind NGF to the high-affinity (slow) receptors. Treatment of cells with insulin or IGF-II causes a return of the capacity to bind and respond to NGF. This observation alone, however, cannot entirely explain the basis for the synergistic potentiation of neurite formation caused by insulin and NGF in SH-SY5Y and PC12 cells (Recio-Pinto et al., 1984).

The case for cooperation is extended because of the finding that NGF has minimal effects on the level of tubulin mRNAs in SH-SY5Y cells, whether serum or insulin is present or not (Mill et al., 1985), despite the observation that NGF is a strong agonist for neurite formation (Sonnenfeld and Ishii, 1982). This suggests that, in SH-SY5Y cells, the synergistic potentiation in neurite formation may arise from elevated tubulin mRNA levels due largely to insulin and/or IGFs, and enhanced assembly and/or stabilization of microtubules mediated largely by NGF. There may be other complementary metabolic events supported to different extents by NGF and the insulin homologs (see Section IV,C). These mechanisms may act in concert with the activation of NGF binding by insulin homologs to explain the synergistic potentiation.

C. Phosphorylation and Neurite Formation

The importance of phosphorylation in the mediation of transmembrane signaling and control of metabolism is well documented (Cohen, 1982). A great deal of interest has been raised in phosphorylation as a mechanism of transmembrane transduction for neurotransmitters (Browning et al., 1985). Phosphorylation may additionally play a critical role in the mediation of neuronal differentiation and neuritic outgrowth.

Early studies which implicated phosphorylation as a mediator of neuronal differentiation were performed by Erlich et al. (1978) in murine neuroblastoma cells. Murine neuroblastoma cells can be induced to morphologically differentiate by a variety of agents, including cAMP and phosphodiesterase inhibitors (Prasad, 1975). Cells induced by phosphodiesterase inhibitors show altered phosphorylation of several proteins (Erlich et al., 1978). These seminal experiments did not include study of specificity for differentiation, nor whether phosphorylation could be altered by physiological regulators of differentiation and neuritogenesis.

1. PHOSPHORYLATION IS ASSOCIATED WITH BINDING TO
   INSULIN, NGF, AND TUMOR-PROMOTER RECEPTORS

   a. NGF-Mediated Phosphorylation. Guroff and colleagues (Yu et al., 1980) were among the first to study the effects of NGF on phosphorylation in superior cervical ganglia. The phosphorylation of a nonhistone nuclear protein is increased by NGF or cAMP treatment. This protein can be distinguished from histone H1 by its lower mobility on acid/urea gels. For this reason it has been called the *slow migrating protein*, or SMP.

SMP is a protein with a molecular weight of 30,000 extracted in the nonhistone fraction of nuclei from PC12, superior cervical ganglia, and sympathetic cells (Yu *et al.*, 1980). The phosphorylation of SMP is increased almost exclusively on serine residues by both cAMP and NGF, and it binds to chromatin.

Triton extracts of nuclear fractions, derived from intact PC12 and sympathetic cells pretreated with NGF or cAMP, show an increase in SMP phosphorylation (Nakanishi and Guroff, 1985). Induction of NGF occurs within 30 minutes and can be maintained for at least 3 days. The kinase does not seem to be cAMP dependent, nor does it seem to belong to the protein kinase C class because diolein and calcium are not observed to enhance phosphorylation of SMP in cell-free extracts.

Another phosphoprotein and kinase sensitive to NGF have been described by Guroff in PC12 cells (End *et al.*, 1983). This phosphoprotein, of 100,000 Da, shows a marked but transient decrease in phosphorylation upon exposure to NGF, and is denoted Nsp100. Nsp100 is a cytosolic protein as is its associated kinase. The kinase associated with Nsp100 is similar in both its cation sensitivity and substrate requirements to the SMP-associated kinase. However, the two appear to have separate cellular locations, and Nsp100 is phosphorylated on threonine residues whereas the SMP kinase is serine specific.

Nsp100 and its kinase have been partially purified (Togari and Guroff, 1985). The substrate and kinase can be resolved, and autophosphorylation is not involved. The kinase has a molecular weight of 110,000–130,000 and is highly heat labile. These proteins are found distributed in brain, adrenals, and superior cervical ganglia (Hama and Guroff, 1985). However, they are found in nonneuronal tissues as well.

Halegoua and Patrick (1980), using intact PC12 cells, identified by gel electrophoresis a number of protein bands phosphorylated in response to NGF treatment. The bands are tentatively identified as tyrosine hydroxylase, ribosomal protein S6, and histones H1, H2A, and H3. Two additional bands are from the nonhistone high-mobility group 17. The putative histone H1 band may potentially be SMP, because the acid/urea gel system was not used to distinguish between the two. The authors suggest that the pattern of phosphorylation caused by dibutyryl cAMP is very similar to that caused by NGF. However, Guroff's studies clearly demonstrate that many of NGF's actions occur through a cAMP-independent mechanism.

   b. Kinase Activity Associated with an NGF-Binding Component in Melanoma Cells. An NGF-binding component of the human melanoma cell line A875 has been purified by NGF–sepharose affinity chromatography (Puma *et al.*, 1983). A875 cells have a large number of NGF binding sites (Fabricant *et al.*, 1977), which appear kinetically similar to the fast sites found in other cell lines. Preliminary evidence was cited that this purified preparation contains a kinase activity. This is particularly interesting, because the NGF-binding protein is located in the plasma membrane, whereas the other identified kinase activities are located within the interior of the cell. The NGF-binding protein is phosphorylated, but the phosphorylation appears to be NGF independent both *in vitro* and in intact cells (Ross *et al.*, 1984; Grob *et al.*, 1985). Some implications of receptor phosphorylation are discussed below.

The relevance of these findings to neuronal differentiation remains to be established because the A875 cells are not known to respond to NGF (an effect on cell survival has been suggested, but has yet to be confirmed). NGF, as previously noted, can readily adsorb to a variety of substances due to its large electrostatic charge. It is difficult to assign receptor status to a binding component in the absence of a physiological response. Nevertheless, the relationship to the fast NGF sites on cells of neuronal origin might eventually be established by immunochemical or molecular biological means.

c. Phosphorylation Mediated by Insulin Homologs. Insulin is well known to modulate intermediary metabolism. Many of the key regulatory enzymes are phosphoproteins, whose state of phosphorylation is altered in the classic kinase cascade following treatment with insulin (Cohen, 1982). It has been suggested that many of these events may be mediated by low-molecular-weight *second messengers* (Cheng *et al.*, 1981).

With regard to microtubule function, the insulin-induced phosphorylation of MAP 2, tubulin and τ proteins is notable (Kadowaki *et al.*, 1985). The observation that phosphorylation of serine and threonine residues on MAP proteins can regulate microtubule assembly (Burns *et al.*, 1984) suggests that insulin may modify microtubule function directly through phosphorylation of its tyrosines.

d. Autophosphorylation of the Insulin Receptor. Insulin induces the phosphorylation of the β-subunit of its own receptor both in intact cells (Kasuga *et al.*, 1982a), and in cell-free extracts (Kasuga *et al.*, 1982b). The phosphorylation of the α-subunit, however, is observed only in the cell-free system, and its consequence is less certain. IGF-II can also induce phosphorylation of the insulin receptor, presumably through cross-occupancy.

Autophosphorylation of the insulin receptor occurs almost exclusively on tyrosine residues (Kasuga *et al.*, 1982a,b). Phosphotyrosine makes up only 0.03% of the total pool of phosphoamino acids. A potentially significant growth regulatory role is suggested. For example, other powerful mitogens, such as epidermal growth factor (EGF) (Cohen *et al.*, 1980b) and IGF-I

(Jacobs *et al.*, 1983) also induce tyrosine kinase activities, and phosphorylation of their own receptors. Moreover, there is a high level of tyrosine-specific phosphorylation during embryogenesis (Dasgupta and Garbers, 1983), and many of the RNA tumor viruses have oncogene products that are tyrosine kinases (Bishop, 1983).

The recent cloning of the cDNA encoding the insulin receptor (Ebina *et al.*, 1985) has served to clarify whether the associated kinase activity is intrinsic to the receptor, or simply associated with it. Several distinct domains have been deduced from the nucleotide sequence of the insulin-receptor cDNA. The domain containing the cytoplasmic region of the β-subunit has homology to the corresponding regions of both the EGF-receptor cDNA, and the viral oncogene *v-ros*. Thus the tyrosine kinase activity seen in insulin-receptor preparations is almost surely intrinsic to the molecule itself.

e. Phosphorylation and Cooperative Interactions. Studies on phosphorylation may shed further light on the biochemical basis for cooperative effects between neuritogenic factors. Insulin promotes a pattern of phosphorylation in PC12 cells which overlaps, but is not identical to, that induced by NGF (Halegoua and Patrick, 1980). This is intriguing in light of the synergistic enhancement of neurite outgrowth caused by the combination of insulin and NGF in these cells (Recio-Pinto *et al.*, 1984). Furthermore, NGF's primary action on MAP 1 appears to be regulation of its synthesis (Greene *et al.*, 1983). On the other hand, insulin promotes the phosphorylation of MAP 2. These separate effects of NGF and insulin may, therefore, provide the biochemical basis for synergistic interactions when these factors are present in combination. However, there remains the task of elucidating more completely the relationship of phosphorylation to neuronal differentiation.

f. Tumor-Promoter Receptor and Phosphorylation. Several years ago it was shown that tumor promoters can modulate the process of morphological differentiation in cultured dorsal root ganglia (Ishii, 1978). Brain was then found to have the highest concentration of tumor-promoter receptor sites of any tissue (Blumberg *et al.*, 1981), and the number of receptor sites is increased during development (Nagle *et al.*, 1981; Murphy *et al.*, 1983). A more complete discussion of the role of tumor promoter receptors in the nervous system is available elsewhere (Ishii *et al.*, 1985). The distribution of tumor-promoter receptors in the nervous system, and their effects on morphological differentiation, suggest a role for these sites in the regulation of neuronal differentiation. In analogy to the opiate receptors, a more precisely defined physiological role for tumor-promoter receptors may eventuall come to light. Whatever that physiological role may be, tumor promoters are powerful agents for studying the mechanism of neuronal differentiation.

Tumor promoters can specifically and reversibly enhance neurite formation in SH-SY5Y cells (Spinelli *et al.*, 1982), and the specific receptors have been described (Spinelli and Ishii, 1983). Tumor-promoting compounds, such as the phorbol esters, mezerein, and teleocidin, reversibly enhance neurite formation and compete for binding to the receptors. On the other hand, nonpromoting structural congeners of the phorbol esters neither enhance neurite formation nor compete for binding. Other neuritogenic agents, such as NGF and insulin, do not cross-occupy the phorbol ester tumor-promoter receptors (Spinelli and Ishii, 1983).

The phorbol ester tumor-promoter receptor appears to be, or is closely associated with, protein kinase C (Castagna *et al.*, 1982; Niedel *et al.*, 1983). Thus, there is the common thread that the three major classes of agents that can cause neurite formation in SH-SY5Y cells can also cause phosphorylation. The phosphorylation of the Nsp100 protein in PC12 cells is modulated by both NGF and tumor promoters (Hama *et al.*, 1986).

2.   INSULIN-MEDIATED PHOSPHORYLATION IN SH-SY5Y CELLS

The foregoing suggested that the neurite outgrowth mediated by insulin in SH-SY5Y cells might also be associated with alterations in protein phosphorylation. This possibility was explored. SH-SY5Y cells were cultured in serum-free medium for 1 day with various concentrations of insulin. Thereafter, the cultures were incubated with $^{32}PO_4$ for 2 hours, then extracted with buffer containing SDS and 2-mercaptoethanol. The extracts were electrophoresed in polyacrylamide gels under reducing conditions. the autoradiograms are shown in Fig. 8. There was, first of all, a dose-dependent increase in total phosphorylation. Insulin helps to maintain the phosphorylated state of the cell. This is evident because the serum-free, insulin-treated cultures retain a pattern of phosphorylation very similar to that of cultures grown in serum-containing medium (compare lanes a, b, and g). There is specific phosphorylation in the 34,000-Da region. The specificity is evident because the phosphorylation in this region is increased relative to other regions on the autoradiograms, such as those marked X and Y.

Densitometric scans were prepared from the autoradiograms shown in Fig. 8. In order to normalize and correct for the increase in total phosphorylation, the relative peak heights of the 34,000-Da region were compared to those of regions X and Y (Fig. 9). There was a dose-dependent increase in the 34,000-Da region when the data were plotted relative to the extent of phosphorylation in either regions X or Y. On the other hand, there was no change in the ratio of X to Y, showing that these regions were both increased in phosphorylation to the same extent by insulin. The dose–response curve for this specific insulin-mediated phosphorylation of the 34,000-Da band follows closely the dose–response curves for enhancement of neurite formation (Recio-Pinto and Ishii, 1984) and tubulin mRNA content (Mill *et al.*, 1985).

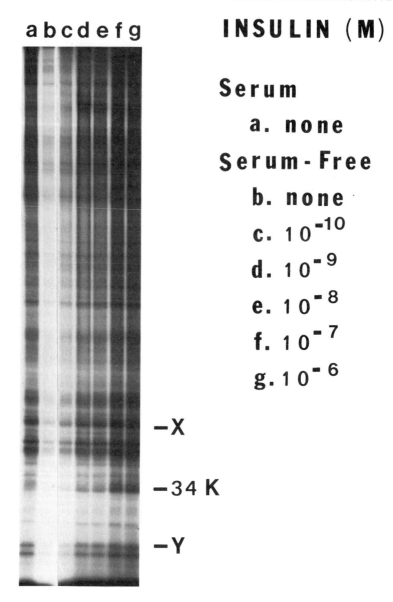

abcdefg

INSULIN (M)

Serum

   a. none

Serum - Free

   b. none

   c. 10^{-10}

   d. 10^{-9}

   e. 10^{-8}

   f. 10^{-7}

   g. 10^{-6}

−X

−34 K

−Y

FIG. 8.   Effect of insulin concentration on phosphorylation in SH-SY5Y cells. The cells (2.5 × 10⁵ per 35-mm dish) were plated for 2 days in RPMI 1640 with serum, then incubated for 1 day in serum-free medium. The cultures were incubated for another day under the conditions indicated. The cells were washed gently three times with phosphate-free RPMI 1640, then incubated for 2 hours at 37 °C in phosphate-free RPMI 1640 medium with 0.1 mCi/ml ³²PO₄ and the ligand concentrations shown. The cells were extracted and boiled for 3 minutes in 1 ml buffer containing 2.3% sodium dodecyl sulfate, 3 m*M* EDTA, 3 m*M* EGTA, 5%

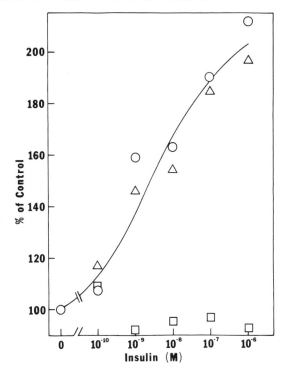

FIG. 9. Insulin dose-dependent phosphorylation of the 34,000-Da band. Densitometric scans were prepared from lanes b (untreated) and f (0.1 μM insulin) of the autoradiograms shown in Fig. 8. The relative peak heights of the 34,000-Da region were compared to those of regions X and Y, in order to normalize and correct for the increase in total phosphorylation. The figure shows the percentage increase relative to untreated cultures for the following: (○) ratio of 34-kDa region to region X; (△) ratio of 34-kDa region to region Y; (□) ratio of region X to region Y. Note that there is no change in the ratio of regions X to Y, showing that they were both increased to the same extent by insulin. This indicates regions X and Y are reasonable standards against which to normalize the effect on the 34-kDa band.

2-mercaptoethanol, 10% glycerol, and 60 mM Tris pH 6.8. The 10,000 g supernatant fractions of the extracts (1.5 μg protein per lane) were electrophoresed in 10% polyacrylamide gels containing 0.1% sodium dodecyl sulfate (Laemmli, 1970). The autoradiograms are shown. Molecular weight markers were used to calibrate the gel. Note that total phosphorylation is increased over the same range of insulin concentrations that enhance neurite outgrowth. Phosphorylation of the 34,000-Da region is specifically increased, when considered relative to the regions marked X or Y.

FIG. 10. Insulin-enhanced phosphorylation of SH-SY5Y cells cultured in serum-containing medium, and reversibility on washout of insulin. Cells were cultured in RPMI 1640 medium containing 12% serum under the following conditions: (U) 2 days without additions; (I) 2 days with 0.1 μM insulin; and (W) 1 day with 0.1 μM insulin followed by washout for 1 day. Then, phosphorylation was studied in all cultures by methods described in Fig. 8. The autoradiograms of untreated, insulin-treated, and withdrawn cultures are shown. The arrow shows the position of the 34,000-Da region.

The phosphorylation of the putative 34,000-Da protein can occur also in cultures grown in serum-containing medium, and is reversible. Following exposure to insulin for 1 day in medium containing 12% serum, specific phosphorylation of the 34,000-Da band was increased (not shown). The increase in total phosphorylation under these conditions is not as marked as under the serum-free conditions. Most likely this is due to the higher background caused by the activity of endogenous neuritogenic factors in serum on the untreated cultures (Recio-Pinto and Ishii, 1984). One day after washout of insulin from some of the cultures, the level of phosphorylation was examined (Fig. 10). The phosphorylation caused by insulin was reversed on washout.

These results serve to show that insulin can cause specific, reversible phosphorylation, which correlates with increased neurite outgrowth. The function of the putative 34,000-Da phosphoprotein, particularly its relationship to enhancement of tubulin mRNA levels, remains a subject for future experimentation.

V.   CONCLUDING REMARKS

Physiological concentrations of insulin and IGF-II can enhance survival and neurite formation in sensory and sympathetic neurons. These are direct effects on the neurons, and do not require the presence of other serum factors. The expression of the prepro-IGF-II mRNA, and its related transcripts, is developmentally regulated in a tissue-specific manner.

Because the prepro-IGF-II transcript is detected in brain and spinal cord, it is probable that IGF-II has a role in the development of the central, as well as the peripheral, nervous sytem. The detection of specific and distinct binding sites for insulin, IGFs, and NGF in SH-SY5Y cells, together with the observation that insulin homologs can synergistically potentiate the response to NGF, supports the view that these polypeptides may cooperate in the construction of neurites. The neuritogenic agents can increase the level of α- and β-tubulin transcripts, a prelude to the assembly of the microtubules necessary for neurite extension. A general consequence, which follows the binding of the various classes of neuritogenic agents to their respective receptors, is the increased phosphorylation of macromolecules, implicating phosphorylation as a common transmembrane event closely involved with morphological differentiation.

ACKNOWLEDGMENTS

We thank Wendy A. Nute for assistance in preparing the manuscript. This work was supported in part by grant RO1 NS24787 from the National Institute of Neurological and Communicative Disorders and Stroke, and an award from the Diabetes Research and Education Foundation. JFM is supported by a postdoctoral fellowhip from the National Institutes of Health.

REFERENCES

Aloe, L., and Levi-Montalcini, R. (1979). Nerve growth factor-induced transformation of immature chromaffin cells *in vito* into sympathetic neurons. Effect of antiserum to nerve growth factor. *Proc. Natl. Acad. Sci. U.S.A.* **76**, 1246–1250.

Angeletti, R. H., and Bradshaw, R. A. (1971). Nerve growth factor from mouse submaxillary gland: amino acid sequence. *Proc. Natl. Acad. Sci. U.S.A.* **68**, 2417–2420.

Banerjee, S. H., Snyder, S. H., Cuatrecasas, P., and Greene, L. A. (1973). Binding of nerve growth factor receptor in sympathetic ganglia. *Proc. Natl. Acad. Sci. U.S.A.* **70**, 2519–2523.

Barde, Y. A., Edgar, D., and Thoenen, H. (1983). New neurotrophic factors. *Annu. Rev. Physiol.* **45**, 601–612.

Ben Ze'ev, A., Farmer, S. R., and Penman, S. (1979). Mechanisms regulating tubulin synthesis in cultured mammalian cells. *Cell* **17**, 319–325.

Biedler, J. L., Tarlov, S. R., Schachner, M., and Freedman, L. S. (1978). Multiple neurotransmitter synthesis by human neuroblastoma cell lines and clones. *Cancer Res.* **38**, 3751–3757.

Bishop, J. M. (1983). Cellular oncogenes and retroviruses. *Annu. Rev. Biochem.* **52**, 301–354.

Bjerre, B. O., Bjorklund, A., Mobley, W., and Rosengren, E. (1975). Short and long-term effects of NGF on the sympathetic nervous system in the adult mouse. *Brain Res.* **94**, 263–277.

Black, M. M., and Greene, L. A. (1982). Changes in the colchicine susceptibility of microtubules associated with neurite outgrowth: Studies with nerve growth factor-responsive PC12 pheochromocytoma cells. *J. Cell Biol.* **95**, 379–386.

Blumberg, P. M., Declos, K. B., and Jaken, S. (1981). Tissue and species specificity for phorbol ester receptors. *In* "Organ and Species Specificity in Chemical Carcino-genesis" (R. Langenbach, S. Nesnow, and J. M. Rice, eds.), pp. 201–227. Plenum, New York.

Blundell, T. L., and Humbel, R. E. (1980). Hormone families: Pancreatic hormones and homologous growth factors. *Nature (London)* **287**, 781–787.

Bond, J. F., and Farmer, S. R. (1983). Regulation of tubulin and actin mRNA production in rat brain: Expression of a new β-tubulin mRNA with development. *Mol. Cell. Biol.* **3**, 1333–1342.

Borisy, G. G., Marcum, J. M., Olmstead, J. B., Murphy, D. B., and Johnson, K. A. (1975). Purification of tubulin and associated high molecular weight proteins from porcine brain and characterization of microtubule assembly *in vitro. Ann. N.Y. Acad. Sci.* **253**, 101–132.

Bottenstein, J. E. (1980). Serum-free culture of neuroblastoma cells. *In* "Advances in Neuroblastoma Research" (A. E. Evans, ed.), pp. 161–170. Raven, New York.

Boyd, F. T., Jr., Deo, R., and Raizada, M. K. (1982). Lack of down regulation of insulin receptors by insulin in neuron-enriched primary cultures of rat brain. *Soc. Neurosci. Abstr.* **8**, 980.

Bradshaw, R. A. (1978). Nerve growth factor. *Annu. Rev. Biochem.* **47**, 191–216.

Bray, D. (1973). Model for membrane movements in the neural growth cone. *Nature (London)* **244**, 93–95.

Brissenden, J. E., Ullrich, A., and Francke, U. (1984). Human chromosomal mapping of genes for insulin-like growth factors I and II and epidermal growth factor. *Nature (London)* **310**, 781–784.

Browning, M. D., Huganir, R., and Greengard, P. (1985). Protein phosphorylation and neuronal function. *J. Neurochem.* **45**, 11–23.

Bunge, M. B. (1973). Fine structure of nerve fibers and growth cones of isolated sympathetic neurons in culture. *J. Cell Biol.* **56**, 713–735.

Burmeister, D. W., and Lyser, K. (1982). Process formation in the human neuroblastoma clone SK-N-SH-SY5Y *in vitro. Diss. Abstr. Int.* **43**, 1334–B.

Burns, R. G., Islam, K., and Chapman, R. (1984). The multiple phosphorylation of the microtubule-associated protein MAP2 controls the MAP2:tubulin interaction. *Eur. J. Biochem.* **141**, 609–615.

Buxser, S. E., Kelleher, D. J., Watson, L., Puma, P., and Johnson, G. L., (1983). Change in state of nerve growth factor receptor. *J. Biol. Chem.* **258**, 3741–3749.

Calissano, P., and Cozzari, C. (1974). Interaction of nerve growth factor with the mousebrain neurotubule protein(s). *Proc. Natl. Acad. Sci. U.S.A.* **74**, 2131–2135.

Campenot, R. B. (1977). Local control of neurite development by nerve growth factor. *Proc. Natl. Acad. Sci. U.S.A.* **74**, 4516–4519.

Carbonetto, S., and Stach, R. W. (1982). Localization of nerve growth factor bound to neurons growing nerve fibers in culture. *Dev. Brain Res.* **3**, 463–473.

Carbonetto, S. F., Gruver, M. M., and Turner, D. C. (1982). Nerve fiber growth on defined hydrogel substrates. *Science* **216**, 897–899.

Castagna, M., Takai, Y. Kaibuchi, K., Sano, K., Kikkawa, U., and Nishizuka, Y. (1982). Direct activation of calcium-activated, phospholipid-dependent protein kinase by tumorpromoting phorbol esters. *J. Biol. Chem.* **257**, 7847–7851.

Chalfie, M., and Thomson, J. N. (1979). Organization of neuronal microtubules in the nematode *Caenorhabditis elegans. J. Cell Biol.* **82**, 278–289.

Chamley, J. H., Goller, I., and Burnstock, G. (1973). Selective growth of sympathetic nerve fibers to explants of normally densely innervated autonomic effector organs in tissue culture. *Dev. Biol.* **31**, 362–379.

Charlwood, K. A., Griffith, M. J., Lamont, M. D., Vernon, C. A., and Wilcock, J. C. (1974). Effects of nerve growth factor from the venom of *Vipera russelli* on sensory and sympathetic ganglia from the embyonic chick in culture. *J. Embryol. Exp. Morphol.* **32**, 239–252.

Cheng, K., Kikuchi, K., Huang, L., Kellog, J., and Larner, J. (1981). Chemical mediators of insulin action on protein phosphorylation and studies on the activity of degraded insulins and insulin fragments. In "Current Views on Insulin Receptors" (D. Andreani, R. De Pirro, R. Lauro, J. Olefsky, and J. Roth, eds.), pp. 235–244. Academic Press, New York.

Claude, P., Hawrot, E., Dunis, D. A., and Campenot, R. B. (1982). Binding, internalization, and retrograde transport of ^{125}I-nerve growth factor in cultured rat sympathetic neurons. J. Neurosci. 2, 431–442.

Cleveland, D. W. (1983). The tubulins: From DNA to RNA to protein and back again. Cell 34, 330–332.

Cleveland, D. W., Lopata, M. A., MacDonald, R. J., Rutter, W. J., and Kirschner, M. W. (1980). Number and evolutionary conservation of alpha- and beta-tubulin and cytoplasmic beta- and gamma-actin genes using specific cloned cDNA probes. Cell 20, 95–105.

Cleveland, D. W., Lopata, M. A., Sherline, P., and Kirschner, M. W. (1981). Unpolymerized tubulin modulates the level of tubulin mRNAs. Cell 25, 537–546.

Cohen, P. (1982). The role of protein phosphorylation in neural and hormonal control of cellular activity. Nature (London) 296, 613–620.

Cohen, P., Sutter, A., Landreth, G., Zimmermann, A., and Shooter, E. M. (1980a). Oxidation of tryptophan-21 alters the biological activity and receptor binding characteristics of mouse nerve growth factor. J. Biol. Chem. 255, 2949–2954.

Cohen, S., Carpenter, G., and King, L. E. (1980b). Epidermal growth factor–receptor–protein kinase interactions. Co-purification of receptor and epidermal growth factor-enhanced phosphorylation activity. J. Biol. Chem. 255, 4834–4842.

Costrini, N. V., Kogan, M., Kukreja, K., and Bradshaw, R. A. (1979). Physical properties of the detergent-extracted nerve growth factor receptor of sympathetic ganglia. J. Biol. Chem. 254, 11242–11246.

Coughlin, M. D., Boyer, D. M., and Black, I. B. (1977). Embryologic development of a mouse sympathetic ganglion in vivo and in vitro.. Proc. Natl. Acad. Sci. U.S.A. 74, 3438–3442.

Cuatrecasas, P., and Hollenberg, M. D. (1975). Binding of insulin and other hormones to non-receptor materials: saturability, specificity and apparent "negative cooperativity." Biochem. Biophys. Res. Commun. 62, 31–41.

Cummings, R., Burgoyne, R. D., and Lytton, N. A. (1982). Differential immunocytochemical localisation of alpha-tubulin and beta-tubulin in cerebellum using monoclonal antibodies. Cell. Biol. Int. Rep. 6, 1047–1053.

Czech, M. P., and Massague, J. (1982). Subunit structure and dynamics of the insulin receptor. Fed. Proc., Fed. Am. Soc. Exp. Biol. 41, 2719–2723.

Dasgupta, J. D., and Garber, D. L. (1983). Tryrosine protein kinase activity during embryogenesis. J. Biol. Chem. 258, 6174–6178.

Daughaday, W. H., Parker, K. A., Borowsky, S., Trivedi, B., and Kapadia, M. (1982). Measurement of somatomedin-related peptides in fetal, neonatal, and maternal rat serum by IGF-I radioimmunoassay, IGF-II radioreceptor assay, and MSA radioreceptor assay after acid–ethanol extraction. Endocrinology 110, 575–581.

De Meyts, P. (1976). Cooperative properties of hormone receptor in cell membranes. J. Supramol. Struct. 4, 241–258.

Dull, T. J., Gray, A., Hayflick, F. S., and Ullrich, A. (1984). Insulin-like growth factor II precursor gene organization in relation to the insulin gene family. Nature (London) 310, 777–781.

Ebina, Y., Ellis, L., Jarnagin, K., Edery, M., Graf, L., Clauser, E., Ou, J., Masiarz, F., Kan, Y. W., Goldfine, I. D., Roth, R. A., and Rutter, W. J. (1985). The human insulin receptor cDNA: The structural basis for hormone-activated transmembrane signalling. Cell 40, 747–758.

Ehrlich, Y. H., Prasad, K. N., Sinha, P. K., Davis, L. G., and Brunngraber, E. G. (1978).

Selective changes in the phosphorylation of endogenous proteins in subcellular fractions from cyclic AMP-induced differentiated neuroblastoma cells. *Neurochem. Res.* **3**, 803–813.

End, D., Tolson, N., Hashimoto, S., and Guroff, G. (1983). Nerve growth factor-induced decrease in the cell-free phosphorylation of a soluble protein in PC12 cells. *J. Biol. Chem.* **258**, 6549–6555.

Fabricant, R. N., De Larco, J. E., and Todaro G. J. (1977). Nerve growth factor receptors on human melanoma cells in culture. *Proc. Natl. Acad. Sci. U.S.A.* **74**, 565–569.

Fernyhough, P., and Ishii, D. N. (1987). Nerve growth factor modulates tubulin transcript levels in pheochromocytoma PC12 cells. *Neurochem. Res.* **12**, 891–899.

Francke, U., De Martinville, B., Coussens, L., and Ullrich, A. (1983). The human gene for the beta subunit of nerve growth factor is located on the proximal short arm of chromosome 1. *Science* **222**, 1248–1251.

Frankel, F. R. (1976). Organization and energy-dependent growth of microtubules in cells. *Proc. Natl. Acad. Sci. U.S.A.* **73**, 2798–2802.

Frazier, W. A., Ohlendorf, C. E., Boyd, L. F., Aloe, L., Johnson, E. M., Ferrendelli, J. A., and Bradshaw, R. A. (1973). Mechanism of action of nerve growth factor and cyclic AMP on neurite outgrowth in embryonic chick sensory ganglia: Demonstration of independent pathways of stimulation. *Proc. Natl. Acad. Sci. U.S.A.* **70**, 2448–2452.

Goldstein, A., Lowney, L. I., and Pal, B. K. (1971). Stereospecific and non-specific interactions of the morphine congener levorphanol in subcellular fractions of mouse brain. *Proc. Natl. Acad. Sci. U.S.A.* **69**, 1742–1747.

Goodyer, C. G., De Stephano, L., Lai, W. H., Guyda, H. J., and Posner, B. I. (1984). Characterization of insulin-like growth factor receptors in rat anterior pituitary, hypothalamus, and brain. *Endrocrinology* **114**, 1187–1195.

Gorin, P. D., and Johnson, E. M. (1980). Effects of long-term nerve growth factor deprivation on the nervous system of the adult rat: An experimental autoimmune approach. *Brain Res.* **198**, 27–42.

Gozes, I., and Sweadner, K. J. (1981). Multiple brain tubulin forms are expressed by a single neuron. *Nature (London)* **276**, 411–413.

Grafstein, B. (1977). *In* "Cellular Biology of Neurons, Part I, Handbook of Physiology, Section 1: The Nervous System, Vol. I" (E. R. Kandel, ed.), pp. 691–717. Amer. Physiol. Soc., Bethesda, Maryland.

Grafstein, B., and McQuarrie, I. G. (1978). Role of the nerve cell body in axonal regeneration. *In* "Neuronal Plasticity" (C.W. Cotman, ed.), pp. 156–195. Raven, New York, New York.

Greene, L. A., Liem, R. K. H., and Shelanski, M. L. (1983). Regulation of a high molecular weight microtubule-associated protein in PC-12 cells by nerve growth factor. *J. Cell. Biol.* **96**, 76–83.

Grob, P. M., Ross, A. H., Koprowski, H., and Bothwell, M. (1985). Characterization of the human melanoma nerve growth factor receptor. *J. Biol. Chem.* **260**, 8044–8049.

Gundersen, R. W., and Barrett, J. N. (1980). Characterization of the turning response of dorsal root neurites toward nerve growth factor. *J. Cell Biol.* **87**, 546–554.

Halegoua, S., and Patrick, J. (1980). Nerve growth factor mediates phosphorylation of specific proteins. *Cell* **22**, 571–581.

Hall, J. L., Dudley, L., Dobner, P. R., Lewis, S. A., and Cowan, N. J. (1983). Identification of two human beta-tubulin isotypes. *Mol. Cell. Biol.* **3**, 854–862.

Hama, T., and Guroff, G. (1985). Distribution of Nsp100 and Nsp100 kinase, a nerve growth factor-sensitive phosphorylation system, in rat tissues. *J. Neurochem.* **45**, 1279–1287.

Hama, T., Huang, K.-P., and Guroff, G. (1986). Protein kinase C as a component of a nerve growth factor-sensitive phosphorylation system in PC12 cells. *Proc. Natl. Acad. Sci. U.S.A.* **83**, 2353–2357.

Hamburger, V., and Oppenheim, R. W. (1982). Naturally occurring neuronal death in vertebrates. *Neurosci. Comment.* **1**, 39–55.

Harper, G. P., and Thoenen, H. (1980). Nerve growth factor: Biological significance, measurement, and distribution. *J. Neurochem.* **34**, 5–16.

Havercroft, J. C., and Cleveland, D. W. (1984). Programmed expression of β-tubulin genes during development and differentiation of the chicken. *J. Cell Biol.* **99**, 1927–1935.

Havrankova, J., Roth, J., and Brownstein, M. (1978a). Insulin receptors are widely distributed in the central nervous system of the rat. *Nature (London)* **272**, 827–829.

Havrankova, J., Schmechel, D., Roth, J., and Brownstein, M. (1978b). Identification of insulin in rat brain. *Proc. Natl. Acad. Sci. U.S.A.* **75**, 5737–5741.

Havrankova, J., Brownstein, M., and Roth, J. (1981). Insulin and insulin receptors in rodent brain. *Diabetologia* **20**, 268–273.

Hendricks, S. A., de Pablo, F., and Roth. J. (1984). Early development and tissue-specific patterns of insulin binding in chick embryo. *Endocrinology* **115**, 1315–1323.

Hendry, I. A. (1977). The effect of the retrograde axonal transport of NGF on the morphology of adrenergic neurons. *Brain Res.* **134**, 213–223.

Hendry, I. A., Stockel, K., Thoenen, H., and Iversen, L. L. (1974). The retrograde axonal transport of nerve growth factor. *Brain Res.* **68**, 103–121.

Herrup, K., and Shooter, E. M. (1973). Properties of the β-nerve growth factor receptor of avian dorsal root ganglia. *Proc. Natl. Acad. Sci. U.S.A.* **70**, 3884–3888.

Herzog, W., and Weber, K. (1978). Microtubule formation by pure brain tubulin *in vitro*: The influence of dextran and poly(ethylene glycol). *Eur. J. Biochem* **92**, 1–8.

Hier, D. B., Arnason, B. G., and Young, M. (1972). Studies on the mechanism of action of nerve growth factor. *Proc. Natl. Acad. Sci. U.S.A.* **69**, 2268–2272.

Hosang, M., and Shooter, E. M. (1985). Molecular characteristics of nerve growth factor receptors on PC12 cells. *J. Biol. Chem.* **260**, 665–662.

Ishii, D. N. (1978). Effect of tumor promoters on the response of cultured embryonic chick ganglia to nerve growth factor. *Cancer Res.* **38**, 3886–3893.

Ishii, D. N. (1982a). Effect of the suspected tumor promoters saccharin, cyclamate, and phenol on nerve growth factor binding and response in cultured embryonic chick ganglia. *Cancer Res.* **42**, 429–432.

Ishii, D. N. (1982b). Inhibition of iodionated nerve growth factor binding by the suspected tumor promoters saccharin and cyclamate. *J. Natl. Cancer Inst.* **68**, 299–303.

Ishii, D. N., and Shooter, E. M. (1975). Regulation of nerve growth factor synthesis in mouse submaxillary gland by testosterone. *J. Neurochem.* **25**, 843–851.

Ishii, D. N., Recio-Pinto, E., Spinelli, W., Mill, J. F., and Sonnenfeld, K. H. (1985). Neurite formation modulated by nerve growth factor, insulin, and tumor promoter receptors. *Int. J. Neurosci.* **26**, 109–127.

Jacobs, S., Schechter, V., Bissell, K., and Cuatrecasas, P. (1977). Purification and properties of insulin receptors from rat liver membranes. *Biochem. Biophys. Res. Commun.* **77**, 981–988.

Jacobs, S., Hazum, E., Schechter, Y., and Cuatrecasas, P. (1979). Insulin receptor: Covalent labeling and identification of subunits. *Proc. Natl. Acad. Sci. U.S.A.* **76**, 4918–4921.

Jacobs, S., Kull, F. C., Earp, H. S., Svoboda, M. E., van Wyk, J. J., and Cuatrecasas, P. (1983). Somatomedin-C stimulates the phosphorylation of the β subunit of its own receptor. *J. Biol. Chem.* **258**, 9581–9584.

Johnson, E. M., Gorin, P. D., Brandeis, L. D., and Pearson, J. (1980). Dorsal root ganglion neurons are destroyed by exposure *in utero* to maternal antibody to nerve growth factor. *Science* **210**, 916–918.

Kadowaki, T., Fujita-Yamaguchi, Y., Nishida, E., Takaku, F., Akiyama, T., Kathuria, S., Akanuma, Y., and Kasuga, M. (1985). Phosphorylation of tubulin and microtubule-associated proteins by the purified insulin receptor kinase. *J. Biol. Chem.* **260**, 4016–4020.

Kappy, M., Sellinger, S., and Raizada, M. (1984). Insulin binding in four regions of the developing rat brain. *J. Neurochem.* **42**, 198–203.

Kasuga, M., Van Obberghens, E., Nissley, S. P., and Rechler, M. M. (1981). Demonstration of two subtypes of insulin-like growth factor receptors by affinity cross-linking. *J. Biol. Chem.* **256**, 5305–5308.

Kasuga, M., Karlsson, F. A., and Kahn, C. R. (1982a). Insulin stimulates phosphorylation of the 95,000-dalton subunit of its own receptor. *Science* **215**, 185–187.

Kasuga, M., Zick, Y. Blithe, D. L., Crettaz, M., and Kahn, R. (1982b). Insulin stimulates tyrosine phosphorylation of the insulin receptor in a cell-free system. *Nature (London)* **298**, 667–669.

Kim, H., Binder, L. I., and Rosenbaum, J. L. (1979). The periodic association of MAP$_2$ with brain microtubules *in vitro*. *J. Cell Biol.* **80**. 266–276.

Kolber, A. R., Goldstein, M. N., and Moore, B. W. (1974). Effect of nerve growth factor on the expression of colchicine-binding activity and 14-3-2 protein in an established line of human neuroblastoma. *Proc. Natl. Acad. Sci. U.S.A.* **71**, 4203–4207.

Laemmli, U. K. (1970). Cleavage of structural proteins during the assembly of the head of Bacteriophage T$_4$. *Nature (London)* **227**, 680–685.

Landau, B. R., Takaoka, Y., Abrams, M. A., Genuth, S. M., van Houten, M., Posner, B. I., White, R. J., Ohgaku, S., Horvat, A., and Hemmelgarn, E. (1983). Binding of insulin by monkey and pig hypothalamus. *Diabetes* **32**, 284–292.

Landreth, G. E., and Shooter, E. M. (1980). Nerve growth factor receptors on PC12 cells: Ligand-induced conversion from low- to high-affinity states. *Proc. Natl. Acad. Sci. U.S.A.* **77**, 4751–4755.

Lemischka, I. R., Farmer, S., Rancaniello, V. R., and Sharp, P. A. (1981). Nucleotide sequence and evolution of a mammalian alpha-tubulin messenger RNA. *J. Mol. Biol.* **151**, 101–120.

Letourneau, P. C. (1975a). Possible roles for cell-to-substratum adhesion in neuronal morphogenesis. *Dev. Biol.* **44**, 77–91.

Letourneau, P. C. (1975b). Cell-to-substratum adhesion and guidance of axonal elongation. *Dev. Biol.* **44**, 92–101.

Levi, A., Cimino, M., Mercanti, D., Chen, J. S., and Calissano, P. (1975). Interaction of nerve growth factor with tubulin. Studies on binding and induced polymerization. *Biochim. Biophys. Acta* **399**, 50–60.

Levi-Montalcini, R., and Angeletti, P. U. (1966). Second symposium on catecholamines. Modification of sympathetic function. Immunosympathectomy. *Pharmacol. Rev.* **18**, 619–628.

Levi-Montalcini, R., and Angeletti, P. U. (1968). Nerve growth factor. *Physiol. Rev.* **48**, 534–569.

Levi-Montalcini, R., and Booker, B. (1960). Excessive growth of the sympathetic ganglia evoked by a protein isolated from mouse salivary glands. *Proc. Natl. Acad. Sci. U.S.A.* **46**, 373–384.

Levi-Montalcini, R., and Cohen, S. (1960). Effects of the extract of the mouse salivary glands on the sympathetic system of mammals. *Ann. N.Y. Acad. Sci.* **85**, 324–341.

Levi-Montalcini, R., and Hamburger, V. (1951). Selective growth-stimulating effects of mouse sarcoma on the sensory and sympathetic nervous system of the chick embryo. *J. Exp. Zool.* **116**, 321–362.

Levi-Montalcini, R., Revoltella, R., and Calissano, P. (1974). Microtubule proteins in the nerve growth factor mediated response: Interction between the nerve growth factor and its target cells. *Recent Prog. Horm. Res.* **30**, 635–699.

Lopata, M. A., Havercroft, J. C., Chow, L. T., and Cleveland, D. W. (1983). Four unique genes required for beta-tubulin expression in vertebrates. *Cell* **32**, 713–724.

Lopresti, V., Macagno, E. R., and Levinthal, C. (1973). Structure and development of neuronal connections in isogenic organisms: Cellular interactions in the development of the optic lamina of *Daphnia. Proc. Natl. Acad. Sci. U.S.A.* **70**, 433–437.

Lyons, C. R., Stach, R. W., and Perez-Polo, J. R. (1983). Binding constants of isolated nerve growth factor receptors from different species. *Biochem. Biophys. Res. Commun.* **115**, 368–374.

Mains, R. E., and Patterson, P. H. (1973). Primary cultures of dissociated sympathetic neurons. I. Establishment of long-term growth in culture and studies of differentiated properties. *J. Cell Biol.* **59**, 329–345.

Marquardt, H., Todaro, G. J., Henderson, L. E., and Oroszlan, S. (1981). Purification and primary structure of a polypeptide with multiplication stimulating activity from rat liver cell cultures. *J. Biol. Chem.* **256**, 6859–6865.

Massague, J., and Czech, M. P. (1982). The subunit structures of two distinct receptors for insulin-like growth factors I and II and their relationship to the insulin receptor. *J. Biol. Chem.* **257**, 5038–5045.

Massague, J., Guillette, B. J., and Czech, M. P. (1981). Affinity labeling of multiplication stimulating activity receptors in membranes from rat and human tissues. *J. Biol. Chem.* **256**, 2122–2125.

Menesini-Chen M. G. M., Chen J. S., and Levi-Montalcini, R. (1978). Sympathetic nerve fibers ingrowth in the central nervous system of neonatal rodent upon intracerebral NGF injections *Arch. Ital. Biol.* **116**, 53–84.

Mill, J. F., Chao, M. V., and Ishii, D. N. (1985). Insulin, insulin-like growth factor II, and nerve growth factor effects on tubulin mRNA levels and neurite formation. *Proc. Natl. Acad. Sci. U.S.A.* **82**, 7126–7130.

Mizel, S. B., and Bamburg, J. R. (1975). Studies on the action of nerve growth factor II. Neurotubule protein levels during neurite outgrowth. *Neurobiology* **5**, 283–290.

Mobley, W. C., Server, A. C., Ishii, D. N., Riopelle, R. J., and Shooter, E. M. (1977). Nerve growth factor. *New Engl. J. Med.* **297**, 1096–1104. (Part I); 1149–1158 (Part II); 1211–1218 (Part III).

Moses, A. C., Nissley, S. P., Short, P. A., Rechler, M. M., White, R. M., Knight, A. B., and Higa, O. Z. (1980). Increased levels of multiplication stimulating activity and insulin-like growth factors in fetal rat serum. *Proc. Natl. Acad. Sci. U.S.A.* **77**, 3649–3653.

Murphy, D. B., Valee, R. B., and Borisy, G. G. (1977). Identity and polymerization-stimulatory activity of the nontubulin proteins associated with microtubules. *Biochemistry* **16**, 2598–2605.

Murphy, K. M. N., Gould, R. J., Oster-Granite, M. L., Gearheart, J. D., and Snyder, S. H. (1983). Phorbol ester receptors: Autoradiographic identification in the developing rat. *Science* **222**, 1036–1038.

Nagle, D. S., Jaken, S., Castagna, M., and Blumberg, P. M. (1981). Variation with embryonic development and regional localization of specific [^{3}H]phorbol 12,13-dibutyrate binding to brain. *Cancer Res.* **41**, 89–93.

Nakanishi, N., and Guroff, G. (1985). Nerve growth factor-induced increase in the cell-free phosphorylation of a nuclear protein in PC12 cells. *J. Biol. Chem.* **260**, 7791–7799.

Neumann, D., Scherson, T., Ginzburg, I., Littauer, U. Z., and Schwartz, M. (1983). Regulation of mRNA levels for microtubule proteins during nerve regeneration. *Fed. Eur. Biochem. Soc. Lett.* **162**, 270–276.

Niedel, J. E., Kuhn, L. J., and Vanderbark, G. R. (1983). Phorbol diester receptor copurifies with protein kinase C. *Proc. Natl. Acad. Sci. U.S.A.* **80**, 36–40.

Olender, E. J., and Stach, R. W. (1980). Sequestration of ^{125}I-labeled nerve growth factor by sympathetic neurons. *J. Biol. Chem.* **255**, 9338–9343.

Olender, E. J., Wagner, B. J., and Stach, R. W. (1981). Sequestration of ^{125}I-labeled nerve growth factor by embryonic sensory neurons. *J. Neurochem.* **37**, 436–442.

Pacold, S. T., and Blackard, W. G. (1979). Central nervous system insulin receptors in normal and diabetic rats. *Endocrinology* **105**, 1452–1457.

Perez-Polo, J. R., Werrbach-Perez, K., and Tiffany-Castiglioni, E. (1979). A human clonal cell line model of differentiating neurons. *Dev. Biol.* **71**,341–355.

Pfenninger, K. H., and Maylie-Pfenninger, M.-F. (1981). Lectin labeling of sprouting neurons. II. Relative movement and appearance of glycoconjugates during plasmalemmal expansion. *J. Cell Biol.* **89**, 547–559.

Posner, B. I., Kelly, P. A., Shiu, R. P. C., and Friesen, H. G. (1974). Studies of insulin, growth hormone and prolactin binding: Tissue distribution, species variation and characterization. *Endocrinology* **95**, 521–531.

Prasad, K. N. (1975). Differentiation of neuroblastoma cells in culture. *Biol. Rev.* **50**, 129–165.

Puma, P., Buxser, S. E., Watson, L., Kelleher, D. J., and Johnson, G. L. (1983). Purification of the receptor for nerve growth factor from A875 melanoma cells by affinity chromatography. *J. Biol. Chem.* **258**, 3370–3375.

Puro, D. G., and Agardh, E. (1984). Insulin-mediated regulation of neuronal maturation. *Science* **225**, 1170–1172.

Puro, D. G., Battelle, B.-A., and Hansmann, K. E. (1982). Development of cholinergic neurons of the rat retina. *Dev. Biol.* **91**, 138–148.

Purves, D., and Lichtman, J. W. (1978). Formation and maintenance of synaptic connections in autonomic ganglia. *Physiol. Rev.* **58**, 821–862.

Raizada, M. K., Stamler, J. F., Quinlan, J. T., Landas, S., and Phillips, M. I. (1982). Identification of insulin receptor-containing cells in primary cultures of rat brain. *Cell. Mol. Neurobiol.* **2**, 47–52.

Rakic, P. (1981). Neuronal–glial interaction during brain development. *Trends Neurosci.* **4**, 184–187.

Rechler, M. M., Kasuga, M., Sasaki, N., de Vroede, M. A., Romanus, J. A., and Nissley, S. P. (1983). Properties of insulin-like growth factor receptor subtype. *In* "Insulin-like growth factors/Somatomedins: Basic Chemistry, Biology, and Clinical Importance" (E. M. Spencer, ed.), pp. 459–490. De Gruyter, New York.

Recio-Pinto, E., and Ishii, D. N. (1984). Effects of insulin, insulinlike growth factor-II, and nerve growth factor on neurite outgrowth in cultured human neuroblastoma cells. *Brain Res.* **302**, 323–334.

Recio-Pinto, E., Lang, F. F., and Ishii, D. N. (1984). Insulin and insulin-like growth factor II permit nerve growth factor binding and the neurite formation response in cultured human neuroblastoma cells. *Proc. Natl. Acad. Sci. U.S.A.* **81**, 2562–2566.

Recio-Pinto, E., Rechler, M. M., and Ishii, D. N. (1986). Effects of insulin, insulinlike growth factor-II and nerve growth factor on neurite formation and survival in cultured sympathetic, and sensory neurons. *J. Neurosci.* **6**, 1211–1219.

Recio-Pinto, E., and Ishii, D. N. (1987). Insulin and insulin-like growth factor receptors regulating neurite outgrowth in cultured human neuroblastoma cells. *J. Neurosci. Res.*, in press.

Rice, B. L., and Stach, R. W. (1976). The dimer–monomer equilibrium constant for [^{125}I]beta nerve growth factor. *Biochem. Biophys. Res. Commun.* **73**, 479–485.

Rinderknecht, E., and Humbel, R. E. (1978a). Primary structure of human insulin-like growth factor-II. *Fed. Eur. Biochem. Soc. Lett.* **89**, 283–286.

Rinderknecht, E., and Humbel, R. E. (1978b). The amino acid sequence of human IGF-I and its structural homology with proinsulin. *J. Biol. Chem.* **253**, 2769–2776.

Riopelle, R. J., Klearman, M., and Sutter, A. (1980). Nerve growth factor receptors: Analysis of the interaction of NGF with membranes of chick embryo dorsal root ganglia. *Brain Res.* **199**, 63–77.

Roederer, E., and Cohen, M. J. (1983). Regeneration of an identified central neuron in the cricket. I. Control of sprouting from soma, dendrites, and axon. *J. Neurosci.* **3**, 1835–1847.

Rosenfeld, R. G., Ceda, G., Wilson, D. M., Dollar, L. A., and Hoffman, A. R. (1984). Characterization of high affinity receptors for insulin-like growth factors I and II in rat anterior pituitary cells. *Endocrinology* **114**, 1571–1575.

Ross, A. H., Grob, P., Bothwell, M., Elder, D. E., Ernst, C. S., Marano, N., Ghrist, B. F. D., Slemp, C. C., Herlyn, M., Atkinson, B., and Koprowski, H. (1984). Characterization of nerve growth factor receptor in neural crest tumors using monoclonal antibodies.. *Proc. Natl. Acad. Sci. U.S.A.* **81**, 6681–6685.

Rutishauser, U., Gall, W. E., and Edelman, G. M. (1978). Adhesion among neural cells of the chick embryo. IV. Role of the cell surface molecule CAM in the formation of neurite bundles in cultures of spinal ganglia. *J. Cell Biol.* **79**, 382–393.

Sara, V.R., Hall, K., Misaki, M., Fryklund, L., Christensen, N., and Wetterberg, L. (1983). Ontogenesis of somatomedin and insulin receptors in the human fetus. *J. Clin. Invest.* **71**, 1084–1094.

Schechter, A. L., and Bothwell, M. A. (1981). Nerve growth factor receptors on PC12 cells: Evidence for two receptor classes with differing cytoskeletal association. *Cell* **24**, 867–874.

Schwartz, J. H. (1979). Axonal transport: Components, mechanisms, and specificity. *Annu. Rev. Neursci.* **2**, 467–504.

Scott, J., Selby, M., Urdea, M., Quiroga, M., Bell, G. I., and Rutter, W. J. (1983). Isolation and nucleotide sequence of a cDNA encoding the precursor of mouse nerve growth factor. *Nature (London)* **302**, 538–540.

Shankland, M., Goodman, C. S. (1982). Development of the dendritic branching pattern of the medial giant interneuron in the grasshopper embryo. *Dev. Biol.* **92**, 489–506.

Shelton, D. L., and Reichardt, L. F. (1984). Expression of the beta nerve growth factor gene correlates with the density of sympathetic innervation in effector organs. *Proc. Natl. Acad. Sci. U.S.A.* **81**, 7951–7955.

Shooter, E. M., Yanker, B. A., Landreth, G. E., and Sutter, A. (1981). Biosynthesis and mechanism of action of nerve growth factor. *Recent Prog. Horm. Res.* **37**, 417–446.

Soares, M. B., Ishii, D. N., and Efstratiadis, A. (1985). Developmental and tissue specific expression of a family of transcripts related to rat insulin-like growth factor II mRNA. *Nucleic Acids Res.* **13**, 1119–1134.

Soares, M. B., Turken, A., Ishii, D., Mills, L., Episkopou, V., Cotter, S., Zeitlin, S., and Efstratiadis, A. (1986). Rat insulin-like growth factor II gene: A single gene with two promoters expressing a multitranscript family. *J. Mol. Biol.* **192**, 737–752.

Sonderegger, P., Fishman, M. C., Bokoum, M., Bauer, H. C., and Nelson, P. G. (1983). Axonal proteins of presynaptic neurons during synaptogenesis. *Science* **221**, 1294–1297.

Sonnenfeld, K. H., and Ishii, D. N. (1982). Nerve growth factor effects and receptors in cultured human neuroblastoma cell lines. *J. Neurosci. Res.* **8**, 375–391.

Sonnenfeld, K. H., and Ishii, D. N. (1985). Fast and slow nerve growth factor binding sites in human neuroblastoma and rat pheochromocytoma cell lines: Relationship of sites to each other and to neurite formation. *J. Neurosci.* **5**, 1717–1728.

Sonnenfeld, K. H., Bernd, P., Sobue, G., Lebwohl, M., and Rubenstein, A. B. (1986). Nerve growth factor receptors on dissociated neurofibroma Schwann-like cells. *Cancer Res.* **46**, 1446–1452.

Spinelli, W., and Ishii, D. N. (1983). Tumor promoter receptors regulating neurite formation in cultured human neuroblastoma cells. *Cancer Res.* **43**, 4119–4125.

Spinelli, W., Sonnenfeld, K. H., and Ishii, D. N. (1982). Paradoxical effect of phorbol ester tumor promoters on neurite outgrowth in cultured human neuroblastoma cells. *Cancer Res.* **42**, 5067–5073.

Stach, R. W., and Shooter, E. M. (1974). The biological activity of cross-linked beta nerve growth factor protein. *J. Biol. Chem.* **249**, 6668–6674.

Stephenson, R. P. (1956). A modification of receptor theory. *Br. J. Pharmacol. Chemother.* **11**, 379–393.

Stoeckel, K., Salomon, F., Paravicini, U., and Thoenen, H. (1974). Dissociation between effects of nerve growth factor on tyrosine hydroxylase and tubulin synthesis in sympathetic ganglia. *Nature (London)* **250**, 150–151.

Sullivan, K. F., and Wilson, L. (1984). Developmental and biochemical analysis of chick brain tubulin heterogeneity. *J. Neurochem.* **42**, 1363–1371.

Sutter, A., Riopelle, R. J., Harris-Warrick, R. M., and Shooter, E. M. (1979). Nerve growth factor receptors: Characterization of two distinct classes of binding sites on chick embryo sensory ganglia cells. *J. Biol. Chem.* **254**, 4972–4982.

Tennyson, V. (1970). The fine structure of the axon and growth cone of the dorsal root neuroblast of the rabbit embryo. *J. Cell Biol.* **44**, 62–70.

Togari, A., and Guroff, G. (1985). Partial purification and characterization of a nerve growth factor-sensitive kinase and its substrate from PC12 cells. *J. Biol. Chem.* **260**, 3804–3811.

Tricoli, J. B., Rall, L. B., Scott, J., Bell, G. I., and Shows, T. B. (1984). Localization of insulin-like growth factor genes to human chromosomes 11 and 12. *Nature (London)* **310**, 784–786. 784–786.

Ullrich, A., Gray, A., Berman, C., and Dull, T. J. (1983). Human beta nerve growth factor gene sequence highly homologous to that of mouse. *Nature (London)* **303**, 821–825.

Van Houten, M., Posner, B. I., Kopriwa, B. M., and Brawer, J. R. (1979). Insulin binding sites in the rat brain: *In vivo* localization to the circumventricular organs by quantitative radioautography. *Endocrinology* **105**, 666–673.

Varon, S. (1975). Nerve growth factor and its mode of action. *Exp. Neurol.* **48**, 75–92.

Varon, S., Nomura, J., and Shooter, E. M. (1967). Isolation of mouse nerve growth factor in high molecular weight form. *Biochemistry* **7**, 1296–1303.

Vinores, S. and Guroff, G. (1980). Nerve growth factor: Mechanism of action. *Annu. Rev. Biophys. Bioeng.* **9**, 223–257.

Whitfield, H. J., Bruni, C. B., Frunzio, R., Terrell, J. E., Nissley, S. P., and Rechler, M. M. (1984). Isolation of a cDNA clone encoding rat insulin-like growth factor-II precursor. *Nature (London)* **312**, 277–280.

Yamada, K. M., and Wessels, N. K. (1971). Axon elongation: Effect of nerve growth factor on microtubule protein. *Exp. Cell Res.* **66**, 346–352.

Yamada, K. M., Spooner, B. S., and Wessells, N. K. (1971). Ultrastructure and function of growth cones and axons of cultured nerve cells. *J. Cell Biol.* **49**, 614–635.

Yip, C. C., Yeung, C. W. T., and Moule, M. L. (1978). Photoaffinity labeling of insulin receptor of rat adipocyte plasma membrane. *J. Biol. Chem.* **253**, 1743–1745.

Young, W. S., III, Kuhar, M. J., Roth, J., and Brownstein, M. J. (1980). Radiohistochemical localization of insulin receptors in the adult and developing rat brain. *Neuropeptides* **1**, 15–22.

Yu, M. W., Tolson, N. W., and Guroff, G. (1980). Increased phosphorylation of specific nuclear proteins in superior cervical ganglia and PC12 cells in response to nerve growth factor. *J. Biol. Chem.* **255**, 10481–10492.

Zahniser, N. R., Goens, M. B., Hanaway, P. J., and Vinych, J. V. (1984). Characterization and regulation of insulin receptors in rat brain. *J. Neurochem.* **42**, 1354–1362.

Zapf, J., Schoenle, E., and Froesch, E. R. (1978). Insulin-like growth factors I and II. Some biological actions and receptor binding characteristics of two purified constituents of non-suppressible insulin-like activity of human serum. *Eur. J. Biochem.* **87**, 285–296.

Zapf, J., Froesh, E. R., and Humbel R. E. (1981). The insulin-like growth factors (IGF) of human serum: Chemical and biological characterization and aspects of their possible physiological role. *Curr. Top. Cell. Regul.* **19**, 257–309.

Zimmermann, A., and Sutter, A. (1983). Beta nerve growth factor (β-NGF) receptors on glial cells. Cell–cell interaction between neurons and Schwann cells in cultures of chick sensory ganglia. *EMBO J.* **2**, 879–885.

Zimmermann, A., Sutter, A., and Shooter, E. M. (1981). Monoclonal antibodies against nerve growth factor and their effects on receptor binding and biological activity. *Proc. Natl. Acad. Sci. U.S.A.* **78**, 4611–4615.

Pituitary Regulation by Gonadotropin-Releasing Hormone (GnRH): Gonadotropin Secretion, Receptor Numbers, and Target Cell Sensitivity

WILLIAM C. GOROSPE AND P. MICHAEL CONN

Department of Pharmacology
University of Iowa
College of Medicine
Iowa City, Iowa 52242

I. INTRODUCTION

Luteinizing hormone (LH) and follicle-stimulating hormone (FSH), the gonadotropins, are glycoprotein hormones secreted by the anterior pituitary. These hormones regulate sperm and ovum maturation as well as steroidogenesis in the gonads. During the past two decades, mainly due to methodological improvements, a great deal has been learned about factors controlling gonadotropin release. Although steroids, opiates, and other neuromodulators have been shown to affect the secretion of LH and FSH either by a direct action at the anterior pituitary gland or by indirect means, the actual physiological stimulus triggering release of the gonadotropins is the hypothalamic decapeptide gonadotropin-releasing hormone (GnRH). Consequently, the secretory patterns of LH and FSH *in vivo* are regulated by complex interactions between GnRH and other modulatory substances. Because these interactions are ultimately deciphered at the level of the gonadotrope, it is of interest to understand the molecular mechanism by which GnRH regulates pituitary function. In this article, the current status of our understanding of the cellular mechanism and regulatory processes involved in the GnRH (stimulus)–gonadotropin release (response) coupling phenomenon will be discussed and a working model proposed.

II. MOLECULAR CHARACTERISTICS OF GnRH

Gonadotropin-releasing hormone is a hypothalamic decapeptide: pyro-Glu^1-His^2-Trp^3-Ser^4-Tyr^5-Gly^6-Leu^7-Arg^8-Pro^9-Gly^{10}-NH_2 . Both termini of the GnRH molecule are blocked; the carboxyl end is amidated and the amino end is internally cyclized by a bond between the α-amino and carbonyl moieties. These features enhance the resistance of the molecule to chemical attack by amino and carboxyl peptidases. Analysis of GnRH has revealed that the favored structure (based on lowest energy state) of the molecule is a configuration in which the $pyroGlu^1$ and Gly^{10}-NH_2 are in close proximity (Fig. 1; Momany, 1976). It has been suggested that in this conformation the pyrrolidone carbonyl group can form a hydrogen bond with the glycinamide group and enhance the stability of this configuration (Coy *et al.*, 1979; Nikolics *et al.*, 1977).

In general, the GnRH molecule may be envisioned as being composed of two physical domains (Conn *et al.*, 1985b). One domain includes that region of the molecule which is responsible for recognition and binding of GnRH to its specific receptor. The other domain is responsible for activation of the receptor once the ligand binds. In the case of a GnRH agonist, both domains are present. A competitive antagonist would contain only a domain that binds to the receptor; the activation domain is absent. Structure–activity studies of

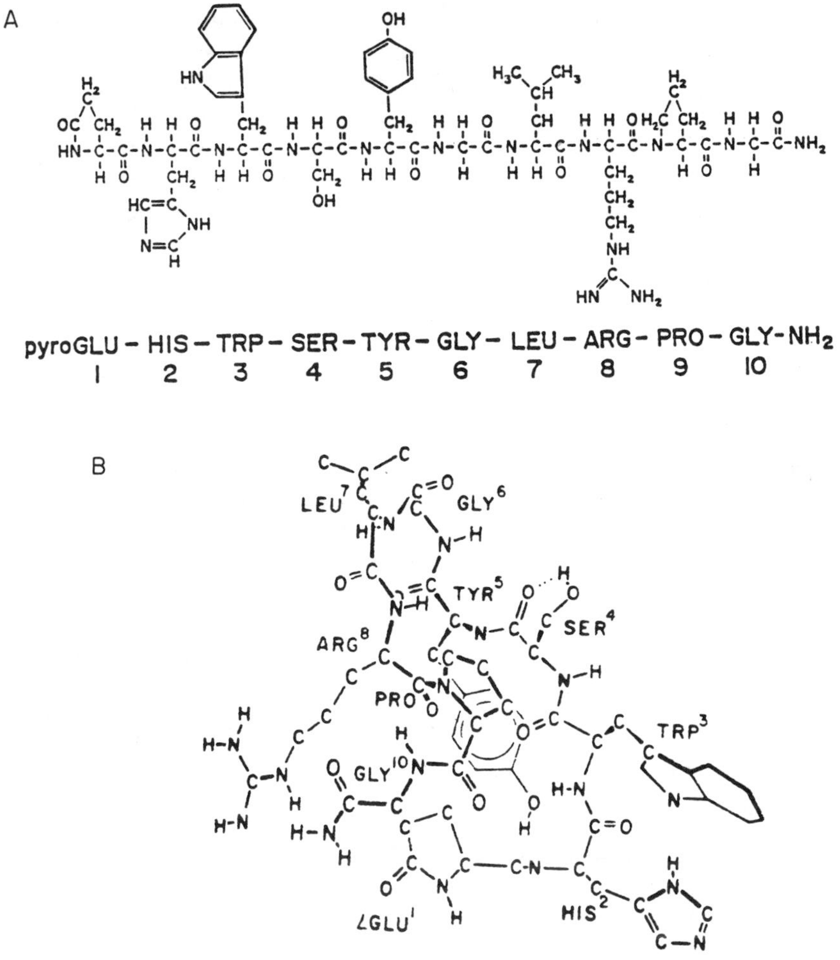
A
pyroGLU – HIS – TRP – SER – TYR – GLY – LEU – ARG – PRO – GLY-NH2
1 2 3 4 5 6 7 8 9 10
B
LEU7
GLY6
TYR5
SER4
ARG8
PRO9
GLY10
TRP3
HIS2
GLU1

FIG. 1. Structure of GnRH. (A) Linear amino acid composition. (B) Schematic representation of the three-dimensional configuration. (From Momany, 1976.)

GnRH have identified individual amino acid residues in the releasing hormone that correspond to these domains. For instance, His^2 and Trp^3 are believed to be principal components in the *activation domain* of the molecule since substitution of these two amino acids results in loss of biological activity although binding still occurs (Conn *et al.,* 1984b). $Pyro\text{-}Glu^1$ and Gly^{10} appear to play major roles in the binding character of the molecule. Modification at either end of GnRH significantly alters the affinity of the peptide for the receptor (Coy *et al.*, 1979). Gly^6 provides a flexible hinge by which the peptide can enter the β-turn needed to maintain the biologically active conformation. Substitution at position 6 with D-amino acids, particularly those containing hydrophobic side chains, results in the production of GnRH derivatives with increased potency (Coy et al., 1976). Thus, D-Ala[6]-GnRH is six- to sevenfold more potent than GnRH (Coy et al., 1976; Fujino et al., 1974). D-Phe[6]-GnRH and D-Trp[6]-GnRH were found to be longer lasting and 10 and 13 times more potent than GnRH, respectively (Coy et al., 1976). Much of the enhanced activity displayed by these analogues is attributed to stabilization of this β-turn by the D-amino acids (Monahan et al., 1973). D-Amino acids at position 6 also stabilize the peptide against biological degradation.

When substituted D-amino acids in the sixth position are combined with substitution of ethylamide for Gly[10], even greater enhancement in binding affinity and potency is observed (Fujino et al., 1974). For example, [D-Ser(t-Bu)[6]-des-Gly[10]]GnRH-ethylamide is 17- to 19-fold more potent than GnRH; [D-Ala[6]-des-Gly[10]]GnRH-ethylamide is 30- to 40-fold more potent than GnRH; and [D-Leu[6]-des-Gly[10]]GnRH-ethylamide is 53 times higher in potency (Coy et al., 1974, 1975; Vilchez-Martinez et al., 1974). The remaining amino acid residues are probably required to maintain proper steric configuration of the molecule for ligand–receptor interactions and biologial response. Indeed, shortening the molecule leads to substantial loss of biological activity (Schally et al., 1972). For greater detail concerning the structure–activity relationships of GnRH and its derivatives, comprehensive reviews are available (Coy et al., 1979; Nestor et al., 1984; Karten and Rivier, 1986).

III. PLASMA MEMBRANE RECEPTOR FOR GnRH

A. Receptor Characterization

Biochemical evidence suggests that the GnRH receptor is a plasma membrane protein (Marian and Conn, 1983). Indeed, pituitary membranes pretreated with proteolytic enzymes such as trypsin or chymotrypsin no longer bind GnRH analogs (Hazum, 1981a; Hazum and Keinan, 1984). Glycosidic enzymes, such as neuraminidase, as well as lectin wheat-germ agglutinin also diminish the binding of GnRH agonists to plasma membranes of the pituitary. This observation suggests that the GnRH receptor may contain sialic acid in the region of the active binding site (Hazum, 1982).

GnRH-binding activity has been recovered from solubilized plasma membranes of bovine gonadotropes using 3[(3-cholamidopropyl)dimethylammonio]-1-propanesulfonate (CHAPS), a zwitterionic detergent (Perrin et al., 1983; Winiger et al., 1983). Chromatographic analysis by gel filtration with Sepharose 6B revealed that the GnRH-binding activity migrates with an apparent molecular weight of 80K (Winiger *et al.*, 1983). Hazum (1981b), using a ^{125}I-labeled photoreactive GnRH analog D-Lys^6-[N^ϵ-azidobenzoyl]-GnRH, identified a major protein band (MW 60K) with sodium dodecyl sulfate–polyacrylamide gel electrophoresis (SDS-PAGE) following solubilization of the covalently coupled antagonist–receptor complex. Using immunoblotting, Eidne *et al.* (1985) confirmed Hazum's size estimation of a 60K GnRH receptor. It should be noted that the estimates by Hazum (1981b) and Eidne *et al.* (1985) are on fractions that had been denatured in sodium dodecyl sulfate. Accordingly, they provide estimates of a GnRH-binding component only; this may represent a subunit of the intact receptor.

Conn and Venter (1985), employing the technique of radiation inactivation and target size analysis, estimated the functional molecular weight of the GnRH receptor to be approximately 136K. A distinction of this methodology is that it measures molecular weight of molecules while still components of the plasma membrane. The determination is based on the observation that there exists an inverse correlation between the size of a macromolecule and the dose-dependent inactivation of that molecule by ionizing radiation (Conn and Venter, 1985; Venter, 1985). Conn and Venter (1985) purified plasma membranes from rat pituitary tissue, layered the fractions to a depth of 0.5 mm in open aluminum trays, and froze the samples in liquid nitrogen. A set of calibration enzymes with known molecular weights was prepared in a similar manner. The trays were fitted in an aluminum, air-tight target chamber and flushed with nitrogen to maintain $-45\,°C$ and to remove oxygen. Irradiation for varying intervals of time was performed with a 0.5-mA beam of 1.5-MeV electrons produced by a Van der Graff electron generator. Following electron bombardment, samples were stored at $-80\,°C$ until the amount of binding activity for the GnRH receptor or enzymatic activity of the standards could be evaluated. The number but not the affinity of GnRH receptors declined proportionally to the amount of radiation exposure. This indicated that the ionizing radiation caused destruction and not partial damage to the GnRH binding site. Since the enzymatic activity of the calibrating standards declined as a simple exponential function of the radiation dosage, it was possible to establish an inverse relationship between the dose of radiation and the degree of enzyme inactivation (*inactivation ratio*) for each standard. This ratio was then related to the known molecular weights of the standards. A direct comparison between the amount of radiation required to inactivate the GnRH receptor and that amount required to inactivate the standards can then be related to the inactivation ratio of the standards and the molecular weight of the GnRH receptor estimated (for details of these calculations, see Venter, 1985). The 136K estimate of the GnRH receptor, significantly higher than the previously reported molecular weight determinations, suggests that in the plasma membrane the receptor may exist as a multisubunit complex.

B. GnRH-Receptor Interactions

Much of what is known about the binding parameters of GnRH to its receptor has arisen from the availability of superactive derivatives of the GnRH molecule as well as the development of reliable radioreceptor and radioligand assays (Clayton *et al.*, 1979). With the use of these superactive analogs, it has been possible to demonstrate the presence of a single class of high-affinity (K_a = 1–3 × 10^{10} M^{-1}) receptors in the anterior pituitary gland which is also specific and saturable (Clayton *et al.*, 1979; Clayton and Catt, 1980, 1981a; Conne *et al.*, 1979; Marian *et al.*, 1981; Marian *et al.*, 1983; Perrin *et al.*, 1980; Wagner *et al.*, 1979).

Recognition and binding of GnRH to its receptor on the gonadotrope plasma membrane appear to constitute the first of the molecular events leading to the release of the pituitary gonadotropins. Following receptor occupancy, there is measurable lateral mobility of the ligand–receptor complexes in the plane of the cell surface. This movement, observed in dispersed pituitary cell cultures with the aid of a bioactive, fluorescently labeled GnRH derivative and image-intensified microscopy, consists of patching, large-scale capping, and eventual internalization of the ligand–receptor unit into endocytotic vesicles (Hazum *et al.*, 1980). This phenomenon of internalization of the GnRH–receptor complex has received further support through morphological and biochemical studies (Duello and Nett, 1980; Duello *et al.*, 1981; Jennes *et al.*, 1983a; Pelletier *et al.*, 1982). The contribution of these receptor-mediated cell-surface phenomena to the stimulatory actions of GnRH on gonadotropin secretion remains uncertain. Indeed, Conn and Hazum (1981) demonstrated that large-scale clustering and internalization of the GnRH–receptor complex can be prevented by vinblastine, without affecting GnRH-stimulated LH release. Additionally, at a time when a significant degree of internalization has occurred, removal of external GnRH results in an immediate, precipitous fall in LH secretion (Conn and Hazum, 1981; Bates and Conn, 1984). Thus it appears unlikely that patching, capping, or internalization plays an essential role in the mechanisms of GnRH-stimulated LH release because they can be effectively uncoupled from the biological response (Conn *et al.*, 1981c; Conn and Hazum, 1981).

The resolution limitations of image-intensification microscopy (50–100 molecules) do not allow for the visual monitoring of the interaction of small numbers of GnRH receptors (Conn and Hazum, 1981; Pastan and Willingham, 1981). It has been suggested (Conn *et al.*, 1982a, 1985b) that once occupied, and possibly preceding large-scale patching, the GnRH receptors undergo a certain degree of localized cross-linking (i.e., dimerization, trimerization, etc.), termed *microaggregation*. This interaction between hormone–receptor complexes may be required to elicit a biological response. Indeed, in a number of biological systems, such cross-linking between ligand–receptor units has been shown to be important to the integrity of the stimulus–response coupling mechanisms. For instance, Jacobs and co-workers (1978) demonstrated that insulin-receptor antibodies can effectively simulate the actions of insulin. However, if Fab ' fragments (antigen-binding fragments produced by enzymatic digestion of an IgG molecule with papain) are derived from these insulin-receptor antibodies, there is a retention of the binding properties (i.e., fragments will displace ^{125}I-labeled insulin from its receptor) but they can no longer mimic the actions of insulin. However, bioactivity is restored if the Fab ' fragments are cross-linked (Kahn *et al.*, 1978). Similarly, B lymphocytes, which contain immunoglobulin in their plasma membranes, require at least a bivalent antigen to proliferate and differentiate (Clark, 1980). Cyanogen bromide cleavage of epidermal growth factor (EGF) produces a derivative (CnBr-EGF) that can bind to target tissues (i.e., 3T3 fibroblasts) but cannot stimulate mitogenesis. Restoration of biological activity is accomplished if this derivative binds in the presence of a bivalent anti-EGF (Schechter *et al.*, 1979). Finally, it has been suggested that the IgG responsible for the long-acting thyroid stimulating (LATS) syndrome associated with Graves ' disease may exert its effect by cross-linking thyroid-stimulating hormone (TSH) receptors, in turn activating the adenylate cyclase system of thyroid cells (Blum and Conn, 1982; Conn *et al.*, 1984a; Mehdi and Kriss, 1978).

To explore the potential involvement of microaggregation in the stimulatory actions of GnRH, Conn *et al.* (1982b) used the GnRH antagonist D-p-Glu^1-D-Phe^{2}D-Trp^3-D-Lys^6-GnRH. The D-p-Glu^1-D-Phe^2-D-Trp^3 sequence is responsible for antagonist activity, whereas the D-Lys^6 residue both stabilizes the molecule from degradation and provides a free ϵ-amine group that can be chemically modified without altering the binding properties of the derivative. To provoke localized cross-linking, or microaggregation, the following steps were performed. First, two GnRH antagonist molecules were dimerized through the reactive amino group on the D-Lys^6 residue with the use of a bifunctional cross-linker, ethylene glycol bissuccinimidylsuccinate (EGS) (Conn, 1983). The resulting dimerized ligand was composed of two antagonist molecules which were free to rotate about a common axis—the EGS cross-linker (EGS is approximately 12–15 Å long). Treatment of dispersed pituitary cells with the dimerized ligand inhibited GnRH-stimulated LH release with the same efficacy as the "parent" antagonist (correcting for change in molecular weight). This finding suggested the possibility that the length of the dimerized molecule was too short to bridge the distance between two receptors. This was tested by preparing purified IgG with binding specificity directed toward the N terminus of the GnRH antagonist dimer. When purified, the resulting antigen–antibody complex was an IgG with a GnRH antagonist dimer at the end of each Fab ' arm; one end was bound by the antibody and the other was free to rotate in space about the EGS bridge. Because the distance between Fab ' arms is approximately 120–150 Å and the calculated maximum length of the dimer does not exceed 45 Å, the formation of an internally cyclic structure is not possible. Addition of the antibody–dimer complex to cultured pituitary cells resulted in enhanced LH release. This surprising switch from potent antagonist to efficacious agonist indicates that the increased bridge length is sufficient to promote dimerization of the occupied GnRH receptors. In turn, this degree of microaggregation is apparently both sufficient and essential for the LH secretory response following GnRH binding to the gonadotrope. On the other hand, if the antibody prior to conjugation with the GnRH antagonist dimer is subjected to papain digestion or enzymatic cleavage by pepsin and subsequently reduced, the monovalent molecule once again inhibits GnRH-stimulated LH release. Neither the bivalent antibody nor the monovalent fragment when incubated alone with cultured pituitary cells affects LH secretion. A mathematical model of hormone action which quantitatively accounts for the release of LH by the antibody conjugate over a wide range of concentrations has been previously described (Blum and Conn, 1982).

In a subsequent series of studies, Gregory et al. (1982) also demonstrated that LH release could be stimulated by an inactive GnRH antagonist–receptor complex following dimerization of the receptors with a divalent protein-specific IgG directed at the N-terminus of the bound antagonist. More specifically, the D-Lys^6 analogue of the octapeptide Z-Gln-Trp-Ser-Tyr-Gly- Leu-Arg-Pro-ethylamide, a GnRH antagonist, was prepared, coupled to bovine serum albumin using 1-ethyl-3(3-dimethylaminopropyl) carbodiimide, and used to immunize rabbits. An IgG from the serum was purified which bound [^{125}I]-labeled GnRH with specificity directed toward the N-terminus of the antagonist. Treatment of pituitary cells with the antagonist alone prevented GnRH-stimulated LH release. However, in the presence of the antibody, the GnRH antagonist caused significant enhancement of LH release. This transformation of antagonist to agonist as a result of dimerization of bound GnRH receptors both confirmed the earlier observations of Conn et al. (1982b) and further supported the involvement of microaggregation in the GnRH-stimulated LH release pathway. Moreover, these studies supported the concept that the amino end of the GnRH molecule contains components important for the activation of microaggregation.

In the case of a GnRH agonist, a similar approach may be employed to demonstrate an enhancing effect on the potency of the agonist. Specifically,

Conn *et al.* (1982a) prepared a dimer of D-Lys[6]-GnRH (a potent GnRH agonist) using the EGS cross-linker as described earlier. Cultured pituitary cells were treated with the agonist dimer at a dose sufficient to produce an ED_{10} response as measured by LH release. When antibody prepared against D-Lys[6]-GnRH to (D-Lys[6])$_2$-EGS dimer was added, there was a significant increase in the efficacy of the agonist. Thus, the ability of the agonist dimer to provoke LH secretion, which in the absence of antibody was minimal, was greatly enhanced in the presence of the antibody. This finding suggests that the agonist can bind to the receptor but at this low concentration cannot evoke LH release. However, when an appropriate concentration of antibody is added, microaggregation of ligand–receptor complexes occurs and LH release is potentiated. Additionally, like native GnRH, the release of LH provoked by the cross-linking of receptors, either by the dimerized antagonists or agonists, requires extracellular Ca^{2+} and is inhibited by calmodulin antagonists (Conn *et al.*, 1982b). Hence, it appears that microaggregation is sufficient to provoke gonadotropin release and this occurs by a mechanism similar to that which is utilized by GnRH itself. The actual mechanisms underlying localized aggregation of receptors remain unknown.

IV. EFFECTOR COUPLING OF GnRH ACTION

A. Calcium

Since the classic studies of Douglas and Rubin (1961), the importance of Ca^{2+} as an intracellular mediator involved in the coupling mechanisms of stimulus to biological response has been demonstrated in numerous tissues (Rasmussen and Barret, 1984), including the pituitary gland (Moriarty, 1978). As early as 1967, Vale and co-workers found that the release of thyroid-stimulating hormone from isolated hemipituitaries in response to various secretagogues was abolished in Ca^{2+}-free media. Subsequently, Wakabayashi *et al.* (1969) demonstrated that LH release in response to hypothalamic extracts could be prevented if extracellular Ca^{2+} was chelated with EDTA. Samli and Geschwind (1968), in addition to reporting that Ca^{2+} deprivation significantly diminished the secretion of LH from hemipituitaries treated with hypothalamic extracts, also observed that incorporation of [14C]leucine into LH was unaffected. Conversely, when metabolic inhibitors such as dinitrophenol or oligomycin were applied to hemipituitaries in culture in the presence of Ca^{2+}, [14C]leucine uptake into LH was significantly reduced but LH secretion was unaltered. These findings indicate that Ca^{2+} is essential for secretion but not for biosynthesis of LH. Although all of these findings suggest that Ca^{2+} is required for the secretion of LH in response to GnRH, they do not necessarily prove that Ca^{2+} is a second messenger for this peptide. Marian and Conn (1979) described a set of criteria to fulfill in order to establish Ca^{2+} as a second messenger for GnRH. First, removal of Ca^{2+} from its site of action must inhibit GnRH-stimulated LH release. Next the movement of Ca^{2+} to this site by any experimental intervention must evoke LH secretion even in the absence of GnRH. Finally, GnRH stimulation of the gonadotrope must cause mobilization of Ca^{2+} from intracellular or extracellular stores to a site linked with LH secretion.

1. CALCIUM IS ESSENTIAL FOR GnRH-STIMULATED LH SECRETION

Efforts to determine whether Ca^{2+} is essential for GnRH-stimulated LH release have partially relied on the strategies of Ca^{2+} removal and blocking Ca^{2+} entry into cells. For example, Marian and Conn (1979) demonstrated that cultured pituitary cells, preincubated in Ca^{2+}-free medium, fail to release LH in response to GnRH. If Ca^{2+} is returned to the medium, gonadotrope responsiveness to GnRH is restored and LH is secreted (Conn and Rogers, 1979). Similar observations were subsequently made by deKoning *et al.* (1982), who showed that chelation of extracellular Ca^{2+} with EGTA blocked GnRH-stimulated LH secretion from cultured pituitaries and that this inhibition could be completely reversed with the addition of $CaCl_2$ to the incubation medium. This Ca^{2+} dependence of GnRH-stimulated LH secretion has also been shown in studies using *in vitro* perfusion of halved rat pituitary glands (Stern and Conn, 1981) or dispersed anterior pituitary cells (Borges *et al.*, 1983). In both models, removal of extracellular Ca^{2+} abolishes the LH secretory response to GnRH. Upon return of Ca^{2+} to the bathing medium, tissue responsiveness is restored. These studies also provide indirect evidence that intracellular concentrations of Ca^{2+} alone are insufficient in mediating the GnRH stimulus.

Blocking the action of Ca^{2+} with $LaCl_3$ (a competitive inhibitor of Ca^{2+} action) or preventing entry of Ca^{2+} into the cells with Ca^{2+} channel blockers such as ruthenium red, verapamil, or methoxyverapamil (D600) also effectively inhibits the release of LH in response to GnRH (Borges *et al.*, 1983; Conn *et al.*, 1983; Marian and Conn, 1979; Stern and Conn, 1981).

Because Ca^{2+} plays a role in the functional and structural integrity of the plasma membrane, it was possible that the Ca^{2+} requirement reflects a permissive effect, allowing GnRH to bind to its receptor, rather than play an intermediary role in GnRH-coupling mechanism for LH release. Consequently, Marian and Conn (1981), using a superactive analog of GnRH, D-Ser(t-Bu)^6-des-Gly^{10}EA-GnRH (Buserelin), coupled with a specific radioligand assay, explored this possibility by determining the effect of Ca^{2+} and other ions including Ba^{2+}, Mg^{2+}, Co^{2+}, Mn^{2+}, and Na^+ on the binding of Buserelin to rat pituitary plasma membrances. In the presence of elevated Ca^{2+} concentrations (>10 mM), the ability of Buserelin to bind to the plasma membranes was significantly diminished. This decreased binding was due to a decrease in the affinity of the GnRH receptor to its ligand and not a decrease in receptor number. The apparent inverse correlation between the Ca^{2+} requirement for LH secretion and its effect on binding affinity at the level of the GnRH receptor suggest that the actions of Ca^{2+} in the secretory mechanisms of LH cannot be viewed as permissive. This ionic effect on ligand binding was not specific for Ca^{2+}. In fact, all the ions tested exerted inhibition over the binding of Buserelin to the GnRH receptor. On the other hand, the effect of these ions on the release of LH in response to GnRH stimulation was differential. LH secretion was enhanced by Ba^{2+}, unaffected by Mg^{2+}, and Mn^{2+} and Co^{2+} were inhibitors. Taken together, these studies indicate that the locus of action of Ca^{2+} occurs subsequent to the recognition and binding of GnRH to its receptor.

2. INCREASED INTRACELLULAR CALCIUM STIMULATES LH RELEASE

In order to satisfy the second criterion, that Ca^{2+} is an internal mediator of GnRH actions, it was necessary to demonstrate in the absence of GnRH receptor occupancy, that alteration in cytosolic Ca^{2+} levels evokes LH release. The most direct method to investigate this possibility is with the use of the carboxylic ionophore A23187 (a Ca^{2+}-selective ionophore). Administration of A23187 in the presence of 1 mM Ca^{2+} to cultured pituitary cells resulted in a significant dose- and time-dependent increase in LH secretion (Conn et al., 1979). The ability to provoke LH release in the absence of GnRH further indicates that the locus of Ca^{2+} action is at a site distal to the GnRH receptor. In contrast to GnRH, treatment of pituitary cells with A23187 in the presence of Ca^{2+} also circumvents the inhibitory effects induced by either D600 or 2-N-butylaminoindene (Ca^{2+} channel blockers) over GnRH-stimulated LH release.

Liposomes, lipid vesicles which fuse with plasma membranes and deposit their contents into the interior of the cell, containing Ca^{2+} stimulated LH release in the absence of GnRH. However, neither Mg^{2+} nor monovalent ion-containing liposomes affected LH secretion. Thus, the insertion of Ca^{2+} into gonadotropes is sufficient to evoke LH release in the absence of GnRH stimulation. The specificity of the ion further supports the contention that Ca^{2+} is an internal mediator for the action of GnRH.

3. GnRH STIMULATES CALCIUM MOBILIZATION

Establishment of the third criterion involved the demonstration that Ca^{2+} is mobilized in response to GnRH stimulation and prior to LH secretion. In earlier studies, Williams (1976), using hemipituitaries in a perfusion system, demonstrated that Ca^{2+} gating did occur following GnRH stimulation. Hemipituitaries were preincubated with $^{45}Ca^{2+}$, exposed to GnRH, and monitored for the presence of isotope in the perfusate. The recovery of radioisotope in the extracellular medium would be an indirect measure of Ca^{2+} channel activity. It was observed that in response to increasing concentrations of GnRH there was a dose-dependent increase in Ca^{2+} efflux. However, the interval between collected samples was too long to precisely determine the temporal relationship between Ca^{2+} flux and LH secretion. Consequently, it was still uncertain whether the mobilization of Ca^{2+} was a specific action of GnRH or a consequence of LH release. In order to obviate the possibility that LH secretion is responsible for Ca^{2+} movement, Conn et al. (1981b), using a similar system but significantly decreasing the time interval between collected fractions, demonstrated that Ca^{2+} efflux occurs subsequent to GnRH binding (within <1.5 minute) and several minutes prior to any detectable LH release. Although these studies effectively indicate the involvement of Ca^{2+} gating in response to GnRH, it should be understood that the appearance of radioisotope in the extracellular medium merely reflects the changes in Ca^{2+} efflux, influx, intracellular turnover, and redistribution of Ca^{2+} in response to GnRH stimulation and not the net movement of the ion.

It was not until later that the status of cytosolic Ca^{2+} in response to GnRH stimulation could be accurately measured. With the development of Ca^{2+}-sensing probes such as Quin 2 (Tsien et al., 1982), intracellular Ca^{2+} perturbations in response to secretagogues could be evaluated. Quin 2, a fluorescent analog of EGTA (a Ca^{2+} chelator), is a water-soluble, tetracarboxylate anion which binds Ca^{2+} with 1:1 stoichiometry (Tsien et al., 1982) but does not readily enter living cells. However, the unreactive, uncharged, lipophilic compound Quin 2 tetraacetoxymethyl ester (Quin 2A/M) freely crosses the plasma membrane (Wollheim and Pozzan, 1984) and upon entry into the cell is converted to the charged, impermeant form, Quin 2, by nonspecific esterases. It is in this form that the trapped Quin 2 is used as an indicator of intracellular Ca^{2+} levels.

Clapper and Conn (1985), using Quin 2, demonstrated a measurable rise in intracellular Ca^{2+} concentrations of purified gonadotropes in response to GnRH stimulation. Neither low-affinity analogs nor GnRH antagonists were able to stimulate an increase in cytosolic Ca^{2+} levels at similar concentrations to GnRH. If Quin 2 levels were increased, causing greater amounts of intracellular Ca^{2+} to be chelated (Quin 2 acts as an intracellular buffer for Ca^{2+}), a blunted LH response to GnRH stimulation resulted. These studies again support the role of Ca^{2+} as an internal mediator of GnRH by demonstrating a functional relationship between intracellular Ca^{2+} levels and LH secretion.

It is most interesting, however, that the elevated cytosolic levels of Ca^{2+} that occur in response to GnRH are most likely derived from extracellular sources rather than intracellular stores. Bates and Conn (1984), while investigating the relative roles of intra- and extracellular Ca^{2+} sources on GnRH-stimulated LH release, noted that stabilization of intracellular Ca^{2+} sources with dantrolene or 8-(N,N-diethylamino)octyl-/3,4,5-trimethoxybenzoate hydroxychloride (TMB-8), which prevents mobilization of Ca^{2+} from within the gonadotrope, had no apparent effect over the LH secretory response. On the other hand, preventing translocation of Ca^{2+} from extracellular pools to the interior of the gonadotrope by either applying Ca^{2+} channel blockers or through experimentally depleting the external milieu of this ion prompted a rapid (<2 minute) extinction of the LH response to GnRH. These studies further emphasize that the integrity of the LH secretory response to GnRH is absolutely dependent on a continually accessible pool of extracellular Ca^{2+}.

B. Calmodulin

Calmodulin (CaM) is a ubiquitous intracellular receptor for Ca^{2+} and has been shown to be responsible for the activation of many Ca^{2+}-dependent enzymes and cellular processes (for reviews see Cheung, 1980; Klee et al., 1980; Means and Dedman, 1980; Means et al., 1982). Among the enzymes dependent on the formation of a CaM–Ca^{2+} complex for their activation are cyclic nucleotide phosphodiesterases, brain adenylate cyclase, ATPase and Ca^{2+} pump of the erythrocyte plasma membrane, myosin light chain kinases, brain membrane kinase, and phosphorylase kinase. Cellular processes that may involve CaM include cyclic nucleotide and glycogen metabolism, regulation of microtubule assembly and diassembly, Ca^{2+} gating, flagellar movement, exocytosis, and endocytosis in hormone action. Thus, CaM was a logical candidate to be considered as an intracellular target for mobilized Ca^{2+} in the mechanism of GnRH-stimulated LH secretion. To this end, a specific and sensitive radioimmunoassay for CaM was used to determine the quantity and distribution of this Ca^{2+}-binding protein before and during GnRH activation of LH secretion (Conn et al., 1981a). In response to GnRH treatment there is an initial increase in CaM associated with the plasma membrane of the gonadotrope and this increase is concomitant with a depletion of cytosolic CaM. In contrast, des[1]-GnRH, an analog of GnRH that binds to the receptor with a 1000-fold lower affinity than native GnRH, was unable to evoke CaM redistribution. It should be emphasized that only redistribution of CaM between intracellular compartments occurred in response to GnRH stimulation; no detectable changes in the total intracellular content of CaM were measured. Finally, the rapid onset of CaM redistribution in response to GnRH stimulation suggests that this is probably an early molecular event following GnRH receptor-mediated Ca^{2+} mobilization.

Additional evidence supporting the involvement of CaM in the mechanism of action of GnRH is exemplified by inhibition studies. For example, a number of chemically different neurotropic agents, including penfluridol, pimozide, chlordiazepoxide, and chlorpromazine, significantly reduce LH release in response to either GnRH or Ca^{2+} ionophores (Conn et al., 1981c; Hart et al., 1983). The ability of these drugs to inhibit CaM-activated nucleotide phosphodiesterases correlates well with the binding of neurotropic agents to CaM (Conn et al., 1984a; Levin and Weiss, 1976, 1978). Finally, the class of naphthalene sulfonamides, the "W" compounds, which specifically bind and inhibit the actions of CaM, have been shown to decrease GnRH-stimulated LH release in pituitary cell culture systems (Hart et al., 1983).

Finally, Jennes et al. (1985), using indirect immunofluorescence with fluorescein isothiocyanate (FITC) and rhodamine-labeled second antibody, demonstrated morphologically an association of CaM with membrane patches containing GnRH–receptor complexes. These findings provide evidence for a possible site of action of CaM and are consistent with the fact that CaM is redistributed from the cytosol to the plasma membrane during GnRH-stimulated LH secretion.

C. Inositol Phospholipids

The early studies of Hokin and Hokin (1953, 1954) demonstrated that stimulation of the exocrine pancreas with acetylcholine resulted in enhanced incorporation of ^{32}P-labeled phosphate into phospholipids, particularly phosphatidylinositol and phosphatidic acid. This observation led to the suggestion that phospholipid turnover may be important in stimulus–response coupling mechanisms. Since then it has been demonstrated, in a number of systems, that membrane inositol phospholipids are rapidly hydrolyzed following binding of biologically active substances with cell-surface receptors (Nishizuka, 1984a).

In 1982, Snyder and Bleasdale reported that treatment of cultured anterior pituitary cells with GnRH, at concentrations that increase LH secretion, stimulates the uptake of ^{32}P as phosphate into phosphoinositol in a dose-related manner. The potential involvement of phospholipid turnover in the mechanism of GnRH action is further supported by the fact that phospholipase C (the enzyme responsible for hydrolysis of inositol phospholipids to produce diacylglycerols) can effectively evoke the release of LH secretion from cultured anterior pituitary cells (Raymond et al., 1982). However, it is difficult to determine in a heterogeneous population of cells whether the responses are restricted to the gonadotrope. Since gonadotropes constitute only 5% of the pituitary cell population studied (Raymond *et al.*, 1984), these findings could reflect the cumulative responses of all cell types in the culture and, therefore, any changes occurring in the gonadotrope (if occurring at all) may be masked. Subsequent studies (Andrews and Conn, 1986), in which enriched populations of gonadotropes were used, have confirmed that GnRH binding to the gonadotrope stimulates a rapid (within seconds) incorporation of ^{32}P into phospholipids, specifically phosphatidylinositol and phosphatidic acid. Andrews and Conn (1986) extended these studies and showed that there is a differential rate of incorporation of ^{32}P between polyphosphatidylinositols and phosphatidic acid upon GnRH stimulation. Following GnRH stimulation there is an immediate increase in ^{32}P incorporation into phosphatidic acid. This ^{32}P labeling of the phosphatidic acid pool precedes measurable changes in labeled polyphosphatidylinositols. There is an immediate decline in the pool of labeled inositol phospholipids. The mechanism underlying this delay in resynthesis of the polyphosphatidylinositols from phosphatidic acid is unknown. Nevertheless, the rapid decline in labeled phosphatidylinositol following GnRH stimulation of the gonadotrope indicates that this may be an early event leading to the release of LH.

Information regarding the mass changes of phospholipid pools during GnRH stimulation of gonadotropes has also been gained through the use of isotopic pulse–chase studies (Andrews and Conn, 1986). Purified gonadotropes were labeled with ^{33}P and allowed to reach isotopic equilibrium. The cells were then exposed to ^{32}P for 1 hour prior to GnRH stimulation. Analysis of the ^{32}P:^{33}P ratio at various time intervals following the onset of GnRH stimulation revealed that there is a rapid loss (within 45 seconds) of labeled phosphatidylinositol, phosphatidylinositol 4-phosphate, and phosphatidylinositol 4,5-bisphosphate. These pools attain a new steady-state level within 5 minutes after initial onset of GnRH stimulation. A transient increase in labeling of the phosphatidic acid pool is detectable immediately following GnRH stimulation. Ten minutes after onset of GnRH stimulation a new steady state of phosphatidic acid is established. Pools of the other phospholipids (phosphatidycholine, phosphatidylserine, and phosphatidylethanolamine) were unaffected by GnRH. These studies further support the potential involvement and specificity of phosphoinositol turnover in the GnRH-stimulated mechanisms of gonadotropin release.

D. Guanine Nucleotide Binding Protein: A Link between the GnRH Receptor and Phospholipid Hydrolysis

The observation that GnRH stimulates phospholipid hydrolysis following activation of phospholipase C type enzymes (Andrew and Conn, 1986) has focussed attention on the means by which receptor occupancy by GnRH regulates this enzyme. In other systems, GTP binding proteins (G-proteins; Gilman, 1984) have been shown to link receptors to effectors (i.e., adenylate cyclase). G_s and G_i are two such G-proteins which stimulate or inhibit adenylate cyclase activity, respectively, following formation of agonist–receptor complexes. Recently, others (Haslam and Davidson, 1984; Hinkle and Phillips, 1984; Wojcikiewicz et al., 1986), have shown that other G-proteins may be responsible for providing the functional link between agonist-occupied receptor and activation of phosphoinositol-4, 5-biphosphate (PIP_2) phosphodiesterase causing hydrolysis of PIP_2 to diacylglycerol and IP_3. Andrews et al. (1986) have demonstrated that stable GTP analogs provoke both release of LH and phosphoinositide turnover in permeabilized pituitary cell cultures. Both effects could be blocked by GnRH antagonists, implicating an intimate association of a G-protein with the recognition site of the GnRH receptor. Further, it was demonstrated that neither pertussis toxin nor cholera toxin (agents which stimulate cyclic AMP production by interacting with G_s or G_i) exerted any effect on either GnRH- or guanine nucleotide-stimulated LH release or phosphoinositide turnover. However, these agents did cause elevated levels of cyclic AMP, indicating that these toxins were fully active. Taken together, it appears that GnRH stimulates phospholipid hydrolysis by activation of a phospholipase C; the functional coupling of the enzyme involves a guanine nucleotide binding protein that appears distinct from the G-proteins which regulate levels of cellular cAMP.

Thus, it appears that the binding of GnRH to the plasma membrane results in both the increase in intracellular Ca^{2+} as well as stimulated turnover in inositol phospholipids. Other evidence (Huckle and Conn, 1987) provided insight into the interaction of these two pathways which promote LH release. By using LiCl (5 mM) to block the recycling of inositol phosphates to inositol, it has been possible to study the effects of GnRH and other various agents that affect LH release on the accumulating pool of inositol phosphates in cultured pituitary cells. Since inositol phosphates are a product of phospholipid hydrolysis, this approach will provide insight into the relationship between phospholipid turnover and GnRH-stimulated LH secretion. Treatment of cultured pituitary cells with GnRH (10^{-8} M) resulted in LH release as well as a severalfold increase in inositol phosphate accumulation. Both LH release and accumulation of inositol phosphates could be prevented with GnRH antagonists and stimulated with agents that cause GnRH receptor aggregation.

The ability of GnRH to stimulate both turnover of phospholipids and increase in intracellular Ca^{2+} levels (both events of which have been shown to be linked to LH release) suggests that the two seemingly independent pathways are integrated by an unknown mechanism. Insight into this possible interaction was provided with the identification and characterization of protein kinase C.

E. Protein Kinase C and Diacylglycerols

Protein kinase C (PKC) is a Ca^{2+}- and phospholipid-dependent enzyme found in virtually every mammalian tissue (Takai *et al.*, 1979; Kuo *et al.*, 1980). Normally inert, PKC is activated by diacylglycerols in the presence of Ca^{2+} and phospholipids (Nishizuka, 1984b; Takai *et al.*, 1984). This requirement for diacylglycerols is specific. Indeed, cholesterol, glycolipids, mono- and triacylglycerols, and free fatty acids are totally ineffective in activating PKC (Takai *et al.*, 1984). However, diacylglycerol is typically absent from the membrane, except for the transient rise that follows receptor-mediated hydrolysis of phosphoinositols (Nishizuka, 1984a).

Of the various phospholipids examined, phosphatidylserine is the most potent for the activation of PKC (Takai *et al.*, 1984). The remaining phospholipids, such as phosphatidylcholine, phosphatidylethanolamine, phosphatidic acid, and phosphatidylinositol, are either inert or show differential effects depending on the presence of Ca^{2+} and phosphatidylserine. Calcium ion is also inextricably linked to the activation mechanism of PKC (Kaibuchi *et al.*, 1983). Finally, kinetic analysis has revealed that, once activated, PKC has a significantly increased affinity for Ca^{2+} and phospholipid (Takai *et al.*, 1984). Despite these observations, the involvement of PKC and phospholipid turnover in the mechanism of action of GnRH remains uncertain.

The role of diacylglycerols and PKC in the stimulatory actions of GnRH on LH release was suggested through the use of phorbol esters. Phorbol esters are tumor-promoting agents that can substitute for diacylglycerols in the activation of PKC (Castagna *et al.*, 1982; Kikkawa *et al.*, 1983). In fact, observations by Niedel *et al.* (1983) and Aschendel *et al.* (1983) indicate that diacylglycerols and phorbol esters compete for similar site(s) on the PKC molecule. Treatment of dispersed pituitary cells in culture with phorbol esters such as 4β-phorbol-12β- myristate-13α-acetate (PMA) and phorbol dibutyrate (PDB) results in the release of LH (Harris *et al.*, 1985; Smith and Vale, 1980; Smith and Conn, 1984). Based on these observations, Conn *et al.* (1985a) examined the possibility that diacylglycerols may play a role in the secretory process of LH. In order to accomplish this, it was first necessary to synthesize diacylglycerols that could enter living cells. By synthesizing a series of diacylglycerols with varying length side-chain acyl constituents (4–10 carbons) in the *sn*-1,2 positions, it was indeed found that when cultured pituitary cells were challenged with the diacylglycerols a positive correlation occurred between PKC activation and LH release. Of the synthesized diacylglycerols, 1,2-dioctanoylglycerol (diC_8) was the most potent in activating PKC and stimulating LH release. Additionally, the mechanism by which diacylglycerols initiate LH release is a Ca^{2+}-*independent* process. Further support for the involvement of PKC in GnRH-stimulated LH release was provided by the recent *in vitro* findings of Hirota *et al.* (1985), as well as the *in vivo* studies of McArdle and Conn (1986), who demonstrated that, in response to GnRH, there is a redistribution of this Ca^{2+}-activated, phospholipid-dependent enzyme between cytosol and membrane compartments of gonadotropes. Finally, Andrews and Conn (1986) demonstrated that the diacylglycerol pool in gonadotropes increases in response to GnRH stimulation.

F. Diacylglycerols and Calcium: Synergism between Intracellular Mediators

Both Ca^{2+} and diacylglycerol are requisite for the activation of PKC. This consideration led to the exploration of a possible interaction between diacylglycerol-activated PKC and Ca^{2+} and what effects this interaction has on LH secretion. To this end, Harris *et al.* (1985) determined the locus of activity of diacylglycerols in relation to the Ca^{2+}-mediated stimulation of LH secretion. It was demonstrated that activators of PKC (PMA, diC_8) evoke LH secretion via a mechanism that is unaltered by a GnRH antagonist [Ac(D-pCl-Phe^{1,2}-D-Trp^3-D-Lys^6-D-Ala^{10}-)GnRH], a Ca^{2+} channel blocker (D600), or a calmodulin inhibitor (pimozide). These findings indicated that PKC activation is not mediated by direct action on the GnRH receptor, and does not require mobilization of extracellular Ca^{2+} or activation of CaM. However, it has been found that, in the presence of the Ca^{2+} ionophore A23187, the stimulatory effects of PMA and diC_8 are significantly enhanced. This finding suggests that there is indeed an interaction between diacylglycerols and Ca^{2+} and it has been suggested that this interaction is synergistic in nature: The diacylglycerol-activated PKC may serve to amplify the signal triggered by the GnRH-receptor-stimulated, Ca^{2+}-activated pathway of LH release.

G. Does Phosphoinositide Hydrolysis and/or PKC Activation Have a Role in GnRH-Stimulated LH Secretion?

The definitive involvement of the diacylglycerols and PKC as intracellular mediators of LH release requires fulfillment of a set of criteria. These criteria include (1) a demonstration that diacylglycerols can evoke LH release in the absence of GnRH, (2) a detectable change in intracellular diacylglycerol concentrations following gonadotrope stimulations by GnRH, and (3) a diminished or inhibited LH secretory response to GnRH when phosphoinositide hydrolysis, diacylglycerol synthesis, or PKC activation is prevented. In the foregoing sections, evidence has been presented which indicates that occupancy of the GnRH receptor by an agonist stimulates phospholipid hydrolysis, leading to the production of diacyglycerols and activation of PKC. Further, experiments have been described which indicate that agents which activate PKC also cause LH release by gonadotropes. These two lines of evidence, along with the observations of Harris *et al.* (1984) that indicate that a synergistic interaction occurs between diacyglycerol-activated PKC and Ca^{2+} to amplify the GnRH-triggered Ca^{2+}–CaM pathway of LH secretion, suggest that GnRH-mediated LH release occurs, in part, through the stimulation of phosphoinositol hydrolysis and PKC activation. Thus, while the first two criteria mentioned above appear to be satisfied, the third has not been met due to the absence of specific inhibitors for GnRH-stimulated phosphoinositide hydrolysis, diacyglycerol production, or PKC activation. Consequently, the physiological relevance of these events has remained unclear.

Recently, studies have been reported which demonstrate that the events associated with GnRH-stimulated turnover of inositol phospholipids may be uncoupled from the Ca^{2+}–CaM-mediated pathway of GnRH-provoked LH release (Huckle and Conn, 1987). Indeed, it was found that substances which increase intracellular Ca^{2+} and evoke LH release (i.e., veratridine, ionomycin, ionophore A23187) do not appreciably increase the pool of inositol phosphates. In contrast, concentrations of D600 (methoxyverapamil) and pimozide that prevent LH release in response to GnRH, do not block the accumulation of inositol phosphates. Hence, these studies indicate that the turnover of inositol phospholipids in response to GnRH may be uncoupled from the essential Ca^{2+}–CaM mediated pathway of GnRH-stimulated LH release. This finding calls to question the necessity of phosphoinositide hydrolysis in the mechanism of action of GnRH on LH release by gonadotropes.

In a recent study by McArdle *et al.* (1987a), evidence was obtained indicating that GnRH stimulation of LH release was unchanged in cells depleted of all measurable PKC. It was shown that pretreatment of pituitary cell cultures with phorbol ester (PMA) caused a significant and sustained reduction in measurable cellular PKC activity. Further, the loss of cellular PKC activity was associated with a reduced ability of PKC-activating stimuli to provoke LH release. However, LH release in response to GnRH or other Ca^{2+}-mobilizing secretagogues such as A23187 or Maitotoxin (a Ca^{2+} ion-channel activator) is not diminished in PMA-pretreated cells. That GnRH can stimulate maximal LH secretory responses in gonadotropes depleted of measurable PKC activity, whereas PKC-activating agents fail to do so, suggests that, like phospholipid hydrolysis, activation of PKC may not be requisite for GnRH-mediated LH release.

Finally, in related studies, McArdle *et al.* (1987b) demonstrated that neither homologous down-regulation of GnRH receptors nor GnRH-mediated gonadotrope desensitization were dependent on PKC. While it is clear that receptor occupancy by GnRH triggers phosphoinositol hydrolysis lending to activation of PKC, the physiological importance of these events in GnRH-mediated LH release awaits further experimentation.

V. REGULATION OF GONADOTROPE RESPONSIVENESS TO GnRH

Numerous physiological factors regulate the sensitivity of the gonadotrope to GnRH and, therefore, its ability to secrete the gonadotropins. At any given moment, the degree of sensitivity with which the gonadotrope responds to GnRH is not only dependent on prior hormonal exposure but is closely linked to the number of high-affinity receptors for GnRH and the coupling of such receptors to the effector system. It is clear that significant changes in receptor number, but not affinity, occur under various endocrine states, including development, the estrous cycle, castration and steroid replacement, lactation, and senescence. In many cases, there is a direct correlation between the availability of high-affinity GnRH receptors and the sensitivity of the pituitary to GnRH. Consequently, it is important to understand the mechanisms through which GnRH receptor numbers are regulated.

In general, the regulation of receptor numbers is controlled by mechanisms that fall into two main categories: homologous regulation and heterologous regulation.

A. Homologous Regulation

In addition to stimulating the secretion of the gonadotropins, the ability of GnRH to exert homologous regulation over its own receptor number is well documented (Conn *et al.*, 1984c; Loumaye and Catt, 1982; Meidan *et al.*, 1982). A preponderance of evidence now suggests that the differential ability of the gonadotrope to respond to GnRH is significantly dependent on the mode of delivery of the GnRH stimulus to the anterior pituitary gland (Badger *et al.*, 1983; Belchetz *et al.*, 1978; deKoning *et al.*, 1978; Edwardson and Gilbert, 1976; Knobil, 1980; Loughlin *et al.*, 1981; Waring and Turgeon, 1983; Zilberstein *et al.*, 1983). Under physiological conditions, GnRH is secreted by the hypothalamus in a pulsatile manner (Carmen *et al.*, 1976; Knobil, 1980;). The importance of this intermittent exposure of the pituitary to GnRH in maintaining the responsiveness of the gonadotrope to

GnRH becomes apparent under conditions in which the pulses are dynamically altered. For instance, it has been repeatedly demonstrated, in numerous species, that the chronic exposure of the pituitary gland to GnRH results in eventual loss of gonadotrope responsiveness (i.e., desensitization), with a consequential diminution of gonadotropin secretion (Badger *et al.*, 1983; deKoning *et al.*, 1978; Keri *et al.* 1983; Koiter *et al.*, 1981; Sandow, 1983). On the other hand, if the exposure of the pituitary to GnRH, either *in vivo* or *in vitro*, is episodic, pituitary sensitivity is increased and there is enhanced release of gonadotropins to subsequent GnRH administration (Evans *et al.*, 1984; Pickering and Fink, 1977; Van Dieten and van Rees, 1983). This latter phenomenon has been termed *self-priming* (Aiyer *et al.* 1974; Pickering and Fink, 1977).

Further evidence supporting the fact that pituitary sensitivity to GnRH is regulated, in part, by endogenous release of GnRH includes the finding that following castration GnRH-binding capacity of the pituitary is increased (Frager *et al.*, 1981; Clayton and Catt, 1981b), hypothalamic GnRH content is decreased reflecting release of the peptide (Root *et al.*, 1975; Rudenstein *et al.*, 1979), stalk concentrations of GnRH are elevated, and gonadotropin secretion is increased (Eskay *et al.*, 1977). The elevation of gonadotropin secretion as well as GnRH receptors per cell can be prevented in castrated animals by administration of antiserum to GnRH or with lesions in the medical basal hypothalamus (Frager *et al.*, 1981; Pieper *et al.*, 1982). Consequently, the state of responsiveness of the gonadotrope to GnRH, either refractory or sensitized, is partially dependent on the parameters (frequency, amplitude, and duration) of endogenous GnRH secretion. Moreover, evidence indicates that the GnRH secretory pattern which results in increased or decreased responsiveness of the pituitary to GnRH may also be coincidental with an increase or decrease, respectively, in GnRH receptor concentration. Indeed, Zilberstein *et al.* (1983), using a superfusion system, observed that dispersed pituitary cells chronically exposed to GnRH are desensitized. Accompanying this refractory state was a concomitant decrease in GnRH receptor numbers. Similary, Nett *et al.* (1981) found, in ewes, that a decreased LH secretory response following prolonged infusion of GnRH corresponded with a decrease in GnRH receptor number. Therefore, it seems reasonable that if a decrease in pituitary sensitivity to GnRH is associated with a decrease in GnRH receptor number, then increases in pituitary sensitivity should be accompanied by increases in GnRH receptor number. In support of this hypothesis, Frager *et al.* (1981) observed that increased numbers of GnRH receptors parallel periods of increased sensitivity of the pituitary to GnRH (i.e., postcastration). Finally, it has been demonstrated that GnRH does indeed exert a biphasic autoregulation over its own receptor (Clayton, 1982; Conn *et al.*, 1984c). GnRH receptor numbers increased in response to constant low-dose infusions of GnRH whereas constant high-dose infusions of GnRH caused a significant decrease in receptor numbers. Thus, it seems that GnRH may play a significant role in the phenomena of pituitary sensitization and desensitization by regulating the concentration of GnRH receptor numbers.

B. Heterologous Regulation

In addition to GnRH, modulation of pituitary responsiveness is significantly affected by numerous other physiological factors, including steroids, opiates, and neurotransmitters. These factors affect sensitivity by acting either directly at the pituitary to alter the number of GnRH receptors and/or at the level of the hypothalamus to modify the secretion of GnRH.

1. STEROIDS

The effects of steroids on pituitary sensitivity to GnRH have been intensively studied and are evident in experimental models in which the steroidal milieu has been altered. For instance, orchidectomy augments and testosterone diminishes gonadotropin secretion following GnRH administration (Debeljuk et al., 1974). In female rats, pituitary responsiveness to endogenous and exogenous GnRH varies throughout the estrous cycle, being highest on the afternoon of proestrus (Aiyer et al., 1974; Cooper et al., 1974; Gordon and Reichlin, 1974). The magnitude of this heightened response has subsequently been shown to be dependent on the circulating levels of 17β-estradiol (Fink et al., 1975; Fink, 1979). Yet, estradiol administration in ovariectomized rats, although initially decreasing pituitary responsiveness to GnRH administration (Libertun et al., 1974; Vilchez-Martinez et al., 1974), over time will eventually enhance pituitary responsiveness. Progesterone has been reported to alter pituitary responsiveness to GnRH administration in both a stimulatory as well as an inhibitory fashion (Lagace et al., 1980; Martin et al., 1974;).

Many of the observed in vivo characteristics of the anterior pituitary gland are retained by dispersed pituitary cells in culture (O'Connor et al., 1980; Waring and Turgeon, 1980). Steroids have been demonstrated to modify GnRH-stimulated gonadotropin release in vitro. For instance, 17β-estradiol increases LH secretory responsiveness to GnRH in cultured anterior pituitary cells (Drouin et al., 1976; Hsueh et al., 1979; Smith and Conn, 1982). On the other hand, testosterone appears to diminish the sensitivity of the gonadotrope to GnRH (Kamel and Krey, 1982). Progesterone exerts a biphasic control over gonadotropin secretion, depending on its temporal exposure to the gonadotrope and the presence or absence of estradiol (Drouin and Labrie, 1981; Lagace et al., 1980).

The mechanisms by which the steroids manifest their effects to alter the sensitivity of the pituitary and the site(s) of action are still unclear. As mentioned, evidence demonstrates that the steroids act at both the level of the hypothalamus as well as the pituitary gland to regulate pituitary responsiveness to GnRH. Heber and Odell (1979) demonstrated that estrogen can modulate pituitary GnRH receptors directly and hence pituitary sensitivity to GnRH. Giguere *et al.* (1981) reported that androgens decrease GnRH binding sites in rat anterior pituitary cells. The possibility that the steroids effect synthesis and/or release of gonadotropins has also been proposed (Hsueh *et al.*, 1979). This is supported by the fact that cycloheximide will block the stimulatory effects of estrogen and the inhibitory effects of testosterone on GnRH-stimulated LH release in culture (Kamel and Krey, 1982). Although the site(s) of action of the steroids remain unknown, the effects of the steroids, particularly estrogen, are exerted prior to Ca^{2+} mobilization in the GnRH-stimulated gonadotrope (Smith and Conn, 1982).

2. OPIATES

The influence of the opiates on reproductive function is well known (Motta and Martini, 1982). Initially in 1955, Barraclough and Sawyer demonstrated that administration of morphine sulfate to cycling animals on the morning of proestrus blocked the ensuing ovulation. Packman and Rothchild (1976) then demonstrated that opiate-blocked ovulation could be reinstated with the administration of Naloxone (a potent opiate antagonist). In 1977, Pang *et al.* observed that morphine sulfate caused a biphasic effect over the release of gonadotropin. At low doses, opiate appeared to cause an enhanced secretory response of the gonadotropins; however, as the dose of opiate was progressively increased, there was a dose-related suppression of both FSH and LH preovulatory surges. This suppression could be reversed with Naloxone. Moreover, Naloxone when administered alone causes an enhancement of gonadotropin secretion, possibly by stimulating GnRH release from the hypothalamus (Bruni *et al.*, 1977; Cicero *et al.*, 1985; Pang *et al.*, 1977). These findings indicate that the opiates may be involved in the regulation of pituitary sensitivity to GnRH. Supporting this notion, Barkan *et al.* (1983) demonstrated that pituitary GnRH receptors may be acutely regulated by opiate-active compounds. Injection of morphine sulfate caused a rapid increase in the GnRH-binding capacity of the pituitary gland. This effect could be reversed with the simultaneous administration of naloxone. However, the influence of the opiates over the sensitivity of the pituitary to GnRH as well as their effects over the release of the gonadotropins is highly dependent on the endocrine status of the animal. Although it is unclear through what mechanism(s) and site(s) of action the opiates manifest their regulatory influences, the role of endogenous opiate activity in the modulation of pituitary sensitivity and GnRH receptor number is probable.

3. NEUROTRANSMITTERS

The involvement of the neurotransmitters, in particular the catecholamines and serotonin, in controlling pituitary activity has been suggested in numerous studies (for reviews, see Meites and Sonntag, 1981; Weiner and Ganong, 1978). With respect to gonadotropin secretion, most studies have failed to demonstrate a direct action of these neurotransmitters at the level of the pituitary gland (Ryu et al., 1980). However, given that the hypothalamus is richly innervated by monoaminergic fiber tracts, and these tracts run in close proximity to GnRH-producing neurons (Jennes et al., 1982), it is probable that the influences imparted by the monoamines over gonadotropin release occur indirectly by effecting the release of GnRH. In fact, through experimental manipulation of central catecholaminergic activity it has been shown that GnRH secretion can be significantly altered (Barraclough and Wise, 1982). Because the pattern of GnRH release has been shown to directly affect the concentration of GnRH receptors in the pituitary, this may be the manner in which the neurotransmitters regulate pituitary sensitivity and hence their influence over gonadotropin secretion. Reviews dealing with specific effects of the neurotransmitters on GnRH secretion from the hypothalamus and its effects on gonadotropin release are available (Barraclough and Wise, 1982; McCann et al., 1982); therefore, this topic will not be covered here.

VI. MECHANISMS UNDERLYING THE REGULATION OF GnRH RECEPTOR NUMBER

Processes and factors that influence the concentration of GnRH receptors on the gonadotropes have been described above. However, the molecular mechanisms underlying down-regulation (decrease in receptor number) and up-regulation (increase in receptor number) and the relationship of these processes to desensitization and sensitization of the gonadotrope to GnRH remain unclear in detail.

Relating to the GnRH-stimulated LH release system, several important relationships have been established. A number of studies (Conn et al., (1985b) suggest that, at least for desensitization, receptor occupancy (binding, but not activation) and LH release do not play significant roles. Neither receptor occupancy by a GnRH antagonist nor stimulation of LH secretion with a Ca^{2+} ionophore provokes desensitization. However, prolonged receptor occupancy by a GnRH agonist (binding and activation) causes loss of gonadotrope responsiveness. Furthermore, production of this refractory state is a Ca^{2+}-independent process. This finding indicates that the mechanism underlying desensitization is fundamentally distinct from the mechanism resulting in the secretion of gonadotropins. Smith and Conn (1984) found that stimulation of LH release with a GnRH antagonist dimer also resulted in desensitization. Thus, it appears that desensitization of the gonadotrope, although linked to the binding and activation of the GnRH receptor, is not the consequence of depleted stores of LH due to secretion. Furthermore, the hypothesis that desensitization follows or is contemporaneous with down-regulation, although attractive, is not necessarily the case. Indeed, periods of cell refractoriness to GnRH are not always accompanied by loss of GnRH receptors (Conn et al., 1985b; Keri et al., 1983; Gorospe and Conn, 1987b).

For example, the time course of development of gonadotrope desensitization (measurable 20 minutes after addition of GnRH and increasing for 6–12 hours; Jinnah and Conn, 1986) differs from the time course of GnRH receptor down-regulation and up-regulation in response to GnRH (the number of GnRH receptors initially lost during GnRH-provoked desensitization is completely restored to pretreatment long before gonadotrope responsiveness has recovered; Conn et al., 1984c). Further, Gorospe and Conn (1987a) have demonstrated, using either chemical or physical approaches, that blockade of GnRH receptor internalization does not prevent GnRH-mediated gonadotrope desensitization. These findings suggest that factors other than altered receptor number may be important in regulation of gonadotrope responsiveness to releasing hormone. Such factors may include altered coupling of the GnRH receptor to its effector molecules in the desensitized versus sensitized state of the gonadotrope. Indeed, Conn et al. (1987) in a recent study have indicated a potential role for the receptor-regulated Ca^{2+} ion channel in the development of desensitization. Additionally, Gorospe and Conn (1987b) have demonstrated using fluorescence polarization that GnRH-mediated gonadotrope desensitization is dependent upon the fluidity state (gel versus liquid) of the gonadotrope plasma membrane.

Pituitary cell culture systems have also been used to examine the mechanism underlying up-regulation of the GnRH receptor. Occupancy of the receptor by a GnRH antagonist, while blocking secretion of LH, is not in itself able to induce down-regulation or up-regulation. Additionally, up-regulation, but not down-regulation, is a Ca^{2+}-dependent process because it can be prevented with the removal of extracellular Ca^{2+} or in the presence of methoxyverapamil (D600; Ca^{2+} channel blocker). Additional evidence supporting the fact that up-regulation is a Ca^{2+}-mediated event comes from the studies of Conn et al. (1979), who demonstrated that A23187 or veratridine, agents that mobilize extracellular Ca^{2+}, can stimulate LH release and increase GnRH receptor concentrations. Increase in GnRH receptors occurred even if the doses of A23187 or verapamil were too low to stimulate LH secretion. Thus, like down-regulation, up-regulation is independent of LH release. However, in contrast to desensitization is the fact that up-regulation can be blocked in the presence of cycloheximide or actinomycin D. Consequently, it appears that RNA and protein synthesis are involved in this latter process (Conn *et al.*, 1984c). The association of up-regulation with increase in cell sensitivity is yet to be established.

VII. CONCLUSIONS: A MODEL FOR GnRH ACTION

Central to the "Three-Step Model" of GnRH action (Conn *et al.*, 1981b) is the view that, after binding to its plasma-membrane receptor, the informational message of GnRH was carried by Ca^{2+}. This model remains useful for describing gonadotropin release, but has been broadened in subsequent years to explain the mechanisms by which GnRH regulates cell responsiveness and its own receptor. The molecular resolution of the model has been increased to add the possibility of GnRH-receptor microaggregation and the demonstration that Ca^{2+} is likely to be mobilized from extracellular spaces. Calmodulin has specifically been identified as an intracellular receptor for Ca^{2+} and information is now available about its intracellular localization. More recently, an apparent modulatory role for the diacylglycerols and protein kinase C has been identified. We have combined the salient features of our current understanding and speculation of GnRH action in Fig. 2.

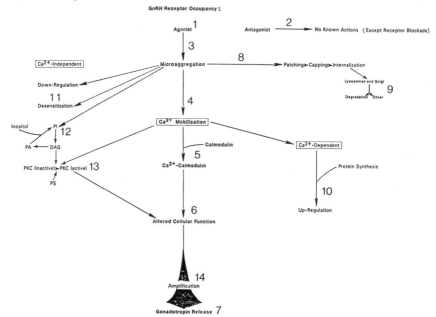

ACKNOWLEDGMENT

Supported by HD06609 (NRSA) and HD19899.

REFERENCES

Andrews, W. V., and Conn, P. M. (1986). Gonadotropin-releasing hormone stimulated mass changes in phosphoinositides and diacyglycerol accumulation in purified gonadotrope cell cultures. *Endocrinology* **118**, 1148–1158.

Andrews, W. V., Staley, D. D., Huckle, W. R., and Conn, P. M. (1986). Stimulation of luteinizing hormone (LH) release and phospholipid breakdown by granosine triphosphate in permeabilized pituitary gonadotropes: Antagonist action suggests association of a G-protein and gonadotropin-releasing hormone receptor. *Endocrinology* **119**, 2537–2546.

Aiyer, M. S., Fink, G., and Grief, F. (1974). Changes in the sensitivity of the pituitary gland to luteinizing hormone releasing factor during the estrous cycle of the rat. *J. Endocrinol.* **60**, 47–64.

Ashendel, C. L., Staller, J. M., and Boutwell, R. K. (1983). Protein kinase activity associated with phorbol ester receptor purified from mouse brain. *Cancer Res.* **43**, 4333–4337.

Badger, T. M., Loughlin, J. S., and Naddaff, P. G. (1983). The luteinizing hormone-releasing hormone (LHRH)-desensitized rat pituitary: Luteinizing hormone responsiveness to LHRH *in vitro*. *Endocrinology* **112**, 793–799.

Barkan, A., Regiani, S., Duncan, J., Papvasiliou, S., and Marshall, J. C. (1983). Opioids modulate pituitary receptors for gonadotropin-releasing hormone. *Endocrinology* **112**, 387–389.

Barraclough, C. A., and Sawyer, C. H. (1955). Inhibition of the release of pituitary ovulatory hormone in the rat by morphine. *Endocrinology* **57**, 329–337.

Barraclough, C. A., and Wise, P. M. (1982). The role of catecholamines in the regulation of pituitary luteinizing hormone and follicle-stimulating hormone secretion. *Endocr. Rev.* **3**, 91–119.

Bates, M. D., and Conn, P. M. (1984). Calcium mobilization in the pituitary gonadotrope: Relative roles of intra- and extracellular sources. *Endocrinology* **115**, 1380–1385.

FIG. 2. Model showing the actions of GnRH at the pituitary gonadotrope. GnRH (or an agonist) recognizes, occupies, and activates the GnRH receptor in the plasma membrane of the gonadotrope (1). Binding of an antagonist to the GnRH receptor produces no known biological actions, although the receptor is blocked (2). Activation of the GnRH receptor provokes localized cross-linking of GnRH receptors (microaggregation) (3). Ca^{2+} is mobilized from extracellular spaces (4), and binds to and stimulates calmodulin redistribution (5). As a result of alterations in intracellular function (6), gonadotropins are secreted via exocytosis (7). Microaggregation also results in patching, capping, and internalization of the GnRH–receptor complex (8). Once internalized, the ligand–receptor complex is routed to either lysosomes or Golgi vesicles. (9). These events lead to GnRH degradation and, although internalization of GnRH is neither necessary nor sufficient for gonadotropin release, internalized hormone may have other unidentified actions on the cell. Mechanisms that regulate gonadotrope sensitivity and GnRH receptor concentrations are also provoked by microaggregation. These include up-regulation (a Ca^{2+}- and protein synthesis-dependent event) (10) as well as the Ca^{2+}-independent processes of down-regulation and desensitization (11). Microaggregation also initiates phosphatidylinositol (PI) turnover (12). Hydrolysis of PI results in increased intracellular diacylglycerols (DAG). In turn, DAG, in the presence of phosphatidylserine (PS) and Ca^{2+}, activate protein kinase C (PKC) (13). PKC may serve as an amplification signal for the intracellular actions stimulated by the calmodulin–Ca^{2+}-mediated pathway of GnRH-stimulated LH release (14).

Belchetz, P. E., Plant, T. M., Nakai, Y., Keogh, E. J., and Knobil, E. (1978). Hypophysial response to continuous and intermittent delivery of hypothalamic gonadotropin-releasing hormone. *Science* **202**, 631–633.

Blum, J. J., and Conn, P. M. (1982). Gonadotropin-releasing hormone stimulation of luteinizing hormone release: A ligand–receptor effector model. *Proc. Natl. Acad. Sci. U.S.A.* **79**, 7307–7311.

Borges, J. L. C., Scott, D., Kaiser, D. L., Evans, W. S., and Thorner, M. O. (1983). Ca^{2+} dependence of gonadotropin-releasing hormone-stimulated luteinizing hormone secretion: *In vitro* studies using continuously perifused dispersed rat anterior pituitary cells. *Endocrinology* **113**, 557–562.

Bruni, J. F., Van Vugt, D. A., Marshall, S., and Meites, J. (1977). Effects of naloxone morphine and methionine enkephalin on serum prolactin, luteinizing hormone, follicle stimulating hormone, thyroid stimulating hormone and growth hormone. *Life Sc.* **21**, 461–466.

Canonico, P. L., and MacLeod, R. (1983). The role of phospholipids in hormonal secretory systems. *In* "Neuroendocrine Perspectives" (E. E. Muller, and R. M. Macleod, eds.), Vol. 2, pp. 123–172. Elsevier, Amsterdam.

Carmen, P. W., Arakai, S., and Ferin, M. (1976). Pituitary stalk portal blood collection in rhesus monkeys: Evidence for pulsatile release of gonadotropin-releasing hormone (GnRH). *Endocrinology* **99**, 243–248.

Castagna, M., Takai, Y., Kaibuchi, K., Sano, K., Kikkawa, U., and Nishizuka, Y. (1982). Direct activation of calcium-activated, phospholipid-dependent protein kinase by tumor-promoting phorbol esters. *J. Biol. Chem.* **257**, 7847–7851.

Cheung, W. Y. (1980). Calmodulin plays a pivotal role in cellular regulation. *Science* **207**, 19–27.

Cicero, T. J., Schmoeker, P. F., Meyer, E. R., and Miller, B. T. (1985). Luteinizing hormone releasing hormone mediates naloxone's effects on serum luteinizing hormone levels in normal and morphine-sensitized male rats. *Life Sci.*. **37**, 467–474.

Clapper, D. L., and Conn, P. M. (1985). Gonadotropin-releasing hormone stimulation of pituitary gonadotrope cells produces an increase in intracellular calcium. *Biol. Reprod.* **32**, 269–278.

Clark, W. R. (1980). Production of antibody in response to antigen. *In* "The Experimental Foundation of Modern Immunology," pp. 190–217. Wiley, New York.

Clayton, R. N. (1982). Gonadrotropin-releasing hormone modulation of its own pituitary receptors: Evidence for biphasic regulation. *Endocrinology* **111**, 152–161.

Clayton, R. N., and Catt, K. J. (1980). Receptor-binding affinity of gonadotropin-releasing hormone analogs: Analysis by radioligand–receptor assay. *Endocrinology* **106**, 1154–1159.

Clayton, R. N., and Catt, K. J. (1981a). Regulation of pituitary gonadotropin-releasing hormone receptors by gonadal hormones. *Endocrinology* **108**, 887–895.

Clayton, R. N., and Catt, K. J. (1981b). Gonadotropin-releasing hormone receptors: Characterization, physiological regulation and relationship to reproductive function. *Endocr. Rev.* **2**, 186–209.

Clayton, R. N., Shakespear, R. A., Duncan, J. A., Marshall, J. C., Munson, P. J., and Rodbard, D. (1979). Radioiodinated nondegradable GnRH analogs: New probes for the investigation of pituitary GnRH receptors. *Endocrinology* **105**, 1369–1376.

Clayton, R. N., Solano, A. R., Garcia-Vela, A., Dufau M. L., and Catt, K. J. (1980). Regulation of pituitary receptors for gonadotropin-releasing hormone during the rat estrous cycle. *Endocrinology* **107**, 699–706.

Conn, P. M. (1982). Gonadotropin-releasing hormone stimulation of pituitary gonadotropin release: A model system for receptor-mediated Ca^{2+} dependent secretion. *In* "Cellular Regulation of Secretion and Release." (P. M. Conn, ed.), pp. 459–493. Academic Press, New York.

Conn, P. M. (1983). Ligand dimerization: A technique for assessing receptor–receptor interactions. In "Methods in Enzymology" (P. M. Conn, ed.), Vol. 103, pp. 49–58. Academic Press, New York.

Conn, P. M., and Hazum, E. (1981). Luteinizing hormone release and gonadotropin-releasing hormone (GnRH) receptor internalization: Independent actions of GnRH. Endocrinology 109, 2040–2045.

Conn, P. M., and Rogers, D. C. (1979). Restoration of responsiveness of gonadotropin-releasing hormone (GnRH) in calcium depleted rat pituitary cells. Life Sci. 24, 2461–2466.

Conn, P. M., and Rogers, D. C. (1980). Gonadotropin release from pituitary cultures following activation of endogenous ion channels. Endocrinology 107, 2133–2134.

Conn, P. M., and Venter, J. C. (1985). Radiation inactivation (target size analysis) of the gonadotropin-releasing hormone receptor: Evidence for a high molecular weight complex. Endocrinology 116, 1324–1328.

Conn, P. M., Rogers, D. C., and Sandhu, F. S. (1979). Alteration of the intracellular calcium level stimulates gonadotropin release from cultured anterior pituitary cells. Endocrinology 105, 1122–1127.

Conn, P. M., Chafouleas, J. G., Rogers, D. C., and Means, A. R. (1981a). Gonadotropin releasing hormone stimulates calmodulin redistribution in the rat pituitary. Nature (London) 292, 264–265.

Conn, P. M., Marian, J., McMillian, M., Stern, J., Rogers, D., Hamby, M., Penna, A., and Grant, E. (1981b). Gonadotropin-releasing hormone action in the pituitary: A three step mechanism. Endocr. Rev. 2, 174–185.

Conn, P. M., Rogers, D. C., and Sheffield, T. (1981c). Inhibition of gonadotropin-releasing hormone-stimulated luteinizing release by pimozide: Evidence for a site of action after calcium mobilization. Endocrinology 109, 1122–1126.

Conn, P. M., Rogers, D. C., and McNeil, R. (1982a). Potency enhancement of a GnRH agonist: GnRH-receptor microaggregation stimulates gonadotropin release. Endocrinology 111, 335–337.

Conn, P. M., Rogers, D. C., Stewart, J. M., Niedel, J., and Sheffield, T. (1982b). Conversion of gonadotropin-releasing hormone antagonist to an agonist. Nature (London) 296 653–55.

Conn, P. M., Rogers, D. C., and Seay, S. G. (1983). Structure–function relationship of calcium ion channel antagonists at the pituitary gonadotrope. Endocrinology 113, 1592–1595.

Conn, P. M., Bates, M. D., Rogers, D. C., Seay, S. G., and Smith, W. A. (1984a). GnRH–receptor–effector–response coupling in the pituitary gonadotrope: A Ca^{2+} mediated system. In "The Role of Drugs and Electrolytes in Hormonogenesis (K. Fotherby, and S. B. Pal, eds.) pp. 85–103. DeGruyter, New York.

Conn, P. M., Hsueh, A. J. W., and Crowley, W. F. (1984b). Gonadotropin-releasing hormone: Molecular and cell biology, physiology and clinical applications. Fed. Proc. Fed. Am. Soc. Exp. Biol. 43, 2351–2361.

Conn, P. M., Rogers, D. C., and Seay, S. G. (1984c). Biphasic regulation of the gonadotropin-releasing hormone receptor by receptor microaggregation and intracellular Ca^{2+} levels. Mol. Pharmacol. 25, 51–55.

Conn, P. M., Ganong, B. R., Ebling, J., Staley, D., Neidel, J. E., and Bell, R. M. (1985a). Diacylglycerols release LH: Structure–activity relations reveal a role for protein kinase C. Biochem. Biophys. Res. Commun. 126, 532–539.

Conn, P. M., Rogers, D. C., Seay, S. G., Jinnah, H., Bates, M., and Luscher, D. (1985b). Regulation of gonadotropin release, GnRH receptors and gonadotrope responsiveness: A role for GnRH receptor microaggregation. J. Cell. Biochem. 27, 13–21.

Conn, P. M., Staley, D. D., Yasumoto, T., Huckle, W. R., and Janovick, J. A. (1987). Homologous desensitization with gonadotropin-releasing hormone (GnRH) also diminishes gonadotrope responsiveness to maitotoxin: A role for the GnRH receptor-regulated calcium ion channel in mediation of desensitization. *Mol. Endocrinol.* **1**, 154–159.

Conne, B. S., Scaglioni, S., Lang, U., Sizonenko, P. C., and Aubert, M. L. (1982). Pituitary receptor sites for gonadotropin-releasing hormone: Effect of castration and substitutive therapy with sex steroids in the male rat. *Endocrinology* **110**, 70–79.

Cooper, K. J., Fawcett, C. P., and McCann, S. M. (1974). Variations in pituitary responsiveness to luteinizing hormone/follicle-stimulating hormone releasing factor (LH-RF/FSH-RF) preparation during the rat estrous cycle. *Endocrinology* **95**, 1293–1299.

Coy, D. H., Coy, E. J., Schally, A. V., Vilchez-Martinez, J. Hirotsu, Y., and Arimura A. (1974). Synthetic and biological properties of [D-ALA-6-Des-GLY-NH$_2$-10]-LH-RH ethylamide, a peptide with greatly enhanced LH- and FSH-releasing activity. *Biochem. Biophys. Res. Commun.* **57**, 335–340.

Coy, D. H., Schally, A. V., Vilchez-Martinez, J. A., Coy, E. J., and Arimura, A. (1975). Stimulatory and inhibitory analogs of LH-RH. *In* "Hypothalamic Hormones" (M. Motta, P. G. Crosignani, and L. Martini, eds.), pp. 1–13. Academic Press, New York.

Coy, D. H., Vilchez-Martinez, J. S., Coy, E. J., and Schally, A. V. (1976). Analogs of LH-RH increase biological activity produced by D-amino acid and substitutions in position 6. *J. Med. Chem.* **19**, 423–425.

Coy, D. H., Seprodi, J., Vilchez-Martinez, J. A., Pedroza, E., Gardrer, J., and Schally, A. V. (1979). Structure–function studies and prediction of conformational requirements for LH-RH. *In* "Central Nervous System Effects of Hypothalamic Hormomes and Other Peptides" (R. Collu, A. Barbeau, J. R. Ducharme, and J.-G. Rochefort, eds.), p. 317. Raven, New York.

Debeljuk, L., Vichez-Martinez, J. A., Arimura, A., and Schally, A. V. (1974). Effect of gonadal steroids on the response to LH-RH in intact and castrated male rats. *Endocrinology* **94**, 1519–1524.

deKoning, J., van Dieten, J. A. M. J., and van Rees, G. P. (1978). Refractoriness of the pituitary gland after continous exposure to luteinizing hormone releasing hormone. *J. Endocrinol.* **79**, 311–318.

deKoning, J., Tijssen, A. M. I., van Dieten, J. A. M. J., and van Rees, G. P. (1982). Effect of Ca^{2+} deprivation on release of luteinizing hormone induced by luteinizing hormone releasing hormone from female rat pituitary glands *in vitro*. *J. Endocrinol.* **94**, 11–20.

Douglas, W. W., and Rubin, R. P. (1961). The role of calcium in the secretory response of the adrenal medulla to acetylcholine. *J. Physiol.* (London) **159**, 40–57.

Drouin, J., and Labrie, F. (1981). Interactions between 17β-estradiol and progesterone in control of luteinizing hormone and follicle-stimulating hormone in rat anterior pituitary cell in culture. *Endocrinology* **108**, 52–57.

Drouin, J., Lagace, L., and Labrie, F. (1976). Estradiol-induced increase of the LH responsiveness to LH releasing hormone (LHRH) in rat anterior pituitary cells in cultures. *Endocrinology* **99**, 1477–1481.

Duello, T. M., and Nett, T. M. (1980). Uptake, localization, and retention of gonadotropin-releasing hormone and gonadotropin releasing hormone analogs in rat gonadotrophs. *Mol. Cell. Endocrinol.* **19**, 101–112.

Duello, T. M., Nett, T. M., and Farquhar, M. G. (1981). Binding and internalization of a GnRH agonist by rat pituitary gonadotrophs. *J. Cell Biol.* **91**, 220–222.

Edwardson, J. A., and Gilbert, D. (1976). Application of *in vitro* perifusion technique to studies of luteinizing hormone releasing factor during the estrous cycle of the rat. *J. Endocrinol.* **68**, 197–207.

Eidne, K. A., Hendricks, D. T., and Millar, R. P. (1985). Demonstration of a 60k molecular weight luteinizing hormone-releasing hormone receptor in solubilized adrenal membrane by ligand–immunoblotting technique. *Endocrinology* **116**, 1792–1795.

Eskay, R. L., Mical, R. S., and Porter, J. C. (1977). Relationship between luteinizing hormone releasing hormone release in intact, castrated and electrochemically-stimulated rats. *Endocrinology* **100**, 263–270.

Evans, W. S., Uskavitch, D. R., Kaiser, D. L., Hellman, P., Borges, J. L. C., and Thorner, M. O. (1984). The self-priming effect of gonadotropin-releasing hormone on luteinizing hormone release: Observations using rat anterior pituitary fragments and dispersed cells continuously perifused in parallel. *Endocrinology* **114**, 861–867.

Fink, G. (1979). Feedback action of target hormones on hypothalamus and pituitary with special reference to gonadal steroids. *Annu. Rev. Physiol.* **41**, 571–585.

Fink, G., Aiyer, M. S., Jamieson, M. G., and Chiappa, S. A. (1975). Factors modulating the responsiveness of the anterior pituitary gland in the rat with special reference to gonadotropin-releasing hormone (GnRH). *In* "Hypothalamic Hormones" (M. Motta, P. G. Crosignani, and L. Martini, eds.), pp. 139–161. Academic Press, London.

Frager, M. S., Pieper, D. R., Tonetta, J. A., and Marshall, J. C. (1981). Pituitary GnRH receptors: Effects of castration, steroid replacement, and the role of GnRH in modulating receptors in the rat. *J. Clin. Invest.* **67**, 615–621.

Fujino, M., Fukuda, S., Shinagawa, S., Kobayashi, S., Yamazaki, I., and Nakayama, R. (1974). Synthetic analogs of luteinizing hormone releasing hormone (LH-RH) substituted in position 6 and 10. *Biochem. Biophys. Res. Commun.* **60**, 406–413.

Giguere, V., Lefebvre, F. A., and Labrie, F. (1981). Androgens decrease LHRH binding sites in rat anterior pituitary cells in culture. *Endocrinology* **108**, 350–352.

Gilman, A. G., (1984). G-proteins and dual control of adenylate cyclase. *Cell* **36**, 577–579.

Gordon, J. H., and Reichlin, S. (1974). Changes in pituitary responsiveness to luteinizing hormone-releasing factor during the rat estrous cycle. *Endocrinology* **94**, 974–978.

Gorospe, W. C., and Conn, P. M. (1987a). Agents that decrease gonadotropin-releasing hormone (GnRH) receptor internalization do not inhibit GnRH-mediated gonadotrope desensitization. *Endocrinology* **120**, 222–229.

Gorospe, W. C., and Conn, P. M. (1987b). Membrane fluidity regulates development of gonadotrope desensitization to GnRH. *Mol. Cell. Endocrinol.*, in press.

Grant, G., Vale, W. W., and Rivier, J. (1973). Pituitary binding sites for (^{3}H)-labelled luteinizing hormone releasing factor. *Biochem. Biophys. Res. Commun.* **50**, 771–778.

Gregory, H., Taylor, C. L., and Hopkins, C. R. (1982). Luteinizing hormone release from dissociated pituitary cells by dimerization of occupied LHRH receptors. *Nature* (London) **300**, 269–271.

Harris, C. E., Staley, D., and Conn, P. M. (1985). Diacylglycerol and protein kinase C: Potential amplifying mechanism for Ca^{2+}-mediated gonadotropin-releasing hormone-stimulated luteinizing hormone release. *Mol. Pharmacol.* **27**, 532–536.

Hart, R. C., Bates, M. D., Cormier, M. J., Rosen, G. M., and Conn, P. M. (1983). Synthesis and characterization of calmodulin antagonistic drugs. *In* "Methods in Enzymology (A. R. Means and B. W. O'Malley, eds.), Vol. 102, pp. 195–204. Academic Press, New York.

Haslam, R. J., and Davidson, M. M. L. (1984). Receptor-induced diacylglycerol formation in permeabilized platelets: Possible role for GTP-binding proteins. *J. Receptor Res.* **4**, 605–629.

Hazum, E. (1981a). Some charateristics of GnRH receptors on rat pituitary membranes: Differences between agonist and antagonist. *Mol. Cell. Endocrinol.* **23**, 275–281

Hazum, E. (1981b). Photoaffinity labeling of luteinizing hormone releasing hormone receptor of rat pituitary membrane preparations. *Endocrinology* **109**, 1281–1283.

Hazum, E. (1982). GnRH-receptor of rat pituitary is a glycoprotein: Differential effect of neuraminidase and lectins on agonists and antagonists binding. *Mol. Cell. Endocrinol.* **26**, 217–222.

Hazum, H., and Keinan, D. (1984). Characterization of GnRH receptors in bovine pituitary membranes. *Mol. Cell. Endocrinol.* **35**, 107–111.

Hazum, E., Cuatrecasas, P., Marian, J., and Conn, P. M. (1980). Receptor mediated internalization of fluorescent gonadotropin releasing hormone by pituitary gonadotropes. *Proc. Natl. Acad. Sci. U.S.A.* **77**, 6692–6695.

Hinkle, P. M. and Phillips, W. J. (1984). Thyrotropin-releasing hormone stimulates GTP hydrolysis by membranes from $G_1H_4C_1$ rat pituitary tumor cells. *Proc. Natl. Acad. Sci. U.S.A.* **81**, 6183-6187.

Hirota, K., Hirota, T., Aguilera, G., and Catt, K. J. (1985). Hormone-induced redistribution of calcium-activated phospholipid-dependent protein kinase in pituitary gonadotrope. *J. Biol. Chem.* **260**, 3243-3246.

Hokin, M. R., and Hokin, L. E. (1953). Enzyme secretion and incorporation of ^{32}P into phospholipids of pancreas slices. *J. Biol. Chem.* **203**, 967-977.

Hokin, M. R., and Hokin, L. E. (1954). Effects of acetylcholine and phospholipids in the pancreas. *J. Biol. Chem.* **209**, 549-558.

Hopkins, C. R., and Walker, A. M. (1978). Calcium as a second messenger in the stimulation of luteinizing hormone secretion. *Mol. Cell. Endocrinol.* **12**, 189-208.

Hsueh, A. J. W., Erickson, G. F., and Yen, S. S. C. (1979). The sensitivity of estrogen on cultured pituitary cells to luteinizing hormone releasing hormone: Its antagonism by progestins. *Endocrinology* **104**, 807-813.

Huckle, W. R., and Conn, P. M. (1987). The relationship between gonadotropin-releasing hormone-stimulated luteinizing hormone release and inositol phosphate production: Studies with calcium antagonists and protein kinase C activators. *Endocrinology* **120**, 160-169.

Jacobs, S., Chang, K.-J., and Cuatrecasas, P. (1978). Antibodies to purified insulin receptor have insulin-like activity. *Science* **200**, 1283-1284.

Jennes, L., Beckman, W. C., Stumpf, W. E., and Grzanna, R. (1982). Anatomical relationship of serotoninergic and noradrenalinergic projections with the GnRH system in the septum and hypothalamus. *Exp. Brain Res.* **46**, 331-338.

Jennes, L., Stumpf, W. E., and Conn, P. M. (1983a). Intracellular pathways of electron-opaque gonadotropin-releasing hormone derivatives bound by cultured gonadotropes. *Endocrinology* **113**, 1683-1689.

Jennes, L., Stumpf, W. E., and Tappaz, M. L. (1983b). Anatomical relationships of dopaminergic and GABAergic systems with GnRH-systems in the septohypothalamic area. *Exp. Brain Res.* **50**, 91-99.

Jennes, L., Bronson, D., Stumpf, W. E., and Conn, P. M. (1985). Evidence for an association between calmodulin and membrane patches containing GnRH-receptors in cultured pituitary gonadotropes *Cell Tissue Res.* **239**, 311-315.

Kahn, C. R., Baird, K. L., Jarrett, D. B., and Flier, J. S. (1978). Direct demonstration that receptor crosslinking or aggregation is important in insulin action. *Proc. Natl. Acad. Sci. U.S.A.* **75**, 4209-4213.

Kaibuchi, K., Takai, Y., Sawamura, M., Hoshijima, M., Fujikura, T., and Nishizuka, Y. (1983). Synergistic functions of protein phosphorylation and calcium mobilization in platelet activation. *J. Biol. Chem.* **258**, 6701-6704.

Kamel, F., and Krey, L. C. (1982). Gonad/steroid modulation of LHRH-stimulated LH secretion by pituitary cell cultures. *Mol. Cell. Endocrinol.* **26**, 151-164.

Karten, M., and Rivier, J. (1986). Overview of GnRH analog development. *Endocr. Rev.* **7**, 44-66.

Keri, G., Nikolics, K., Teplan, I., and Molnar, J. (1983). Desensitization of luteinizing release in cultured pituitary cells by gonadotropin releasing hormone. *Mol. Cell Endocrinol.* **30**, 109-120.

Kikkawa, U., Takai, Y., Tanaka, Y., Miyake, R., and Nishizuka, Y. (1983). Protein kinase C as a possible receptor protein of tumor-promoting phorbol esters. *J. Biol. Chem.* **258**, 11442-11445.

Klee, C. B., Crouch, T. H., and Richman, P. G. (1980). Calmodulin. *Annu. Rev. Biochem.* **49**, 489-515.

Knobil, E. (1980). The neuroendocrine control of the menstrual cycle. *Recent Prog. Horm. Res.* **36**, 53-88.

Koiter, T. R., Pols-Valkhof, N., and Schuiling, G. A. (1981). Long-lasting desensitizing effect of short-term LRH exposure on pituitary responsiveness to LRH in the ovariectomized rat. *Acta Endocrinol.* **99**, 200-205.

Kuo, J. F., Anderson, R. G. G., Wise, B. C., Mackerlova, L., Salomonsson, I., Brackett, N. L., Katoh, N., Shoji, M., and Wrenn, R. W. (1980). Calcium-dependent protein kinase: Widespread occurrence in various tissues and phyla of the animal kingdom and comparison of effects of phospholipid, calmodulin, and trifluoperazine. *Proc. Natl. Acad. Sci. U.S.A.* **77**, 7039–7043.

Lagace, L., Massicotte, J., and Labrie, F. (1980). Acute stimulatory effects of progesterone on luteinizing hormone and follicle-stimulating hormone release in rat anterior pituitary cells in culture. *Endocrinology* **106**, 684–689.

Levin, R. M., and Weiss, B. (1976). Mechanism by which psychotropic drugs inhibit cyclic AMP phosphodiesterase. *Mol. Pharmacol.* **12**, 581–589.

Levin, R. M., and Weiss, B. (1978). Selective binding of antipsychotics and psychoactive agents to calcium-dependent activator of cyclic nucleotide phosphodiesterase. *J. Pharmacol. Exp. Ther.* **208**, 454–459.

Libertun, C., Orias, R., and McCann, S. M. (1974). Biphasic effect of estrogen on the sensitivity of the pituitary to luteinizing hormone releasing factor. *Endocrinology* **94**, 1094–1100.

Loughlin, J. S., Badger, T. M., and Crowley, W. F. (1981). Perifused pituitary cultures: A model for the LHRH regulation of LH release. *Am. J. Physiol.* **240**, E591–596.

Loumaye, E., and Catt, K. J. (1982). Homologues regulation of gonadotropin-releasing hormone receptors in cultured pituitary cells. *Science* **215**, 983–985.

McArdle, C. A., and Conn, P. M. (1986). Hormone stimulated redistribution of gonadotrope protein kinase C *in vivo*: Dependence on calcium mobilization. *Mol. Pharmacol.* **29**, 570–576.

McArdle, C. A., Huckle, W. R., and Conn, P. M. (1987a). Phorbol esters reduce gonadotrope responsiveness to protein kinase C activators but not to Ca^{2+}-mobilizing secretagogues. *J. Biol. Chem.* **262**, 5028–5035.

McArdle, C. A., Gorospe, W. C., Huckle, W. R., and Conn P.M. (1987b). Homologous down-regulation of gonadotropin-releasing hormone receptors and desensitization of gonadotropes: lack of dependence on protein kinase C. *Mol. Endocrinol.* **1**, 420–429.

McCann, S. M., Lumpkin, M. D., Mizunuma, H., Samson, W. K., Steele, M. K., Ojeda, S. R., and Negro-Vilar, A. (1982). Control of luteinizing hormone releasing hormone (LH-RH) release by neurotransmitters. *In* "The Gonadotropins: Basic Science and Clinical Aspects in Females" (C. Flamigni and J. R. Givens, eds.), pp. 107–116. Academic Press, New York.

Marian, J., and Conn, P. M. (1979). GnRH stimulation of cultured pituitary cells requires calcium. *Mol Pharmacol.* **16**, 196–201.

Marian, J., and Conn, P. M. (1980). The calcium requirement in GnRH-stimulated LH release is not mediated through a specific action on the receptor. *Life Sci.* **27**, 87–92.

Marian, J., and Conn, P. M. (1983). Subcellular localization of the receptor for gonadotropin-releasing hormone in pituitary and ovarian tissue. *Endocrinology* **112**, 104–112.

Marian, J., Cooper, R., and Conn, P. M., (1981). Regulation of the rat pituitary GnRH-receptor. *Mol. Pharmacol.* **19**, 399–405.

Marshall, J. C, Shakespear, R. A., and Odell, W. D. (1976). LHRH-pituitary plasma membrane binding: The presence of specific binding sites in other tissues. *Clin. Endocrinol.* **5**, 671–677.

Martin, J. E., Tyrey, L., Everett, J. W., and Fellows, R. E. (1974). Estrogen and progesterone modulation of the pituitary response to LRF in cyclic rat. *Endocrinology* **95**, 1664–1673.

Means, A. R., and Dedman, J. R. (1980). Calmodulin—an intracellular calcium receptor. *Nature* (London) **285**, 73–77.

Means, A. R., Tash, J. S., and Chafouleas, J. G. (1982). Physiological implication of the presence, distribution and regulation of calmodulin in eukaryotic cells. *Physiol. Rev.* **62**, 1–39.

Mehdi, S. Q., and Kriss, J. P. (1978). Preparation of thyroid-stimulating immunoglobulins (TSI) by recombining TSI heavy chains with ^{125}I-labeled light chains: Direct evidence that the product binds to the membrane thyrotropin receptor and stimulates adenylate cyclase. *Endocrinology* **103**, 296–301.

Meidan, R., and Koch, Y. (1981). Binding of luteinizing hormone-releasing hormone analogues to dispersed rat pituitary cells. *Life Sci.* **28**, 1961–1967.

Meidan, R., Aroya, N. B., and Koch, Y. (1982). Variation in the number of LHRH receptors correlated with altered responsiveness to LHRH. *Life Sci.* **30**, 535–541.

Meites, J., and Sonntag, W. E. (1981). Hypothalamic hypophysiotropic hormones and neuro-transmitter regulation: Current views. *Annu. Rev. Pharmacol. Toxicol.* **21**, 295–322.

Momany, F. A. (1976). Conformational energy analysis of the molecule LH-RH 1. Native decapeptide. *J. Am. Chem. Soc.* **98**, 2990–2996.

Monahan, M. W., Amoss, M. S., Anderson, H. A., and Vale, W. (1973). Synthetic analogs of hypothalamic LRF with increased agonist or antagonist properties. *Biochemistry* **12**, 4616–4620.

Moriarity, C. M. (1978). Role of calcium in the regulation of adenohypophysical hormone secretion. *Life Sci.* **23**, 185–194.

Motta, M., and Martini, L. (1982). Effect of opioid peptides on gonadotropin secretion. *Acta Endocrinol.* **99**, 321–325.

Naor, Z., Clayton, R. N., and Catt, K. J. (1980). Characterization of gonadotropin-releasing hormone receptors in cultured rat pituitary cells. *Endocrinology* **107**, 1144–1152.

Naor, Z., Atlas, A., Clayton, R. N., Forman, D. S., Amsterdam, A., and Catt, K. J. (1981). Interaction of fluorescent GnRH with receptors in pituitary cells. *J. Biol. Chem.* **256**, 3049–3052.

Nestor, Jr., J. J., Ho, T. L., Tahilramani, R., McRae, G. I., and Vickery, B. H. (1984). Long acting LHRH agonists and antagonists. *In* "LHRH and its Analogues: Basic and Clinical Aspects" (F. Labire, A. Belanger, and A. Dupont, eds.), pp. 24–35. Academic Press, New York.

Nett, T. M., Crowder, M. E., Moss, G. E., and Duello, T. M. (1981). GnRH–receptor inter-action. V. Down-regulation of pituitary receptors for GnRH in ovariectomized ewes by infu-sion of homologous hormone. *Biol. Reprod.* **24**, 1145.

Niedel, J. E., Kuhn, L. J., and Vanderbark, G. R. (1983). Phorbol receptor copurifies with pro-tein kinase C. *Proc. Natl. Acad. Sci. U.S.A.* **80**, 36–40.

Nikolics, K., Coy, D. H., Vilchez-Martinez, J. A., Coy, E. J., and Schally, A. V. (1977). Synthesis and biological activity of position 1 analogs pf LH-RH. *Int. J. Peptide Protein Res.* **9**, 57–62.

Nishizuka, Y. (1984a). Turnover of inositol phospholipids and signal transduction. *Science* **225**, 1365–1370.

Nishizuka, Y. (1984b). The role of protein kinase C in cell surface transduction and tumor promotion. *Nature* (London) **308**, 693–698.

O'Connor, J. L., Allen, M. B., and Mahesh, V. B. (1980). Castration effects on the response of rat pituitary cells to luteinizing hormone-releasing hormone: Retention in dispersed cell cultures. *Endocrinology* **106**, 1706–1714.

Packman, P. M., and Rothchild, J. A. (1976). Morphine inhibition of ovulation and reversal by naloxone. *Endocrinology* **99**, 7–10.

Pang, C. N., Zimmerman, E., and Sawyer, C. H. (1977). Morphine inhibition of the preovula-tory surges of plasma luteinizing hormone and follicle stimulating hormone. *Endocrinology* **101**, 1726–1732.

Pastan, I. H., and Willingham, M. C. (1981). Receptor-mediated endocytosis of hormones in cultured cells. *Annu. Rev. Physiol.* **43**, 239–250.

Pedroza, E., Vilchez-Martinez, J. A., Fishback, J., Arimura, A., and Schally, A. V. (1977). Binding capacity of luteinizing hormone-releasing hormone and its analogues for pituitary receptor sites. *Biochem. Biophys. Res. Commun.* **79**, 234–246.

Pelletier, G., Dube, D., Guy, J., Seguin, C., and Lefebvre, F. A. (1982). Binding and inter-nalization of a luteinizing hormone-releasing hormone agonist by rat gonadotrophic cells: A radioautographic study. *Endocrinology* **111**, 1068–1076.

Perrin, M. H., Rivier, J., and Vale, W. W. (1980). Radioligand assay for gonadotropin-releasing hormone: Relative potencies of agonists and antagonists. *Endocrinology* **106**, 1289–1296.

Perrin, M. H., Haas, Y., Rivier, J., and Vale, W. (1983). Solubilization of the GnRH receptor from bovine pituitary plasma membranes. *Endocrinology* **112**, 1538–1541.

Pickering, A. J. M. C., and Fink, G. (1977). A priming effect of luteinizing hormone releasing factor with respect to the release of follicle-stimulating hormone *in vitro* and in *vivo*. *J. Endocrinol.* **75**, 155–159.

Pieper, D. R., Gala, R. R., and Marshall, J. C. (1982). Dependence of pituitary gonadotropin-releasing hormone (GnRH) receptors on GnRH secretion from the hypothalamus. *Endocrinology* **110**, 749–753.

Rasmussen, H., and Barrett, P. Q. (1984). Calcium messenger system: An integrated view. *Physiol. Rev.* **64**, 938–984.

Raymond, V., Veilleux, R., and Leung, P. C. K. (1982). Early stimulation of phosphatidylinositol response by LHRH in an enriched population of gonadotropes in primary culture. *Endocrinology* **110**, (Suppl.), 285.

Raymond, V., Leung, P. C. K., Veilleux, R., Lefevre, G., and Labrie, F. (1984). LHRH rapidly stimulates phosphatidylinositol metabolism in enriched gonadotrophs. *Mol. Cell. Endocrinol.* **36**, 157–164.

Root, A. W., Reiter, E. O., Duchett, G. E., and Sweetland, M. L. (1975). Effect of short-term castration and starvation upon hypothalamic content of luteinizing hormone-releasing hormone in adult male rats. *Proc. Soc. Exp. Biol. Med.* **150**, 602–605.

Rudenstein, R. S., Bigdeli, H., McDonald, M. H., and Snyder, P. J. (1979). Administration of gonadal steroids to the castrated male rat prevents a decrease in the release of gonadotropin-releasing hormone from incubated hypothalamus. *J. Clin. Invest.* **63**, 262–267.

Ryu, K., Williams, J. A., and Gallo, R. V. (1980). Studies on a possible pituitary effect of monoamines on luteinizing hormone release in ovariectomized rats. *Life Sci.* **27**. 1083–1087.

Samli, M. H., and Geschwind, I. I. (1968). Some effects of energy-transfer inhibitors and of Ca^{2+}-free and of K^+-enhanced media on the release of LH from the rat pituitary gland *in vitro*. *Endocrinology* **82**, 225–231.

Sandow, J. (1983). The regulation of LHRH action at the pituitary and gonadal receptor level: A review. *Psychoneuroendocrinology* **8**, 277–297.

Sarkar, D. K., and Fink, G. (1979). Effects of gonadal steroids on output of luteinizing hormone releasing factor into pituitary stalk blood in the female rat. *J. Endocrinol.* **80**, 303–313.

Savoy-Moore, R. T., Schwartz, N. B., Duncan, J. A., and Marshall, J. C. (1980). Pituitary gonadotrophin-releasing hormone receptors during the rat estrous cycle. *Science* **209**, 942–944.

Schally, A. V., Arimura, A., Carter, W. H., Redding, T. W., Geiger, R., Konig, R., Wissman, H., Jaeger, G., Sandow, J., Yanihara, N., Yanihara, C., Hashimoto, T., and Sakagami, M. (1972). Luteinizing hormone-releasing hormone (LH-RH) activity of some synthetic polypeptides. I. Fragments shorter than decapeptide. *Biochem. Biophs. Res. Commun.* **48**, 366–375.

Schechter, Y., Hernaez, L., Schlessinger, J., and Cuatrecasas, P. (1979). Localization aggregation of hormone-receptor complexes is required for activation by epidermal growth factor. *Nature* (London) **278**, 835–838.

Smith, D. E., and Conn, P. M. (1982). The site of estradiol sensitization of the gonadotrope is prior to calcium mobilization. *Life Sci.* **30**, 1495–1498.

Smith, M. A., and Vale, W. W. (1980). Superfusion of rat anterior pituitary cells attached to cytodex beads: Validation of a technique. *Endocrinology* **107**, 1425–1431.

Smith, W. A., and Conn, P. M. (1983a). Causes and consequences of altered gonadotropin secretion in the aging rat. *In* "Clinical and Experimental Intervention in the Pituitary During Aging" (R. Walker and R. L. Cooper, eds.), pp. 3–26. Dekker, New York.

Smith, W. A., and Conn, P. M. (1983b). GnRH-mediated desensitization of pituitary gonadotrope is not calcium-dependent. *Endocrinology* **112**, 408–410.

Smith, W. A., and Conn, M., (1984). Microaggregation of the gonadotropin-releasing hormone receptor: Relation to gonadotrope desensitization. *Endocrinology* **114**, 553–559.

Smith, W. A., Cooper, R. L., and Conn, P. M. (1982). Altered pituitary responsiveness to gonadotropin-releasing hormone in middle-aged rats with 4-day estrous cycles. *Endocrinology* **111**, 1843-1848.

Snyder, G. D., and Bleasdale, J. E. (1982). Effect of LHRH on incorporation of [^{32}P]-orthophosphate into phosphatidylinositol by dispersed pituitary cells. *Mol. Cell. Endocrinol.* **28**, 55-63.

Stern, J. E., and Conn, P. M. (1981). Requirements for GnRH stimulated LH release from perifused rat hemipituitaries. *Am. J. Physiol.* **240**, E504-509.

Takai, Y., Kishimoto, A., Iwasa, Y., Kawahara, Y., Mori, T., and Nishizuka, Y. (1979). Calcium-dependent activation of a multifunctional protein kinase by membrane phospholipids. *J. Biol. Chem.* **254**, 3692-3695.

Takai, Y., Kikkawa, U., Kaibudi, K., and Nishizuka, Y. (1984). Membrane phospholipid metabolism and signal transduction for protein phosphorylation. *Adv. Cyclic Nucleotide Protein Phosphorylation* **18**, 119-157.

Tsien, R. Y., Pozzan, T., and Rink, T. J. (1982). Calcium homeostasis in intact lymphocytes: Cytoplasmic free calcium monitored with a new intracellularly trapped fluorescent indicator. *J. Cell Biol.* **94**, 325-334.

Vale, W., Burgus, R., and Guillemin, R. (1967). Presence of calcium ions as a requisite for the *in vitro* stimulation of TSH release by hypothalaminc TRF. *Experientia* **23**, 853-859.

van Dieten, J. A. M. J., and van Rees, G. P. (1983). Influence of pulsatile administration of LRH on the development of the augmentative (positive) effect of oestradiol on the pituitary response to LRH. *Acta Endocrinol.* **102**, 337-342.

Venter, J. C. (1985). Size of neurotransmitter receptors as determimed by radiation inactivation/target size analysis. *In* "The Receptors" Vol. II., (P. M. Conn, ed.) Academic Press, New York, pp. 245-280.

Vilchez-Martinez, J. A., Coy, D. H., Arimura, A., Coy, E. J., Hirotsu, Y., and Schally, A. V. (1974). Synthesis and biological properties of [Leu-6]-Lh-RH and [D-Leu-6, Des Gly-NH$_2$-10]-LH-RH ethylamide. *Biochem. Biophys. Res. Commun.* **59**, 1226-1232.

Wagner, T. O. F., Adams, T. E., and Nett, T. M. (1979). GnRH interaction with anterior pituitary. I. Determination of the affinity and number of receptors for GnRH in bovine anterior pituitary. *Biol. Reprod.* **20**, 140-149.

Wakabayashi, K., Kamberi, I. A., and McCann, S. M. (1969). *In vitro* response of the rat pituitary to gonadotropin releasing factors and to ions. *Endocrinology* **85**, 1046-1056.

Waring, D. W., and Turgeon, J. L. (1980). Luteinizing hormone-releasing hormone-induced luteinizing hormone secretion *in vitro*: Cyclic changes in responsiveness and self-priming. *Endocrinology* **106**, 1430-1436.

Waring, D. W., and Turgeon, J. L. (1983). LHRH self priming of gonadotropin secretion: Course of development. *Am. J. Physiol.* **244**, C410-418.

Weiner, R. I., and Ganong W. F. (1978). Role of brain monoamines and histamine in regulation of anterior pituitary secretion. *Physiol. Rev.* **58**, 905-976.

Williams, J. A. (1976). Stimulation of ^{45}Ca^{2+} efflux from rat pituitary by LHRH and other pituitary stimulators. *J. Physiol.* (London) **260**, 105-115.

Winiger, B. P., Birabeau, M. A., Lang, U., Capponi, A. M., Sizonenko, P. C., and Aubert, M. L. (1983). Solubilization of pituitary GnRH binding sites by means of a zwitterionic detergent. *Mol. Cell. Endocrinol.* **31**, 77-91.

Wojcikiewicz, J. H., Kent, P. A., and Fain, J. (1986). Evidence that thyrotropin-releasing hormone-induced increases in GTPase activity and phosphoinositide metabolism in GH$_3$ cells are mediated by a guanine nucleotide-binding protein after than G$_s$ or G$_i$. *Biochem. Biophys. Res. Commun.* **138**, 1383-1389.

Wollheim, C. B., and Pozzan, T. (1984). Correlation between cytosolic free Ca^{2+} and insulin release in an insulin-secreting cell line. *J. Biol. Chem.* **259**, 2262-2267.

Zilberstein, M., Zakut, H., and Naor, Z. (1983). Coincidence of down-regulation and desensitization in pituitary gonadotrophs stimulated by gonadotropin-releasing hormone. *Life Sci.* **32**, 663-669.

Thyrotropin-Releasing Hormone: Role of Polyphosphoinositides in Stimulation of Prolactin Secretion

RICHARD N. KOLESNICK

Department of Endocrinology, Department of Medicine
Memorial Sloan-Kettering Cancer Center
New York, New York 10021

AND

MARVIN C. GERSHENGORN

Department of Endocrinology and Metabolism, Department of Medicine
Cornell University Medical College and The New York Hospital
New York, New York 10021

I. INTRODUCTION

Recent investigations into proximal events after thyrotropin-releasing hormone (TRH)–receptor interaction have helped establish a biochemical basis for transduction of the hormonal signal at the cell surface into prolactin secretion from rat mammotropic (GH)[1] pituitary cells (for review, see Gershengorn, 1986). Although GH cells are a neoplastic cell line, they have been used extensively to investigate TRH action as they appear to be valid models of physiologic function. It has become clear that TRH-induced hydrolysis of a minor plasma membrane lipid, phosphatidylinositol 4,5-bisphosphate [PtdIns(4,5)P$_2$], via a phospholipase C results in the generation of two second messengers, inositol trisphosphate (InsP$_3$) and

[1]GH cells are cells in long-term culture that were originally derived from a rat pituitary tumor. Several clonal strains of these cells, such as GH$_3$ and GH$_4$C$_1$ clones, have been used by different groups of investigators.

1,2-diacylglycerol (1,2-DG). These messengers mediate and amplify the hormonal signal intracellularly. Inositol trisphosphate directly releases calcium from cellular, nonmitochondrial stores to elevate the concentration of calcium ion free within the cytoplasm ($[Ca^{2+}]_i$). The elevation of $[Ca^{2+}]_i$ serves to couple, at least in part, receptor binding to activation of specific cellular processes. 1,2-DG appears to exert its influence by enhancing the activity of a Ca^{2+}- and phospholipid-dependent protein kinase (protein kinase C). This action of 1,2-DG leads to phosphorylation of a set of cellular proteins presumed to regulate a series of specific cellular responses. It is generally agreed that the coordinate interaction of these bifurcating pathways serves to transduce the TRH signal into an optimal biologic response, such as stimulation of prolactin secretion. This schema is similar to those being elucidated for signal transduction by calcium-mobilizing stimuli in a number of cell types (for reviews, see Berridge, 1984; Nishizuka, 1984; Rasmussen and Barrett, 1984).

II. CALCIUM AS MEDIATOR OF TRH-INDUCED BURST PROLACTIN SECRETION

The proposal that calcium ion might couple stimulation by secretagogues that interact with cell-surface receptors to secretion was initially made to explain neurotransmitter release from chromaffin cells (see Douglas, 1978). Evidence that calcium was necessary for TRH action was first shown directly by Vale *et al.* (1967). These investigators demonstrated that incubation of rat hemipituitary glands in medium containing no added Ca^{2+} inhibited TRH-stimulated secretion of thyrotropin (thyroid-stimulating hormone, TSH) and that this effect was reversed by Ca^{2+} supplementation. Thereafter, Tashjian *et al.*(1978) utilized a similar design in GH cells to show a calcium requirement for TRH-induced prolactin secretion. Based on these observations, and by analogy to neurotransmitter action, these investigators suggested that TRH acted to cause calcium influx that elevated $[Ca^{2+}]_i$ to couple stimulation to secretion. This proposal was supported by the demonstration first by Kidokoro (1975) and later by others (Biales *et al.*, 1977; Dufy *et al.*, Ozawa and Kimura, 1979; Sand *et al.*, 1980) that GH cells manifested Ca^{2+}-dependent action potentials that were increased in frequency by TRH. However, the observation that removal of medium calcium prevented TRH-stimulated secretion was not sufficient to ascribe a role for enhanced influx of extracellular calcium as mediator of stimulated secretion, as it became evident that such manipulation dramatically depressed cellular calcium levels in GH cells (Gershengorn *et al.*, 1981). Thus, it was possible that depletion of extracellular calcium led to loss of Ca^{2+} from critical intracellular pools. We (Gershengorn, 1982), therefore, interpreted these early data to support a role for Ca^{2+} in stimulated secretion, but not necessarily via increased influx from extracellular sources. Additionally, the association of TRH-induced Ca^{2+}-dependent action potentials and stimulation of secretion only suggested but did not prove a cause and effect relationship between these events.

Further evidence to support the hypothesis that acute alterations in calcium metabolism were involved in stimulated prolactin secretion was derived from experiments in which the intracellular calcium pools of GH cells were labeled with radiocalcium (^{45}Ca) (Gershengorn et al., 1981; Tan and Tashjian, 1981). When these cells were resuspended in medium without radiocalcium, TRH stimulated ^{45}Ca efflux from the cells simultaneously with prolactin secretion. Both events were measurable within 1 minute of stimulation. These findings were, therefore, consistent with the notion that calcium was a coupling factor between stimulation by TRH and hormone secretion. However, these studies did not demonstrate whether the enhanced efflux of ^{45}Ca was secondary to enhanced influx of extracellular calcium or mobilization (or redistribution) of intracellular Ca^{2+}, or both.

To determine whether enhanced ^{45}Ca efflux and prolactin secretion stimulated by TRH were dependent on enhanced Ca^{2+} influx, studies were performed in which GH cells were incubated in low calcium medium for only very brief periods prior to TRH addition. The rationale being that if TRH was still able to cause ^{45}Ca efflux and prolactin secretion even after the Ca^{2+} concentration was lowered in the medium so that the Ca^{2+} gradient had been reversed, that is under conditions favoring outflow of Ca^{2+} from the cells, it would strongly suggest that Ca^{2+} influx was not necessary for the TRH effects. In these experiments, extracellular calcium was set using calcium-chelator buffering systems. Figure 1 demonstrates an experiment directed at the question of whether Ca^{2+} efflux and prolactin secretion stimulated by TRH occurred under conditions that partially or completely prevented calcium influx. At this time, it was only possible to estimate the resting $[Ca^{2+}]_i$ to be between 0.1 and 1.0 μM as direct measurement, which was difficult because of the small size of these cells, had not yet been made; the intracellularly trapped calcium probe Quin 2 was not yet available (see below). Thus, extracellular Ca^{2+} concentrations significantly below 0.1 μM were necessary to ensure reversal of the Ca^{2+} gradient. The effect of TRH was compared to a depolarizing concentration (50 mM) of K^+ because high extracellular K^+ was known to cause prolactin secretion specifically via opening of voltage-dependent calcium channels. In the experiment shown here, cells were labeled with radiocalcium by incubating them in medium containing 1500 μM ^{45}Ca^{2+}. The effects of 1-minute pulses of TRH and 50 mM K^+ were compared during perfusion of the cells in medium containing 1500 or 2.8 μM total calcium, or 0.02 μM free Ca^{2+}. ^{45}Ca efflux and prolactin secretion stimulated by TRH and high K^+ appeared to be simultaneous events when corrections were made for an artifact of the perfusion system. The increment in prolactin release caused by K^+ depolarization in medium containing 2.8

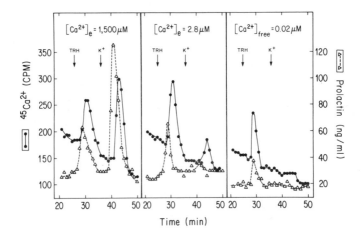

FIG. 1. Effects of TRH and K^+ depolarization on $^{45}Ca^{2+}$ efflux and prolactin release ($\triangle$——$\triangle$) from GH cells incubated in medium containing 1500 or 2.8 μM total Ca^{2+} ($[Ca^{2+}]_e$), or 0.02 μM free Ca^{2+} ($[Ca^{2+}]_{free}$). Reproduced with permission from Gershengorn *et al.* (1981).

μM total calcium was only 3% of that in medium containing 1500 μM calcium (control) and was abolished in medium containing 0.02 μM free Ca^{2+}. In contrast, prolactin release caused by TRH in medium containing 2.8 μM total calcium and 0.02 μM free Ca^{2+} was 50 and 35% of that of control, respectively.

The most parsimonious explanation of these data is that (1) calcium serves as a coupling factor for prolacting secretion stimulated by TRH and high extracellular K^+ and (2) at least part of TRH-induced prolactin secretion is mediated by redistribution of cellular calcium stores. Direct demonstration of elevation of $[Ca^{2+}]_i$ during TRH stimulation of prolactin secretion and the role of mobilization of Ca^{2+} from intracellular stores in contributing to this elevation were not possible until fluorescent and luminescent probes of cytoplasmic Ca^{2+} were used.

The intracellularly trapped fluorescent probe Quin 2 and the photoprotein aequorin have made it possible to measure $[Ca^{2+}]_i$ directly. Most laboratories have reported resting $[Ca^{2+}]_i$ to be in the range of 120 nM in suspensions of GH cells (Snowdowne and Borle, 1984; Snowdowne, 1984; Gershengorn *et al.*, 1984; but see Albert and Tashjian, 1984a); a resting $[Ca^{2+}]_i$ of 37 nM was measured in individual GH cells in monolayer culture (Kruskal *et al.*, 1984). TRH causes an immediate elevation of $[Ca^{2+}]_i$ to a level severalfold greater than the resting concentration, followed by a fall toward the basal level, and then a more sustained elevation to a level approximately twice basal. The upper tracing of Fig. 2 illustrates the typical biphasic elevation of $[Ca^{2+}]_i$ caused by TRH. There was a rapid increase (less than 3 seconds to peak) from a resting level of 140 ± 8.6 to 520 ± 29 nM that returned toward the resting level by 1.5 minute and was followed by a secondary elevation to 260 ± 14 nM. This secondary elevation lasted for at least 20 minutes and was abolished when cells were incubated in medium containing very low calcium. In contrast, the rapid elevation of $[Ca^{2+}]_i$ was only partially inhibited when the cells were incubated in medium with very low Ca^{2+} (Fig. 2, lower tracing). Depolarization by 50 mM K^+ caused a monophasic elevation of $[Ca^{2+}]_i$ in cells incubated in control medium containing 1500 μM Ca^{2+} that was abolished when the cells were incubated in medium containing 10 μM Ca^{2+}. The lower panel of Fig. 2 illustrates a compilation of the data of the effects of lowering the concentration of Ca^{2+} in the medium on the elevation of $[Ca^{2+}]_i$ caused by TRH and K^+ depolarization.

These data demonstrate that the first-phase elevation of $[Ca^{2+}]_i$ stimulated by TRH was only minimally inhibited by lowering the Ca^{2+} level in the medium. This finding was consistent with the notion presented originally by our laboratory (Gershengorn et al., 1981; Gershengorn and Thaw, 1983) that the first-phase elevation of $[Ca^{2+}]_i$ in GH cells induced by TRH was caused in large part by mobilization of intracellular calcium. In contrast, the second-phase elevation of $[Ca^{2+}]_i$ caused by TRH and the monophasic elevation of $[Ca^{2+}]_i$ induced by high extracellular K^+ were abolished when cells were incubated in medium with low Ca^{2+}; hence, these effects appeared to be dependent on enhanced influx of extracellular Ca^{2+}.

To demonstrate further that the second-phase elevation of $[Ca^{2+}]_i$ caused by TRH was dependent on Ca^{2+} influx, the effects of the organic Ca^{2+}-channel blocking agents nifedipine and verapamil were tested. These drugs abolished the second-phase elevation of $[Ca^{2+}]_i$ caused by TRH, as they did the elevation caused by K^+ depolarization, but had only a small effect on the first-phase elevation induced by TRH (Gershengorn and Thaw, 1985; but see Albert and Tashjian, 1984b).

It was then important to determine the intracellular pool(s) that was (were) mobilized during TRH stimulation to cause the rapid elevation of $[Ca^{2+}]_i$. This was assessed indirectly in intact GH cells that were labeled with radiocalcium by using (1) agents that uncouple mitochondrial oxidative phosphorylation, such as cabonyl cyanid chlorophenylhydrazone and bis(hexafluoroacetonyl)acetone, that would specifically release Ca^{2+} from mitochondria and (2) calcium ionophores, such as A23187, that would release Ca^{2+} from all intracellular pools. In these experiments (Gershengorn et al., 1984), the mitochondrial pool was, therefore, defined as the pool released by the mitochondrial uncoupling agents and accounted for approximately 18% of total cell ^{45}Ca. The nonmitochondrial pool(s) was the pool(s) released by the calcium ionophore after depletion of the mitochondrial pool and accounted

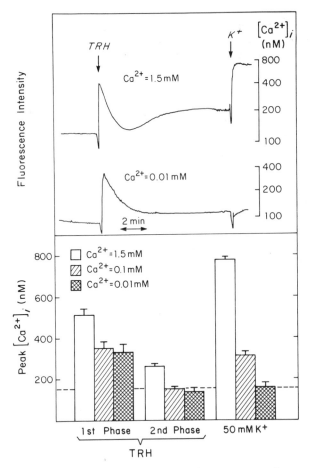

Fɪɢ. 2. Effects of TRH and K^+ depolarization on cytoplasmic free Ca^{2+} concentration ($[Ca^{2+}]_i$) in GH cells incubated in medium containing 1.5, 0.1, or 0.01 mM Ca^{2+}. Reproduced with permission from Gershengorn and Thaw (1985).

for 50% of total cell ^{45}Ca. TRH was shown to deplete approximately 50% of the nonmitochondrial pool(s) and to have no effect on the mitochondrial pool. Hence, TRH appears to cause a rapid elevation of $[Ca^{2+}]_i$ by releasing Ca^{2+} specifically from a nonmitochondrial, intracellular calcium pool, perhaps within the endoplasmic reticulum.

In summary, these data demonstrate that TRH induces a rapid (first-phase) elevation of $[Ca^{2+}]_i$ that is caused in large part, if not solely, by releasing calcium from an intracellular pool and a sustained (second-phase) elevation that is secondary to enhanced influx of extracellular Ca^{2+}. Moreover, these data provide evidence for the hypothesis that elevation of $[Ca^{2+}]_i$ may serve to couple TRH–receptor interaction to stimulation of pro-lactin secretion in GH cells.

III. TRH EFFECTS ON PHOSPHOINOSITIDE METABOLISM

In 1975, Michell noted the relationship of enhanced turnover (or metabolism) of cellular phospholipids, specifically phosphatidylinositol (PtdIns) and phosphatidic acid, to calcium mobilization in many cells stimulated by hormones and neurotransmitters that utilized cell-surface receptors to initiate biological responses, such as secretion. At that time, he proposed that hormone–receptor interaction caused hydrolysis of membrane PtdIns that, via an unknown mechanism, led to opening of calcium channels or redistribution of cellular calcium stores, or both. Early studies from several laboratories (Rebecchi et al., 1981; Drummond and Macphee, 1981; Schlegel et al., 1981; Sutton and Martin, 1982) confirmed that accelerated PtdIns metabolism occurred within 1 minute of TRH stimulation of GH cells. Generally, these studies employed cells incubated for 30–60 minutes with [^{32}P]orthophosphate to label the ATP pool prior to TRH stimulation. TRH stimulated specific labeling of phosphatidic acid within 15 seconds and of PtdIns by 1 minute, but not, for example, of phosphatidylcholine or phosphatidylethanolamine, presumably by stimulated transfer of [^{32}P]phosphate from ATP to lipid.

Because these original studies were performed under conditions in which the labeling of the lipids with [^{32}P]phosphate had not attained an isotopic steady state, that is, with the specific activity of the radiolabel in lipid continuously changing, the observed changes could not be directly correlated to changes in lipid mass. However, since phosphatidic acid was known to be an intermediate in PtdIns synthesis, it was postulated that the initiating event might be hydrolysis of unlabeled membrane PtdIns by a phospholipase C with subsequent phosphorylation of the product of that reaction, 1,2-DG, to phosphatidic acid and then to PtdIns. Additional experiments were designed to prove this contention. In these later studies, cells were labeled to isotopic steady state with [^{32}P]phosphate by incubating cells in medium with [^{32}P]orthophosphate for 48 hours prior to TRH stimulation. Under these conditions, changes in lipid radioactivity directly reflect changes in mass. In experiments performed in our laboratory (Rebecchi et al., 1983), TRH caused a rapid fall (in less than 15 seconds) in [^{32}P]PtdIns to levels that reached 77% of control by 1 minute and caused an increase in phosphatidic acid by 15 seconds, the level of which was 170% of control by 2 minutes. Phosphatidylcholine, phosphatidylethanolamine, sphingomyelin, and phosphatidylglycerol were not affected. These studies were interpreted as supporting the concept that TRH–receptor interaction caused a rapid loss of membrane PtdIns and resynthesis via a phosphatidic acid intermediate.

In an attempt to demonstrate that PtdIns hydrolysis was via a phospholipase C, measurements were made of the products of such a reaction: 1,2-DG and inositol monophosphate (InsP) after TRH. Utilizing an approach similar to that above, before TRH stimulation cells were labeled to isotopic steady state with [³H]myoinositol, which specifically labels inositol-containing lipids, or with [³H]arachidonic acid, which predominantly labels the second position of phospholipids. Figure 3 demonstrates the effect of TRH on [³H]arachidonyl-1,2-DG in GH cells. TRH causes a rapid elevation in the level of 1,2-DG which returned toward the unstimulated level by 1 minute. Similarly, in cells labeled with [³H]myoinositol, TRH caused a very rapid increase in InsP. Hence, by utilizing this approach it was possible to demonstrate that TRH increased the levels of both hydrolysis products of a hypothetical phospholipase C action on PtdIns within seconds in GH cells. However, it was appreciated that these findings did not prove that PtdIns itself was hydrolyzed in GH cells because 1,2-DG and InsP can be formed by several degradative pathways (see below).

Michell *et al.* (1981) suggested that the phosphorylated derivatives of PtdIns, phosphatidylinositol 4-monophosphate (PtdIns4P) and PtdIns(4,5)P$_2$, may actually be the lipids acted upon by the phospholipase C during cellular stimulation and that the loss of membrane PtdIns may represent conversion to these polyphosphoinositides (see Figs. 4 and 5 for the structures of these lipids and inositol sugars and of their metabolic cycle). In this schema, PtdIns(4,5)P$_2$ (and perhaps PtdIns4P) would be hydrolyzed to 1,2 -DG and InsP$_3$ (and 1,2-DG and InsP$_2$). Conversion of these water-soluble inositol sugars by dephosphorylation to InsP and inositol would

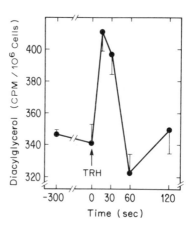

FIG. 3. Effect of TRH on the level of 1,2-diacylglycerol in GH cells prelabeled with [³H]arachidonic acid. Reproduced with permission from Rebecchi *et al.* (1983).

Inositol Lipids

Metabolic Intermediates

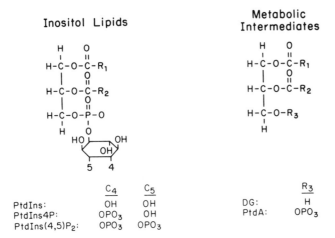

	C_4	C_5
PtdIns:	OH	OH
PtdIns4P:	OPO_3	OH
PtdIns(4,5)P_2:	OPO_3	OPO_3

	R_3
DG:	H
PtdA:	OPO_3

Inositol Sugars

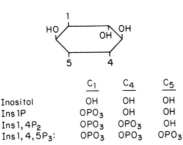

	C_1	C_4	C_5
Inositol	OH	OH	OH
Ins 1P	OPO_3	OH	OH
Ins 1,4P_2	OPO_3	OPO_3	OH
Ins 1,4,5P_3:	OPO_3	OPO_3	OPO_3

FIG. 4. Structures of the inositol phospholipids, sugars, and metabolic intermediates. R_1 and R_2 are fatty acyl moieties. The inositol sugar ring is represented schematically. In both the free and the phospholipid form the inositol ring assumes a stable chair conformation with all substituents oriented equatorially to the ring except the 2-position, which is axial. PtdIns, Phosphatidylinositol; PtdIns4P, phosphatidylinositol 4-phosphate; PtdIns(4,5)P_2, phosphatidylinositol 4,5-bisphosphate; DG, diacylglycerol; PtdA, phosphatidic acid; Ins1P, inositol 1-monophosphate; Ins(1,4)P_2, inositol 1,4-bisphosphate; Ins(1,4,5)P_3, inositol 1,4,5-trisphosphate. Reproduced with permission from Rebecchi and Gershengorn (1985).

then occur by the action of specific phosphatases. Thus, the previously observed increases in 1,2-DG and InsP would not be inconsistent with this paradigm. That such a series of events occurred after TRH stimulation of GH cells was confirmed utilizing cells labeled to isotopic steady state with [^{3}H]myoinositol (Rebecchi and Gershengorn, 1983). In Fig. 6, it can be seen that TRH caused a very rapid (less than 15 second) decrease in the levels of

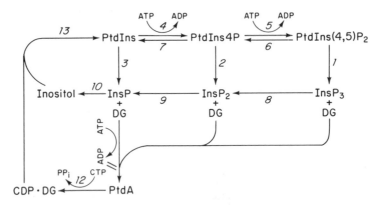

FIG. 5. Pathways of inositol phospholipid metabolism. Enzymes: 1, PtdIns(4,5)P$_2$ phospholipase C; 2, PtdIns4P phospholipase C; 3, PtdIns phospholipase C; 4, PtdIns kinase; 5, PtdIns4P kinase; 6, PtdIns(4,5)P$_2$ phosphomonoesterase; 7, PtdIns4P phosphomonoesterase; 8, InsP$_3$ phosphatase; 9, InsP$_2$ phosphatase; 10, InsP phosphatase; 11, DG kinase; 12, cytidine diphosphate–diacylglycerol (CDP-DG) synthetase; 13, CDP-DG: inositol transferase. Reproduced with permission from Rebecchi and Gershengorn (1985).

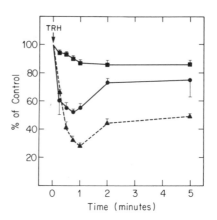

FIG. 6. Time course of the effect of TRH on the levels of PtdIns(4,5)P$_2$ (●——●), PtdIns4P (▲---▲), and PtdIns (■——■) in GH cells prelabeled with [^{3}H]myoinositol. Reproduced with permission from Rebecchi and Gershengorn (1983).

PtdIns(4,5)P$_2$ and PtdIns4P to 60 and 65% of control, respectively. The nadirs for this response occurred at between 45 and 60 seconds. Thereafter, the levels increased but remained below control levels for at least 5 minutes. A slower decrease in PtdIns was again observed that reached a level of 86% of control after 2 minutes. This is consistent with the notion that loss of PtdIns was secondary to PtdIns(4,5)P$_2$ and/or PtdIns4P hydrolysis. However, to prove that PtdIns(4,5)P$_2$ was hydrolyzed by a phospholipase

C, it was necessary to demonstrate a complementary increase in the reaction product InsP$_3$. A stimulated increase in InsP$_3$ can be taken as proof of phospholipase C action on PtdIns(4,5)P$_2$ as InsP$_3$ appears only to be formed by the hydrolysis of PtdIns(4,5)P$_2$ in mammalian cells. Figure 7 illustrates the effect of TRH on inositol sugar metabolism. TRH caused a rapid increase in InsP$_3$ and InsP$_2$ to a peak of 410 (at 15 seconds) and 450% of control (at 30 seconds), respectively. An increase in InsP and inositol was observed only after 0.5 and 1 minute, respectively, in GH cells. The kinetics of these responses are consistent with conversion of InsP$_3$ to InsP$_2$ to InsP to inositol via sequential dephosphorylation. However, these data cannot exclude the possibility that at least part of the elevation of InsP$_2$ and InsP is derived from hydrolysis of PtdIns4P and PtdIns, respectively, via a separate phospholipase(s) C action. Two further pieces of evidence, however, suggest that the decrease in PtdIns is secondary, at least in part, to its conversion to PtdIns(4,5)P$_2$. First, as shown in Figs. 6 and 7, a significant decrease in the level of PtdIns occurs at a time prior to InsP accumulation. Second, and more important, at 15 seconds of TRH stimulation the increase in InsP$_3$ is three times the fall in its precursor, PtdIns(4,5)P$_2$ (Rebechhi and Gershengorn, 1985). These data strongly support the concept that the early decrease in PtIns is via its conversion to PtdIns(4,5)P$_2$. Indeed, PtdIns and PtdIns4P kinases capable of performing this function have been described in plasma membranes isolated from GH cells (A. Imai, M. J. Rebecchi, and M. C. Gershengorn, unpublished observations). Similar observations of the effects of TRH on phosphoinositide metabolism in GH cells have been made by Martin (1983), Drummond and colleagues (Drummond and Raeburn, 1984; Drummond et al., 1984), and Schlegel and Wollheim (1984). These effects are similar to those of other stimulants in many different cell types (for reviews, see Berridge, 1984; Nishizuka, 1984).

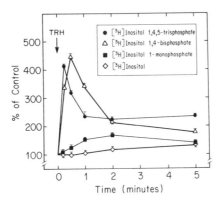

FIG. 7. Time course of the effect of TRH on the levels of radiolabeled derivatives of inositol in GH cells. Reproduced with permission from Rebecchi and Gershengorn (1983).

IV. COUPLING OF TRH RECEPTOR TO POLYPHOSPHOINOSITIDE METABOLISM

If the hydrolysis of membrane polyphosphoinositides were to prove to be the mechanism by which TRH stimulated elevation of $[Ca^{2+}]_i$, this event should be independent of such elevation. Hence, studies were performed to determine whether an elevation of $[Ca^{2+}]_i$ caused by calcium ionophores, such as A23187 and ionomycin, or by K^+ depolarization could mimic the effect of TRH on polyphosphoinositide hydrolysis. In initial studies (Rebecchi and Gershengorn, 1983), we demonstrated that depolarization of GH cells by 50 mM K^+ did not decrease the level of the phosphoinositides or cause an increase in $InsP_3$ or $InsP_2$. A small increase in InsP to 126% of control was observed with K^+ depolarization, but this was abolished if calcium was removed from the medium prior to stimulation with 50 mM K^+. The mechanism of the small increase in InsP, whether it was via inhibition of InsP conversion to inositol or via stimulation of a phospholipase C that hydrolyzed PtdIns, was not determined.

To study this question further, calcium ionophores were utilized to raise $[Ca^{2+}]_i$ directly in a manner that bypasses physiologic mechanisms for calcium mobilization. Thus, the effect of A23187 and ionomycin on polyphoshoinositide metabolism was assessed in cells labeled with [³H]myoinositol to isotopic steady state (Kolesnick and Gershengorn, 1984). A23187 caused a small decrease in $PtdIns(4,5)P_2$, PtdIns4P, and PtdIns. Figure 8 shows the accompanying analysis of the inositol sugars. At 0.5 minutes A23187 caused an increase only in $InsP_2$ to 200% of control; at 5 minutes $InsP_2$ was 230% of control and InsP was elevated to 140% of control. At no time point was $InsP_3$ elevated in these studies.

These data were interpreted to mean that hydrolysis of $PtdIns(4,5)P_2$ to $InsP_3$ via a phospholipase C was independent of elevation of $[Ca^{2+}]_i$ and that hydrolysis of PtdIns4P and PtdIns via a phospholipase(s) C may be caused under some circumstances by elevation of $[Ca^{2+}]_i$. Furthermore, simultaneous stimulation of GH cells by TRH and K^+ depolarization (Rebecchi and Gershengorn, 1983) or TRH and A23187 (Kolesnick and Gershengorn, 1984) failed to affect the rate or level of TRH-induced $InsP_3$ formation. In summary, these data strongly support the contention that TRH-induced hydrolysis of at least $PtdIns(4,5)P_2$ via a phospholipase C does not depend on elevation of $[Ca^{2+}]_i$.

In order to demonstrate that the TRH receptor may be closely coupled to the phospholipase C that hydrolyzes $PtdIns(4,5)P_2$, the effect of decreasing the concentration of receptors for TRH on TRH-stimulated $InsP_3$ formation was investigated (Imai and Gershengorn, 1985). We showed that incubation of cells with dibutyryl cAMP (Bt$_2$cAMP) for 16 hours caused a decrease in the TRH receptor concentration without affecting its affinity

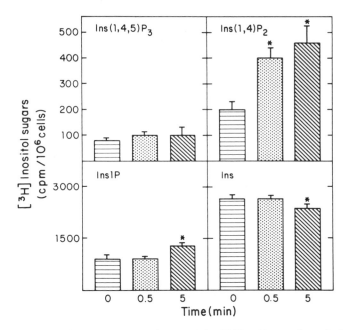

Fig. 8. Effects of A23187, TRH, and A23187 plus TRH at 30 seconds on the levels of inositol sugars in GH cells prelabeled with [³H]myoinositol. Reproduced with permission from Kolesnick and Gershengorn (1984).

for TRH. Similar decreases in receptor level were found in cells incubated for prolonged periods with 8-bromo-cAMP, cholera toxin, sodium butyrate, and TRH. In cells incubated with 1 mM Bt$_2$cAMP for 16 hours, but not for 1 hour, the maximum TRH-induced increase in InsP$_3$ was inhibited to 25 ± 3.2% of that in control cells. Inhibition of TRH-induced InsP$_3$ formation was also observed in cells treated for 16 hours with 8-bromo-cAMP, cholera toxin, and sodium butyrate, and for 48 hours with TRH. Figure 9 illustrates that inhibition of TRH-induced InsP$_3$ formation and lowering of TRH receptor concentration caused by Bt$_2$cAMP occurred in parallel with increasing doses of Bt$_2$cAMP. The concentration dependency of TRH-induced InsP$_3$ formation was the same in control and Bt$_2$cAMP-treated cells; half-maximal effects occurred with 10 nM TRH. These data demonstrated that decreases in TRH receptor concentration caused by several agents that act via different mechanisms are associated with reduced stimulation of InsP$_3$ formation and suggested further that the TRH receptor is tightly coupled to stimulation of hydrolysis of PtdIns(4,5)P$_2$ by a phospholipase C.

Another line of evidence derived from experiments with intact GH cells also supports this notion. It was first shown by Drummond (1985) that

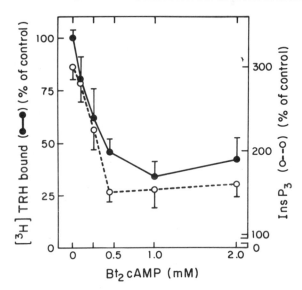

FIG. 9. Effects of dibutyryl cAMP on TRH receptor concentration and TRH-induced InsP₃ formation in GH cells. Reproduced with permission from Imai and Gershengorn (1985).

chlordiazepoxide (CDE) is an antagonist of TRH action in GH cells. We (Gershengorn and Paul, 1986) showed that the effect of CDE is caused by its ability to compete with TRH for binding to receptors on GH cells. CDE decreased the affinity of TRH binding to intact GH cells without affecting the maximum binding capacity. We then used CDE to explore whether continued receptor occupancy by TRH is required for prolonged stimulation of phosphoinositide hydrolysis. CDE itself had no effect on phosphoinositide metabolism but when added simultaneously with TRH caused a dose-dependent inhibition of TRH-induced phosphoinositide metabolism. CDE added to cells 2.5 or 5 minutes after TRH caused a rapid inhibition of TRH-induced phosphoinositide hydrolysis. Most importantly, when cells were stimulated with 50 nM TRH, exposed to 100 μM CDE, and then to 1000 nM TRH, phosphoinositide metabolism was stimulated then inhibited and then restimulated. These data demonstrate that CDE acts as a competitive antagonist of TRH action in GH cells by competing with TRH for binding to its receptor and that continued stimulation by TRH of phospholipase C-mediated hydrolysis of phosphoinositides is tightly coupled to receptor occupancy.

The most direct evidence that the TRH–receptor complex is directly coupled to activation of the polyphosphoinositide-specific phospholipase C comes from *in vitro* studies using membranes isolated from GH cells (Lucas *et al.*, 1985; Straub and Gershengorn, 1986). We (Straub and Gershengorn, 1986) showed that in suspensions of membranes isolated from GH cells that

were prelabeled to isotopic steady state with [³H]inositol and incubated with
ATP, PtdIns(4,5)P₂ and PtdIns4P, and InsP₃ and InsP₂ accumulated. As illustrated in Fig. 10, TRH and GTP stimulated the accumulation of InsP₃ and
InsP₂ in a time-dependent manner. Accumulation of the inositol
polyphosphates stimulated by TRH and GTP was concentration dependent;
half-maximal effects occurred with 10–30 nM TRH and with 3 μM GTP. A
nonhydrolyzable analog of GTP also stimulated inositol polyphosphate accumulation. Morever, when TRH and GTP were added together, their effects
were more than additive. Hence, these data demonstrate that TRH and GTP
act synergistically to stimulate the accumulation of InsP₃ in suspensions of
GH membranes. These findings suggest that the TRH receptor is directly
coupled to a phospholipase C in the plasma membrane that hydrolyzes
PtdIns(4,5)P₂ and that a guanine nucleotide-binding regulatory protein is involved in this coupling. Similar conclusions regarding the involvement of a
guanine nucleotide-binding regulatory protein in coupling several different
receptors to phosphoinositide-specific phospholipases C have been made.

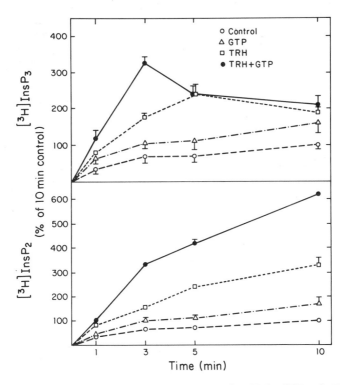

FIG. 10.   Time course of the effects of TRH, GTP, and TRH plus GTP on InsP₃ and InsP₂
accumulation in membrane suspensions from GH cells prelabeled with [³H]myoinositol.
Reproduced with permission from Straub and Gershengorn (1986).

In conclusion, these findings demonstrate that the TRH–receptor complex is tightly coupled to a phospholipase C within the plasma membrane of GH cells and that continued receptor occupancy by TRH is necessary for continued activation of phosphoinositide hydrolysis.

V.　InsP$_3$ AND 1,2-DG AS INTRACELLULAR SIGNALS

Because it appeared that elevation of $[Ca^{2+}]_i$ did not cause the phospholipase C-mediated hydrolysis of PtdIns(4,5)P$_2$ after TRH, it was proposed that this event occurred prior to and was causative in stimulated calcium mobilization and elevation of $[Ca^{2+}]_i$. Based on the findings in other cell systems (Berridge, 1984), it was believed that Ca^{2+} mobilization was mediated by InsP$_3$. In order to test this hypothesis, it was necessary to develop a system in which the plasma membrane of GH cells was made permeable to InsP$_3$ and yet the major intracellular calcium storage pools remained intact (Gershengorn et al., 1984). The saponin-permeabilized preparation of GH cells sequestered Ca^{2+} into mitochondrial and nonmitochondrial pools in an energy-dependent manner. The nonmitochondrial pool had a high affinity and low capacity for Ca^{2+}. In contrast, the mitochondrial pool had a lower affinity and larger capacity for calcium. The notion that this preparation retained its physiologic Ca^{2+}-sequestering activity was strongly suggested by our findings that this preparation was capable of buffering medium Ca^{2+} to a level identical to that found in the cytoplasm of resting intact cells. When saponin-permeabilized cells that had accumulated radiocalcium were incubated with InsP$_3$, a very rapid (less than 0.5 minute) release of calcium was found that was followed by Ca^{2+} reuptake to pretreatment levels (Fig. 11). Similar results were obtained in cells incubated with a mitochondrial uncoupler which released ^{45}Ca from the mitochondrial pool. The half-maximal and maximal doses of InsP$_3$ for this effect were 1 and 10 μM, respectively. InsP$_2$, InsP, inositol, or 1,2-DG did not release $^{45}Ca^{2+}$. These results were interpreted to mean that, of the potential second messengers generated at the plasma membrane by the action of a phospholipase C, only InsP$_3$ was capable of releasing intracellular calcium. Additionally, the transient nature of the response in the permeabilized preparation was similar to the first-phase elevation of $[Ca^{2+}]_i$ after TRH in intact GH cells. These findings provided strong support for the suggestion that InsP$_3$ might serve as coupling factor between TRH–receptor interaction at the cell surface and mobilization of Ca^{2+} from a nonmitochondrial store. Furthermore, it appears that sufficient InsP$_3$ is generated during TRH stimulation of GH cells to achieve intracellular concentrations that exceed that required to mobilize nonmitochondrial calcium in the permeabilized preparation.

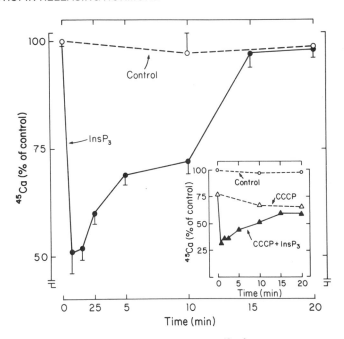

Fig. 11. Time course of the effect of InsP₃ on $^{45}Ca^{2+}$ accumulated by saponin-permeabilized GH cells. Reproduced with permission from Gershengorn *et al.* (1984).

The role of 1,2-DG as a second messenger of TRH action is less certain. 1,2-DG is known to activate the Ca^{2+}- and phospholipid-dependent protein kinase (protein kinase C) *in vitro*; and in studies with a variety of intact cells, exogenously added synthetic 1,2-DGs have been shown to partially mimic the protein phosphorylation pattern of physiologic agonists (for review, see Nishizuka, 1984). The precise role of activation of protein kinase C in regulating cellular processes in most circumstances remains obscure but the resultant phosphorylations are thought to affect the activity of important regulatory proteins. TRH, which rapidly increases the content of 1,2-DG (see above) causes phosphorylation of proteins in cytosolic, microsomal, and nuclear fractions of GH cells. Sobel and Tashjian (1983) and Drust and Martin (1982, 1984; Drust *et al.*, 1982) have presented evidence that protein kinase C mediates phosphorylation of a subset of these cytosolic proteins during stimulation by TRH. These investigations showed that activators of protein kinase C, such as phorbol esters and ex-ogenous phospholipase C (which increases the cellular level of 1,2-DG), cause phosphorylation of some of the same proteins as TRH. Additionally, 1,2-DG and phorbol esters caused phosphorylation of two of these proteins *in vitro*. In an attempt to begin to define a physiological role for these phosphorylations, Sobel and Tashjian (1983) categorized the protein phosphorylations stimulated by TRH as being associated with stimulation of prolactin secretion or synthesis by monitoring the effects of various other factors that stimulate either prolactin secretion of synthesis on these phosphoproteins. Although these investigators were able to show an association between the phosphorylation of certain proteins and stimulation of secretion or synthesis of prolactin, no cause and effect relationship could be established.

VI. Ca^{2+} AND LIPID SECOND MESSENGERS MEDIATE BIPHASIC PROLACTIN SECRETION

A number of laboratories (Aizawa and Hinkle, 1985a,b; Albert and Tashjian, 1984a; Delbeke *et al.*, 1984; Gershengorn and Thaw, 1985; Kolesnick and Gershengorn, 1985b; Martin and Kowalchyk, 1984a,b) have demonstrated that TRH stimulates a biphasic prolactin secretory response; a rapid increase in the secretory rate that lasts for less than 2 minutes (burst phase) is followed by a prolonged stimulation of secretion at a rate that is lower than that attained during the first 2 minutes (sustained phase). Because optimal cellular responses to Ca^{2+}-mobilizing agonists appear to involve activation of both the Ca^{2+} and lipid second messenger pathways, the effect of independent activation of each of these pathways on the two phases of prolactin secretion was studied. Elevation of $[Ca^{2+}]_i$ by calcium ionophores (Delbeke *et al.*, 1984; Aizawa and Hinkle, 1985a; Martin and Kowalchyk, 1984b; but see Albert and Tashjian, 1984b), by depolarization by high extracellular K^+ (Kolesnick and Gershengorn, 1985b; Aizawa and Hinkle, 1985a, Martin and Kowalchyk, 1984b), and by the Ca^{2+} channel agonist Bay K8644 (Enyeart and Hinkle, 1984) was found to cause only burst-phase prolactin secretion from GH cells. In complementary studies, agents that activate protein kinase C, such as phorbol esters and exogenous addition of phospholipase C (by generating endogenous 1,2-DG), caused only sustained-phase secretion.

Based on these observations, a majority of investigators have suggested that, in large part, burst-phase prolactin secretion is caused by elevation of $[Ca^{2+}]_i$ and sustained-phase secretion by activation of protein kinase C. In contrast, based on a series of interesting experiments in which they treated GH cells with calcium ionophores, channel-blocking agents, and phorbol esters as well as TRH, Albert and Tashjian (1984b, 1985) have suggested that elevation of $[Ca^{2+}]_i$ is necessary but not sufficient to cause burst-phase secretion and that elevation of $[Ca^{2+}]_i$ is needed also for the sustained secretory phase. Hence, these studies provided conflicting data as to the precise role of the elevation of cytoplasmic Ca^{2+} in the two phases of prolactin secretion.

Because we were concerned that the use of pharmacologic agents that activate the calcium and lipid pathways independently by bypassing the physiologic receptor for TRH might not produce a physiologic pattern of secretion, we attempted to determine the role of cytoplasmic Ca^{2+} elevation in the two phases of secretion using another approach. We (Kolesnick and Gershengorn, 1985b) studied this process by specifically inhibiting the TRH-induced elevation of $[Ca^{2+}]_i$ by pretreating the cells with arachidonic acid. We had shown that arachidonic acid pretreatment of GH cells stimulated Ca^{2+} extrusion from the cells, and caused depletion of the TRH-responsive intracellular pool of calcium (Kolesnick and Gershengorn, 1985a; Kolesnick *et al.*, 1984) without inhibiting TRH-induced $InsP_3$ formation (unpublished observations). Thus, arachidonic acid pretreatment inhibits the elevation of $[Ca^{2+}]_i$ usually induced by TRH without inhibiting the generation of $InsP_3$ or 1,2-DG and, therefore, can be used in experiments to determine the role of these second messengers in TRH-induced secretion. Figure 12 illustrates the effect of arachidonic acid pretreatment on the elevation of $[Ca^{2+}]_i$ and prolactin secretion stimulated by TRH. Arachidonic acid pretreatment did not affect the resting level of $[Ca^{2+}]_i$ but abolished both phases of the elevation of cytoplasmic Ca^{2+} usually induced by TRH (Fig. 12, insets).

In parallel incubations, arachidonic acid pretreatment altered the usual effect of TRH to cause a biphasic pattern of stimulated prolactin secretion to a monophasic secretory response to TRH. The rate of prolactin secretion stimulated by TRH in arachidonic acid-pretreated cells was indistinguishable from the sustained secretory rate stimulated by TRH in control cells. The burst phase of secretion usually caused by TRH was abolished by arachidonic acid pretreatment. Thus, inhibition of the rise in cytoplasmic Ca^{2+} abolished TRH-induced burst-phase secretion but did not affect sustained secretion stimulated by TRH.

We have interpreted these findings as demonstrating that the elevation of $[Ca^{2+}]_i$ is necessary to cause the burst of prolactin secretion stimulated by TRH. We believe, moreover, that the elevation of cytoplasmic Ca^{2+} is sufficient for this effect because K^+ depolarization, and Bay K8644 and calcium ionophores, agents that act predominantly to affect cellular calcium metabolism, stimulate burst secretion. The sustained phase of secretion induced by TRH does not require any increase in cytoplasmic Ca^{2+}; however, it is likely that Ca^{2+} is needed for this phase of secretion also because if $[Ca^{2+}]_i$ is lowered to below the resting level sustained secretion is inhibited. Hence, it appears that the biphasic prolactin secretory response to TRH in GH cells is mediated in a coordinated, but not synergistic, fashion by the calcium and lipid messengers of this pathway.

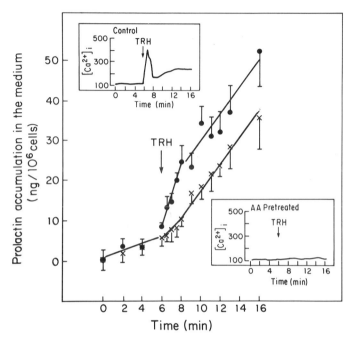

FIG. 12.   Effect of arachidonic acid (AA) on prolactin secretion from and cytoplasmic free calcium concentration ($[Ca^{2+}]_i$) within (inset) GH_3 cells after TRH treatment. Reproduced with permission from Kolesnick and Gershengorn (1985a).

VII. MODEL OF TRH ACTION

The following may, therefore, be proposed as the sequence of intracellular events involved in the mechanism of TRH stimulation of prolactin secretion from GH cells. After TRH binds to its plasma membrane receptor, the TRH–receptor complex activates the hydrolysis of $PtdIns(4,5)P_2$ to yield $InsP_3$ and 1,2-DG using a guanine nucleotide-binding regulatory protein as a coupling factor. There is a simultaneous increase in the conversion of PtdIns to PtdIns4P to $PtdIns(4,5)P_2$. The three phosphoinositides thereby contribute to the formation of $InsP_3$ and 1,2-DG. $InsP_3$, the water-soluble product of $PtdIns(4,5)P_2$ hydrolysis, diffuses from the plasma membrane to a nonmitochondrial pool of calcium (perhaps within the endoplasmic reticulum) and releases previously sequestered Ca^{2+}. The movement of Ca^{2+} from the nonmitochondrial pool into the cytoplasm results in the rapid elevation of $[Ca^{2+}]_i$, which couples stimulus to burst secretion. The elevation of $[Ca^{2+}]_i$ may activate exocytosis directly or through phosphorylation of proteins involved in the exocytotic process via activation of a Ca^{2+}- and calmodulin-dependent protein kinase(s), or both. The elevation of $[Ca^{2+}]_i$ is extended by a delayed, but prolonged, TRH-induced enhancement of influx of extracellular Ca^{2+}. Concomitant with the effects of $InsP_3$ and of elevation of $[Ca^{2+}]_i$, phosphorylation of proteins involved in the exocytotic process may be stimulated through 1,2-DG activation of protein kinase C leading to the sustained phase of secretion. The sustained phase of secretion does not require an elevation of $[Ca^{2+}]_i$ but the effect of submaximal increases in 1,2-DG content may be enhanced by simultaneous elevation of $[Ca^{2+}]_i$. Hence, this model proposes a coordinate regulation of the two phases of prolactin secretion stimulated by TRH by $InsP_3$ and 1,2-DG, the two products of $PtdIns(4,5)P_2$ hydrolysis.

ACKNOWLEDGMENTS

We thank Mario Rebecchi, Atsushi Imai, Elizabeth Geras, and Colette Thaw for many helpful discussions and Robin Richardson for assistance in preparing the manuscript. The work from the authors' laboratory was supported by grants AM33468 and AM33469 to MCG and a posdoctoral fellowship AM07273 to RNK from the National Institutes of Health.

REFERENCES

Aizawa, T., and Hinkle, P. M. (1985a). Thyrotropin-releasing hormone rapidly stimulates a biphasic secretion of prolactin and growth hormone in Gh_4C_1 rat pituitary tumor cells. *Endocrinology* **116**, 73–82.

Aizawa, T., and Hinkle, P. M. (1985b). Differential effects of thyrotropin-releasing hormone, vasoactive intestinal peptide, phorbol ester, and depolarization in GH_4C_1 rat pituitary cells. *Endocrinology* **116**, 909–919.

Albert, P. R., and Tashjian, A. H., Jr. (1984a). Thyrotropin-releasing hormone-induced spike and plateau in cytosolic free Ca^{2+} concentrations in pituitary cells. *J. Biol. Chem.* **259**, 5827–5832.

Albert, P. R., and Tashjian, A. H., (1984b). Relationship of thyrotropin-releasing hormone-induced spike and plateau phases in cytosolic free Ca^{2+} concentrations to hormone secretion. *J. Biol. Chem.* **259**, 15350–15363.

Albert, P. R., and Tashjian, A. H., Jr. (1985). Dual actions of phorbol esters on cytosolic free Ca^{2+} concentrations and reconstitution with ionomycin of acute thyrotropin-releasing hormone responses. *J. Biol. Chem.* **259**, 5827–5832.

Berridge, M. J. (1984). Inositol trisphosphate and diacylglycerol as second messengers. *Biochem. J.* **220**, 345–360.

Biales, B., Dichter, M. A., and Tischler, A. (1977). Sodium and calcium action potentials in pituitary cells. *Nature (London)* **267**, 172–173.

Delbeke, D., Kojima, I., Dannies, P. S., and Rasmussen, H. (1984). Synergistic stimulation of prolactin release by phorbol ester, A23187 and forskolin. *Biochem. Biophys. Res. Commun.* **123**, 735–741.

Douglas, W. W. (1978). Stimulus-secretion coupling: Variations on the theme of calcium-activated exocytosis involving cellular and extracellular sources of calcium. *Ciba Found. Symp.* **54**, 61–90.

Drummond, A. H. (1985). Chlordiazepoxide is a competitive thyrotropin-releasing hormone receptor antagonist in GH₃ pituitary tumor cells. *Biochem. Biophys. Res. Commun.* **127**, 63–70.

Drummond, A. H., and Macphee, C. H. (1981). Phosphatidylinositol metabolism in GH₃ pituitary tumour cells stimulated by TRH. *Br. J. Pharmacol.* **74**, 967P–968P.

Drummond, A. H., and Raeburn, C. A. (1984). The interaction of lithium with thyrotropin-releasing hormone-stimulated lipid metabolism in GH₃ pituitary tumour cells. *Biochem. J.* **224**, 129–136.

Drummond, A. H., Bushfield, M., and Macphee, C. H. (1984). Thyrotropin-releasing hormone-stimulated [³H]inositol metabolism in GH₃ pituitary tumor cells. *Mol. Pharmacol.* **25**, 201–208.

Drust, D. S., and Martin, T. F. J. (1982). Thyrotropin-releasing hormone rapidly and transiently stimulates cytosolic calcium-dependent protein phosphorylation in GH₃ pituitary cells. *J. Biol. Chem.* **257**, 7566–7573.

Drust, D. S., and Martin, T. F. J. (1984). Thyrotropin-releasing hormone rapidly activates protein phosphorylation in GH₃ pituitary cells by a lipid-linked, protein kinase C-mediated pathway. *J. Biol. Chem.* **259**, 14520–14530.

Drust, D. S., Sutton, C. A., and Martin, T. F. J. (1982). Thyrotropin-releasing hormone and cyclic AMP activate distinctive pathways of protein phosphorylation in GH pituitary cells. *J. Biol. Chem.* **257**, 3306–3312.

Dufy, B., Vincent, J. D., Fleury, H., Pasquier, P. D., Gourdji, D., and Tixier-Vidal, A. (1979). Membrane effects of thyrotropin-releasing hormone and estrogen shown by intracellular recording from pituitary cells. *Science* **204**, 509–511.

Enyeart, J. J., and Hinkle, P. M. (1984). The calcium agonist Bay K 8644 stimulates secretion from a pituitary cell line. *Biochem. Biophys. Res. Commun.* **122**, 991–996.

Gershengorn, M. C. (1982). Thyrotropin releasing hormone. A review of the mechanisms of acute stimulation of pituitary hormone release. *Mol. Cell. Biochem.* **45**, 163–179.

Gershengorn, M. C. (1986). Mechanism of thyrotropin releasing hormone stimulation of pituitary hormone secretion. *Annu. Rev. Physiol.* **48**, 515–526.

Gershengorn, M. C., and Paul, M. E. (1986). Evidence for tight coupling of receptor occupancy by thyrotropin-releasing hormone to phospholipase C-mediated phosphoinositide hydrolysis in rat pituitary cells: Use of chlordiazepoxide as a competitive antagonist. *Endocrinology* **119**, 833–839.

Gershengorn, M. C., and Thaw, C. (1983). Calcium influx is not required for TRH to elevate free cytoplasmic calcium in GH₃ cells. *Endocrinology* **113**, 1522–1524.

Gershengorn, M. C., and Thaw, C. (1985). Thyrotropin-releasing hormone (TRH) stimulates biphasic elevation of cytoplasmic free calcium in GH₃ cells. Further evidence that TRH mobilizes cellular and extracellular Ca^{2+}. *Endocrinology* **116**, 591–596.

Gershengorn, M. C., Hoffstein, S. T., Rebecchi, M. J., Geras, E., and Rubin, B. G. (1981). Thyrotropin-releasing hormone stimulation of prolactin release from clonal rat pituitary cells. Evidence for action independent of extracellular calcium. *J. Clin. Invest.* **67**, 1769–1776.

Gershengorn, M. C., Geras, E., Purello, V. S., and Rebecchi, M. J. (1984). Inositol triphosphate mediates thyrotropin-releasing hormone mobilization of nonmitochondrial calcium in rat mammotropic pituitary cells. *J. Biol. Chem.* **259**, 10675–10681.

Hinkle, P. M., and Kinsella, P. A. (1984). Regulation of thyrotropin-releasing hormone binding by monovalent cations and guanyl nucleotides. *J. Biol. Chem.* **259**, 3445–3449.

Hinkle, P. M., and Phillips, W. J. (1984). Thyrotropin-releasing hormone stimulates GTP hydrolysis by membranes from GH₄C₁ rat pituitary tumor cells. *Proc. Natl. Acad. Sci. U.S.A.* **81**, 6183–6187.

Imain, A., and Gershengorn, M. C. (1985). Evidence for tight coupling of thyrotropin-releasing hormone receptors to stimulated inositol trisphosphate formation in rat pituitary cells. *J. Biol. Chem.* **260**, 10536–10540.

Kidokoro, Y. (1975). Spontaneous calcium action potentials in a clonal pituitary cell line and their relationship to prolactin secretion. *Nature (London)* **258**, 741-742.

Kolesnick, R. N., and Gershengorn, M. C. (1984). Ca^{2+} ionophores affect phosphoinositide metabolism differently than thyrotropin-releasing hormone in GH_3 pituitary cells. *J. Biol. Chem.* **259**, 9514-9519.

Kolesnick, R. N., and Gershengorn, M. C. (1985a). Arachidonic acid inhibits thyrotropin-releasing hormone-induced elevation of cytoplasmic free calcium in GH_3 pituitary cells. *J. Biol. Chem.* **260**, 707-713.

Kolesnick, R. N., and Gershengorn, M. C. (1985b). Direct evidence that burst but not sustained secretion of prolactin stimulated by thyrotropin-releasing hormone is dependent on elevation of cytoplasmic calcium. *J. Biol. Chem.* **260**, 5217-5220.

Kolesnick, R. N., Musacchio, I., Thaw, C., and Gershengorn, M. C. (1984). Arachidonic acid mobilizes calcium and stimulates prolactin secretion from GH_3 cells. *Am. J. Physiol.* **246** (*Endocrinol. Metab.* 9), E458-E462.

Kruskla, B. A., Keith, C. H., and Maxfield, F. R. (1984). Thyrotropin-releasing hormone-induced changes in intracellular $[Ca^{2+}]$ measured by microspectrofluorometry on individual quin 2-loaded cells. *J. Cell Biol.* **99**, 1167-1172.

Lucas, D. O., Bajjalieh, S. M., Kowalchyk, J. A., and Martin, T. F. J. (1985). Direct stimulation by thyrotropin-releasing hormone (TRH) of polyphosphoinositide hydrolysis in GH_3 cell membranes by a guanine nucleotide-modulated mechanism. *Biochem. Biophys. Res. Commun.* **132**, 721-728.

MacPhee, C. H., and Drummond, A. H. (1984). Thyrotropin-releasing hormone stimulates rapid breakdown of phosphatidylinositol 4,5-biophosphate and phosphatidylinositol 4-phosphate in GH_3 pituitary tumor cells. *Mol. Pharmacol.* **25**, 193-200.

Martin, T. F. J. (1983). Thyrotropin-releasing hormone rapidly activated the phosphodiester hydrolysis of polyphosphoinositides in GH_3 pituitary cells. *J. Biol. Chem.* **258**, 14816-14822.

Martin, T. F. J., and Kowalchyk, J. A. (1984a). Evidence for the role of calcium and diacylglycerol as dual second messengers in thyrotropin-releasing hormone action: Involvement of diacylglycerol. *Endocrinology* **115**, 1517-1526.

Martin, T. F. J., and Kowalchyk, J. A. (1984b). Evidence for the role of calcium and diacylglycerol as dual second messengers in thyrotropin-releasing hormone action: Involvement of Ca^{2+}. *Endocrinology* **115**, 1527-1536.

Michell, R. H. (1975). Inositol phospholipids and cell surface receptor function. *Biochim. Biophys. Acta* **415**, 81-147.

Michell, R. H., Kirk, C. J., Jones, L. M., Downes, C. P., and Creba, J. A. (1981). The stimulation of inositol lipid metabolism that accompanies calcium mobilization in stimulated cells: Defined characteristics and unanswered questions. *Philos. Trans. R. Soc. London Ser. B* **296**, 123-137.

Nishizuka, Y., (1984). The role of protein kinase C in cell surface signal transduction and tumour promotion. *Nature (London)* **308**, 693-698.

Ozawa, S., and Kimura, N. (1979). Membrane potential changes caused by thyrotropin-releasing hormone in the clonal GH_3 cells and their relationship to secretion of pituitary hormone. *Proc. Natl. Acad. Sci. U.S.A.* **76**, 6017-6020.

Rasmussen, H., and Barrett, P. Q. (1984). Calcium messenger system: An integrated view. *Physiol. Rev.* **64**, 938-984.

Rebecchi, M. J., and Gershengorn, M. C. (1983). Thyroliberin stimulates rapid hydrolysis of phosphatidylinositol 4,5-biphosphate by a phosphodiesterase in rat mammotropic pituitary cells. *Biochem. J.* **216**, 287-294.

Rebecchi, M. J., and Gershengorn, M. C. (1985). Receptor regulation of phosphoinositides and calcium: A mechanism for thyrotropin-releasing hormone action. *In* "The Receptors," (P. M. Conn. ed.), Vol. III, pp. 173-212. Academic Press, New York.

Rebecchi, M. J., Monaco, M. E., and Gershengorn, M. C. (1981). Thyrotropin releasing hormone rapidly enhances [^{32}P]orthophosphate incorporation into phosphatidic acid in cloned GH$_3$ cells. *Biochem. Biophys. Res. Commun.* **101**, 124–130.

Rebecchi, M. J., Kolesnick, R. N., and Gershengorn, M. C. (1983). Thyrotropin-releasing hormone stimulates rapid loss of phosphatidylinositol and its conversion to 1,2-diacylglycerol and phosphatidic acid in rat mammotropic pituitary cells. *J. Biol. Chem.* **258**, 227–234.

Ronning, S. A., Heatley, G. A., and Martin, T. F. J. (1982). Thyrotropin-releasing hormone mobilizes Ca^{2+} from endoplasmic reticulum and mitochondria of GH$_3$ pituitary cells: Characterization of cellular Ca^{2+} pools by a method based on digitonin permeabilization. *Proc. Natl. Acad. Sci. U.S.A.* **79**, 6294–6298.

Sand, O., Haug, E., and Gautvic, K. M. (1980). Effects of thyroliberin and 4-aminopyridine on action potentials and prolactin release and synthesis in rat pituitary cells in culture. *Acta Physiol. Scand.* **108**, 247–252.

Schlegel, W., and Wollheim, C. A. (1984). Thyrotropin-releasing hormone increases cytosolic free Ca^{2+} in clonal pituitary cells (GH$_3$ cells): Direct evidence for the mobilization of cellular calcium. *J. Cell Biol.* **99**, 83–87.

Schlegel, W., Roduit, C., and Zahnd, G. (1981). Thyrotropin releasing hormone stimulates metabolism of phosphatidylinositol in GH$_3$ cells. *FEBS Lett.* **134**, 47–49.

Schlegel, W., Roduit, C., and Zahnd, G. (1984). Polyphosphoinositide hydrolysis by phospholipase C is accelerated by thyrotropin releasing hormone (TRH) in clonal rat pituitary cells (GH$_3$ cells). *FEBS Lett.* **168**, 54–60.

Snowdowne, K. W. (1984). Estimates for cytosolic concentration. *Am. J. Physiol.* **247** (*Endocrinol. Metab.* **10**), E837.

Snowdowne, K. W., and Borle, A. B. (1984). Changes in cytosolic ionized calcium induced by activators of secretion in GH$_3$ cells. *Am. J. Physiol.* **246** (*Endocrinol. Metab.* **9**), E198–E201.

Sobel, A., and Tashjian, A. H., Jr. (1983). Distinct patterns of cytoplasmic protein phosphorylation related to regulation of synthesis and release of prolactin by GH cells. *J. Biol. Chem.* **258**, 10312–10324.

Straub, R. E., and Gershengorn, M. C. (1986). Thyrotropin-releasing hormone and GTP activate inositol trisphosphate formation in membranes isolated from rat pituitary cells. *J. Biol. Chem.* **261**, 2712–2717.

Sutton, C. A., and Martin, T. F. J. (1982). Thyrotropin-releasing hormone (TRH) selectively and rapidly stimulates phosphatidylinositol turnover in GH pituitary cells: A possible second step of TRH action. *Endocrinology* **110**, 1273–1280.

Tan, K.-N., and Tashjian, A. H., Jr. (1981). Receptor-mediated release of plasma membrane-associated calcium and stimulation of calcium uptake by thyrotropin-releasing hormone in pituitary cells in culture. *J. Biol. Chem.* **256**, 8994–9002.

Tan, K.-N., and Tashjian, A. H., Jr. (1984). Voltage-dependent calcium channels in pituitary cells in culture. *J. Biol. Chem.* **259**, 418–426.

Tan, K.-N., and Tashjian, A. H., Jr. (1984). Voltage-dependent calcium channels in pituitary cells in culture. *J. Biol. Chem.* **259**, 427–434.

Tashjian, A. H., Jr., Lomedico, M. E., and Maina, D. (1978). Role of calcium in the thyrotropin-releasing hormone-stimulated release of prolactin from pituitary cells in culture. *Biochem. Biophys. Res. Commun.* **81**, 798–806.

Vale, W., Burgus, R., and Guillemin, R. (1967). Presence of calcium ions as a requisite for the in vitro stimulation of TSH release by hypothalamic TRF. *Experientia* **23**, 853–855.

Part II
Steroid Hormone Action

Steroid Effects on Excitable Membranes

S. D. ERULKAR AND D. M. WETZEL[1]

Department of Pharmacology
University of Pennsylvania Medical School
Philadelphia, Pennsylvania 19104

I. INTRODUCTION

Reproductive behavior in vertebrates is modulated by gonadal steroids (Beach, 1948; Young, 1961). These steroids act at specific and multiple sites on neurons within the central nervous system and at effector organs at the periphery. Their actions are mediated through molecular mechanisms

---

[1]Present address: Section of Neurobiology and Behavior, Cornell University, Ithaca, New York 14853.

(McEwen and Parsons, 1982; Pfaff, 1983). The understanding of these mechanisms, therefore, may provide some insight into the steroid regulations of these behaviors.

The changes induced by steroids are complex and, although they are ultimately chemical, can be decribed more easily as encompassing morphological, biophysical, and biochemical alterations of cell function. Steroids may also play different roles during development than in the adult. Due to the large number of changes that steroids induce and their multiple sites of action it is difficult to say with certainty how a given steroid may be involved in modulating or altering a given cellular process. An understanding of the molecular basis of such changes requires the isolation of appropriate elements under controlled conditions, where the actions of a steroid on a given element can be studied independently of other influences. For example, many lines of evidence suggest changes in cell excitability as being a molecular endpoint of steroid action in neurons and muscle (McEwen and Parsons, 1982; Pfaff, 1983; Moss and Dudley, 1984). Cell excitability is largely a function of the distribution and transfer of ions across the cell membrane and is mediated by voltage-dependent and neurotransmitter-activated ionic channels as well as by transport processes. Therefore, if steroids are affecting cell excitability, they may do so by altering the properties, density, and/or distribution of these ionic channels or the receptors that activate them. In addition, steroids may have direct effects on the physicochemical properties of the membrane itself. We have developed a cell culture preparation of androgen-sensitive muscle to study androgen influences on ionic channels and membrane receptors and we have used patch-clamp techniques (Hamill *et al.*, 1981) to study changes in single-channel properties induced in these cells by steroids.

In this article we shall describe a number of steroid-induced effects at excitable cells that have been found to occur *in vivo* and relate these changes to changes that may occur in various processes involved in synaptic transmission and membrane excitability. We will examine steroid interactions with lipid membranes and present our experimental results concerning androgen modulation of ionic channels in androgen-sensitive muscles. Finally, we will discuss the advantages and disadvantages of a *cell culture* approach in understanding steroid hormone action on excitable cells, and the future directions that this field may take, particularly with regard to the application of techniques of molecular biology for unraveling the changes that steroids may induce in excitable cells.

II. ORGANIZATION OF STEROID-CONCENTRATING EXCITABLE CELLS

Localization

A steroid-concentrating cell will be defined as a cell that has been shown to contain high levels of high affinity cytoplasmic or nuclear steroid receptors.

For some tissues [e. g., the central nervous system (CNS)], this has been determined primarily by steroid autoradiography, whereas in others (e.g., uterus, muscle), receptor binding or exchange assays have been used. In neural tissue the localization that autoradiography provides is necessary due to the heterogeneity of the tissue. However, autoradiography is less quantitative than biochemical assay. For those cases where it has been examined, steroid receptors found in neural tissue appear to be biochemically equivalent to those found in peripheral endocrine tissues (McEwen *et al.*, 1979). Biochemical techniques provide a more precise description of the characteristics of these receptors; however, localization data are often lost. Recently, the techniques of immunocytochemistry and photoaffinity labeling have been useful in analyzing steroid receptors (King and Greene, 1984; Welshons *et al.*, 1984; Ringold, 1985). The former procedure in particular has provided controversial results as to the role of cytoplasmic versus nuclear steroid receptors. For this article the question of their intracellular localization is not crucial and will not be addressed.

Absolute values of high-affinity steroid receptors vary considerably within tissues and species. Therefore, what is considered a high level of receptor in one tissue of a given species may not be so considered in similar tissue for another species.

There are numerous articles describing steroid-concentrating cells in different tissues and species. The following is a brief summary of generalizations drawn from these reports. Further detailed information can be found in the reviews by Morrell *et al.* (1975), Zigmond (1975), Kelley and Pfaff (1978), and McEwen *et al.* (1979).

1. STEROID RECEPTORS IN MUSCLE

Most skeletal muscles have low levels of androgen receptors, whereas only a few skeletal muscles appear to have estrogen receptors. The majority of skeletal muscle from frogs, rats, rabbits, pigs, and humans has low levels of the androgen receptor (Segil *et al.*, 1983, 1987; Saartok *et al.*, 1984). For example, quadriceps femoris muscle from male or female *Xenopus laevis* contains 1–3 fmol androgen receptor/mg protein (Segil *et al.*, 1983). Androgen receptors in muscle have properties similar to those purified from the prostate (Dahlberg *et al.*, 1980; Snochowski *et al.*, 1980). Perineal muscles (levator ani and bulbocavernosus) of rats and mice, as well as muscles associated with male reproductive behaviors in frogs (laryngeal, flexor carpi radialis) and syringeal muscles of some male birds, have high levels of androgen receptor (Krieg *et al.*, 1974, 1980; Michel and Baulieu, 1975, 1980; Dubé *et al.*, 1976; Tremblay *et al.*, 1977; Max *et al.*, 1981; Segil *et al.*, 1983). For example, male *X. laevis* laryngeal muscle has 21 fmol androgen receptor/mg protein, while the female equivalent has 6.3 fmol androgen receptor/mg protein. High levels of androgen receptors may be a prerequisite of androgen-induced changes in muscle. Estrogen receptors in skeletal muscle have been found, but are less extensively studied (Dubé *et al.*, 1976; Dionne *et al.*, 1979; Dahlberg, 1982).

There is little evidence for the presence of androgen receptors in smooth muscle. There is one report of androgen receptors localized in smooth muscle cells of the aortic arch (Lin *et al.*, 1981). On the other hand, the presence of high levels of estrogen receptors in uterus and oviduct (Jensen and Jacobson, 1962; Eisenfeld and Axelrod, 1965; Gorski *et al.*, 1968; Jensen *et al.*, 1968) as well as in coronary vascular smooth muscle (Harder and Coulson, 1979) has been well documented.

Heart muscle appears to have receptors for androgens, estrogens, and glucocorticoids (Stumpf *et al.*, 1977; Harder and Coulson, 1979; McGill *et al.*, 1980).

2. PERIPHERAL NERVOUS SYSTEM

The short adrenergic postganglionic neurons innervating the muscle of the uterus and oviduct appear to be sensitive to estrogen and to progesterone (Owman and Sjöberg, 1973; Marshall, 1981). It has not been shown whether these cells contain high-affinity estrogen receptors. Estrogen and androgen receptors in other elements of the peripheral nervous system have not yet been shown.

3. CENTRAL NERVOUS SYSTEM

There are several excellent reviews of steroid-concentrating cells in the brain. These include Morrell *et al.* (1975), Zigmond (1975), Kelley and Pfaff (1978), and McEwen *et al.* (1979). Table I (from McEwen *et al.*, 1979) summarizes the distribution of androgen and estrogen-concentrating neurons in the brain, as determined by steroid autoradiography. The localization of hormone-concentrating cells in these areas is remarkably conserved across species (Kelley and Pfaff, 1978; Kelley, 1981; Pfaff, 1983). In general, androgens are concentrated in the septal region, the preoptic area, hypothalamus, amygdala, some thalamic nuclei, hippocampus, pituitary, ventral infundibulum, and some motor nuclei of the brain stem and spinal cord. Figure 1 shows examples of steroid autoradiography for androgen cells in the preoptic area of the frog. Estrogens are concentrated in the same regions, with the exception of brain stem, spinal motor nuclei, and hippocampus (Table I). It has not yet been determined whether there is a sexual dimorphism of receptor levels within the cells within these regions. Many of these steroid-concentrating neurons are arranged in neural circuits subserving specific reproductive behaviors (Pfaff and McEwen, 1983; Pfaff and Modianos, 1985), examples of which are lordosis in the rat (Morrell and

TABLE I

LOCALIZATION OF STEROID SENSITIVE CELLS IN BRAIN[a]

Steroid	Septum			Preoptic	Hypothalamus				Amygdala		Hippocampus Ammon's horn	Anterior pituitary
	nsl	ndb	nst	pom	pv	vm	arc	vpm	aco	am		
Estradiol	+	+	+	++	++	++	++	++	++	++	−	++
5α-DHT	++	−	+	+	++	+	+	++	−	+	+	+

[a]After McEwen et al. (1979). Reproduced, with permission, from the *Annual Review of Newoscience*, Vol. 2. © 1979 by Annual Reviews Inc. Structures are nsl, nucleus septi lateralis; ndb, nucleus of the diagonal band of Broca; nst, nucleus of the stria terminalis; pom, nucleus preopticus medialis; pv, nucleus periventricularis hypothalami; vm, nucleus ventromedialis hypothalami; arc, nucleus arcuatus; vpm, nucleus premammillaris ventralis; aco, cortical nucleus of the amygdala; am, medial nucleus of the amygdala. Estradiol data from Pfaff and Keiner (1973); 5α-DHT data from Sar and Stumpf (1977).

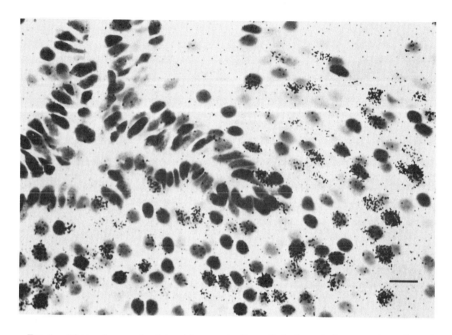

FIG. 1.   Photomicrograph of steroid-concentrating cells in the anterior preoptic nucleus of a castrated male *Xenopus laevis* after injection of [³H]5α-DHT. Darkly staining ependymal cells line the ventricle (kindly provided by Kelley, 1980). Bar, 10 μm.

Pfaff, 1982; Pfaff, 1983), mate calling in the frog (Kelley, 1980; Wetzel *et al.*, 1986), and singing in zebra finches (Arnold *et al.*, 1976).

III. STEROID HORMONE EFFECTS ON EXCITABLE CELLS

Steroid action in adult tissues is in many ways inseparable from its action during development. Three points should be kept in mind. First, there is some evidence that steroids may influence the survival and/or growth of neurons and muscles (Toran-Allerand, 1980; Breedlove, 1984). Thus steroids may be a factor in cell death. Second, sexual dimorphism in size (Arnold and Gorski, 1984) and connectivity (Wetzel *et al.*, 1985) of steroid-sensitive brain areas suggests that developmental interactions between steroid-sensitive cells may help in establishing this connectivity. Finally, many cells are exposed to "waxing and waning" steroid levels throughout their lives. In the CNS, where neurons do not divide, exposure of cells to a given steroid level may predispose them to a particular fate upon subsequent exposure to the same or different steroids. Mechanisms for these effects through gene regulation have been proposed (Yamamoto, 1985a,b).

This is particularly important to keep in mind when studying steroid effects on cells in culture. When possible, primary cultures of adult tissues should be used. The use of cell lines may provide misleading results unless cells are exposed to media that mimic hormonal fluctuations during development.

A. Hormones and Morphology

Steroids, particularly androgens, appear to have pronounced effects on cellular ultrastructure and overall morphology. These effects must be considered when examining steroid-induced effects on cell excitability, synaptic transmission, and impulse propagation. Changes in cell diameter, branching patterns, dendritic length, and synapse size will have profound influences on the ability of a cell to receive and propagate signals. Although much progress has been made in understanding these interactions in non-steroid-sensitive cells (Rall, 1977; Koch *et al.*, 1982), there have been few studies that have investigated steroid-induced changes in these terms.

1. MUSCLE

A number of muscles are known to respond to androgens by increasing in mass. These include the levator ani, bulbocavernosus of the rat and mouse, the flexor carpi radialis and laryngeal muscles of *X. laevis*, and the syrinx of the songbird (Wainman and Shipounoff, 1941; Eisenberg and Gordan, 1950; Segil *et al.*, 1983, 1987; Bleisch *et al.*, 1984). As mentioned previously, these muscles contain high levels of high-affinity androgen receptors. The best described example of morphological changes in muscle due to androgens is found in the levator ani muscle of the rat. Castration of male rats results in a decrease in size of individual levator ani muscle cells through a dissolution of myofilaments from myofibrils and a loss of sarcoplasm (Venable, 1966a,b). These smaller fibers are identical to normal fibers in their ultrastructure. Treatment with testosterone results in enlargement of the fibers with few other morphological changes. The increase in size is accompanied by a faster contraction time and increase in force of contraction (Bass *et al.*, 1969). Castration does not appear to result in significant qualitative changes in the ultrastructure of the neuromuscular junction of the levator ani (Hanzlikcova and Gutmann, 1978) despite muscle atrophy. Treatment of castrates with testosterone results in several changes in pre- and postsynaptic morphology of this muscle; most notable of these is a depletion of synaptic vesicles and the appearance of coated vesicles in the presynaptic terminal. Collateral sprouting of terminal axons may also occur (Hanzlikova and Gutmann, 1978). Increased protein production in the end-plate region is suggested by the increased numbers of ribosomes. Castration decreases levels of acetylcholine receptor (AChR) while testosterone treatment of castrates restores AChR to levels 27% of normal (Bleisch *et al.*,

1982). Dihydrotestosterone treatment of castrates results in increases in levels of AChR to 68% of normal. Levels of acetylcholinesterase (AChE) also decrease with castration and can be increased by treatment of castrates with androgen (Tucek et al., 1976). The above changes in AChR and AChE might be expected with increasing and decreasing size of the neuromuscular junction. There is some evidence that the site density of AChR at the neuromuscular junction is influenced by androgens in these muscles (Bleisch and Harrelson, 1984). Confirmation of these results awaits the application of higher resolution techniques. The electrophysiological consequences of these changes in muscle size and neuromuscular junction morphology have yet to be carefully examined (see, however, Vyskocil and Gutmann, 1977; Souccar et al., 1982).

Rubinstein et al. (1983) found by immunohistochemical techniques that slow tonic muscle fibers in the clasp muscle, the sternoradialis, were found in greater numbers in adult male X. laevis than in the female. Furthermore, castration of the males reduced the number of these fibers to less than those in the female muscle (see also Schneider et al., 1980; Erulkar et al., 1981). Thibert (1986), on the other hand, used a histochemical technique (myosin-ATPase staining) to differentiate slow tonic muscle fibers from fast twitch muscle fibers in Rana temporaria, but failed to see changes in numbers of these by either denervation or hormones.

Estrogens have an interesting effect on uterine muscle cells. Gap junctions between muscle cells, which may help synchronize uterine contractions (Garfield et al., 1979), are increased by estrogens (Garfield et al., 1980). In fact, development of gap junctions in the uterus may be regulated by estrogen and progesterone.

2. NEURONS

Differences in neuronal morphology and ultrastructure due to steroids have been studied extensively and reviewed (Arnold and Gorski, 1984). In the neuronal systems examined, most differences appear to be due to the developmental effects of steroids. There are marked differences in the volume of a given nucleus, neuron number, soma size, dendritic arborizations, and synaptic connections in a number of brain areas containing steroid-concentrating neurons and associated with reproductive behavior. For the most part these dimorphisms are between sexes of a given species and appear to be the result of different hormonal mechanisms during critical periods of development (Arnold and Gorski, 1984).

Changes in neuronal morphology in adult animals due to steroids have been described by a few investigators. Estrogen can "masculinize" a subset of neurons in nucleus robustus archistriatalis (RA) of the songbird brain, increasing somal size but having no effect on total number of neurons (Gurney, 1981). Commins and Yahr (1982) found that the volume of the preoptic area in gerbil brains can be increased by estrogen in adults.

More recently, Clower and DeVoogd (1985) have found that testosterone treatment of female canaries increases the volume and soma size of neurons in the hypoglossal nucleus, a nucleus essential for song production. In addition, testosterone treatment may increase the number of large synaptic vesicles in Gray Type I synapses to dendrites of hypoglossal neurons. Pfaff (1983) has found that estrogen treatment of ovariectomized female rats results in ultrastructual changes in hypothalamic neurons, which may be indicative of increased protein synthesis. These ultrastructural changes include increases in rough stacked endoplasmic reticulum, number of dense core vesicles, and nuclear changes.

B. Steroids and Synaptic Transmission

One area of research which has received an increasing amount of study in recent years is the role that steroids play in the modulation of synaptic transmission in adult animals. A review by McEwen and Parsons (1982) highlights much of this work, particularly with respect to estrogen effects on female reproductive behavior and ovulation. The results of several investigations in a number of laboratories strongly suggest that steroids modulate synaptic transmission at pre- and postsynaptic levels. Tables II, III and IV outline some of the observed effects of steroids at these sites. Presynaptic effects include alterations in neurotransmitter synthesis/metabolism, release, uptake, and degradation. At presumed postsynaptic sites, the most common observation is an increase in neurotransmitter receptor number. Whether this change is due to (1) increased receptor density, (2) increased synaptic size, (3) alterations in receptor degradation or synthesis, and/or (4) changes in receptor affinity is not clear. These changes will be difficult to study in the CNS in vivo and may be more easily studied in culture or at peripheral synapses (e.g., neuromuscular junction). In addition, changes in elements of the basal lamina, such as AChE levels, have been reported with steroid treatment (Luine et al., 1980b). Changes in the density of AChE would be an efficient means of regulating synaptic efficacy, as is suggested by the work of Land et al. (1984).

In view of the facts that (1) estrogens have been shown to influence the numbers of gap junctions beween muscle cells in the uterus (Garfield et al., 1979, 1980); (2) spinal motoneurons in the frog can be electrically coupled by gap junctions (Sotelo and Grofova, 1976; Collins, 1983); and (3) neuromodulators such as dopamine have been found to influence the conductance of gap junctions between retinal horizontal cells (Lasater and Dowling, 1985), we wonder if gonadal steroids may also modulate neuronal function by alterations in gap junction properties and number.

There is no reason to suspect that in any particular brain region one mechanism of modulation prevails. Indeed, there is some evidence that

TABLE II

POSTSYNAPTIC EFFECTS OF ESTROGENS AND ANDROGENS

Receptor	Location	Steroid[a]	Effect	Reference
Adrenergic				
α_1	Uterus	E	Increase number	Roberts et al. (1977); Marshall (1981)
	CNS	E	Increase number	Wilkinson et al. (1981); Johnson et al. (1985)
α_2	Amygdala	E	Increase number	Agnati et al. (1981)
	Uterus	E	Increase number	Hoffman et al. (1979)
β_2	Medial basal hypothalamus	E	Increase number	Vacas and Cardinali (1980)
	Cortex	E	Decrease number	Wagner et al. (1979)
	Hypothalamus	E	Increase number	Wilkinson et al. (1979)
	Lung	E	Increase number	Moawad et al. (1985)
Cholinergic				
Nicotinic	Hypothalamus	E	Increase number	Morley et al. (1983); Miller et al. (1982a); Miller et al. (1982b)
	Levator ani	T	Increase number	Bleisch et al. (1982)
	Syrinx	T	Increase number	Bleisch et al. (1984)
Muscarinic	Preoptic area	E	Increase number	McEwen et al. (1981)
	Hypothalamus	E	Increase number	Rainbow et al. (1980); Block and Billiar (1979); Dohanich et al. (1982)
Dopaminergic	Striatum	E	Increase number	DiPaolo et al. (1979, 1981); Hruska and Silbergeld (1980a,b)
GABAergic	Hippocampus, striatum, cortex	E	Decrease number	Hamon et al. (1983); O'Connor et al. (1985)
Histaminergic	Hypothalamus	E	Decrease number	Portaleone et al. (1980)
Serotinergic	Brain hypothalamus	E	Early reduction, increase later	Biegon and McEwen (1982)

[a]E, Estrogen; T, testosterone.

multiple mechanisms can occur simultaneously (McEwen and Parsons, 1982). This presents perhaps the most difficult problem to those wishing to study the molecular basis of steroid action at synapses. In view of the multiple sites and mechanisms of steroid action, it is useful to develop preparations in which individual components can be isolated and studied under controlled conditions.

C. Advantages and Disadvantages of the Cell Culture Approach

One advantage of using isolated cells in culture is that the cells from which measurements are obtained and to which drugs may be applied can be visualized and the morphology of the cell is known. This allows biophysical, physiological, and some biochemical manipulations to be made which are not possible *in vivo*. Furthermore, as stated earlier, interactions between cells are eliminated or minimized.

There are, however, some disadvantages that should be kept in mind.

First, the techniques of isolation of the cells, including treatment by trypsin or other enzymes, may change cellular properties.

Second, since the cells are grown in culture, they may be at different stages of development and not fully differentiated. Therefore, they may display different characteristics than their adult counterparts *in vivo*.

Third, identification of neurons in culture remains a persistent problem. Although through the use of retrograde labeling techniques (Okun, 1981) and their combination with cell-sorting techniques (Calof and Reichardt, 1984) it has been possible to identify and separate neurons in culture, the application of these techniques has been problematic. An additional problem is encountered in identifying neurons in primary culture as being steroid sensitive. Until now we have been unsuccessful in applying techniques of steroid autoradiography for identifying steroid-concentrating neurons in culture. However, as will be shown below, autoradiography can be used successfully to show the presence of androgen receptors in certain muscle cells in culture.

Fourth, through the use of cell lines, a great deal has been learned about the biochemistry of steroid receptors and their roles in gene regulation. Unfortunately, no cell lines of excitable cells are currently available which are sensitive to gonadal steroids. This does not means that such cell lines do not exist. We can find no evidence that excitable cell lines currently in use have been examined for estrogen or testosterone receptors. In the event that the available lines do not have steroid receptors, it should be relatively easy to generate these cell lines using existing technology. Their use will be necessary for detailed biochemical and molecular experiments where sufficient quantities of starting material are required.

TABLE III

PRESYNAPTIC EFFECTS OF ESTROGENS AND ANDROGENS ON NEUROTRANSMITTERS

Neurotransmitter	Location	Steroid[a]	Effect	Reference
Norepinephrine	Uterus	E	Increase levels	Thorbert (1979)
	Lateral septum, interstitial nucleus of stria terminalis, central grey	E	Increase turnover	Crowley et al. (1978)
	Paraventricular hypothalamus	E	Increase turnover	Fuxe et al. (1981)
	Amygdalaentorhinal cortex	E	Decrease turnover	Fuxe et al. (1981)
	Medial basal and posterior hypothalamus	E	Increase turnover	Cardinali and Gomez (1977)
	Heart	E, T	Inhibits uptake	Iverson and Salt (1970)
	Hypothalamus	E	Increase release	Paul et al. (1979)
Serotonin	Hypothalamus	E	Decrease levels	Kendall and Narayana (1978)
	Hypothalamus	T (after castration)	Increase levels	van de Kar et al. (1978)
	Whole brain	E	Increase/decrease turnover	Everitt et al. (1975)
	Midbrain and pons	E	Decrease uptake	Kendall and Tonge (1977)
	Anterior hypothalamus	E	Increase uptake	Cardinali and Gomez (1977)

152

GABA	Substantia nigra, caudate	E	Decrease levels	Nicoletti and Meek (1985)
Dopamine	Septum accumbens septi	T, E, DHT	Increase levels	Alderson and Baum (1981)
	Tractus diagonalis	E	Increase levels	Crowley et al. (1978)
	n. accumbens striatum, median eminence, anterior hypothalamus suprachiasmatic nucleus, arcuate nucleus, area ventralis tegmenti interpeduncular nucleus	E	Decrease levels	Dupont et al. (1981)
	Hypothalamus	E	Decrease levels	Kendall and Narayana (1978)
	Median eminence	T	Increase levels	Simpkins et al. (1983)
	Mesostriatal and mesolimbic	E	Decrease turnover	Fuxe et al. (1983)
	Anterior hypothalamus	E	Increase uptake	Cardinali and Gomez (1977)
	Hypothalamus	E(+P)	Decrease uptake	Endersby and Wilson (1974)
	Thalamus	E	Decrease uptake	Wirz-Justice et al. (1974)
	Hypothalamus	E	Increase release	Paul et al. (1979)
	Median eminence, tuberoinfundibular neurons	E	Increase turnover	Wiesel et al. (1978)

[a]E, Estrogen; T, testosterone; DHT, dihydrotestosterone; P, progesterone.

TABLE IV
Presynaptic Effects of Steroids on Enzymes

Enzymes	Location	Steroid[a]	Effect	Reference
Choline acetyltransferase	Preoptic area	E	Increase levels	Luine et al. (1980b)
Glutamic-acid decarboxylase	Ventral tegmental substantia nigra	E	Decrease levels	McGinnis et al. (1980); Nicoletti and Meek (1985)
	Arcuate nucleus ventral medial hypothalamus	E	Decrease levels	Wallis and Sutge (1980)
Tyrosine hydroxylase	Median eminence	T	Decrease levels	Kizer et al. (1974)
Type A monoamine oxidase	Basomedial hypothalamus corticomedial amygdala	E	Decrease levels	Luine and McEwen (1977)

[a]E, Estrogen; T, testosterone.

D. The Acetylcholine Receptor as a Model System

Because of the complexity of central neuronal systems studied both *in vivo* and *in vitro* and the multiple sites of action of steroids within these systems, we have decided to examine the properties of a simple preparation in which detailed molecular questions can be addressed. We have, therefore, used steroid-sensitive myotubes derived from adult male *X. laevis* laryngeal muscle to study androgen influences on the properties of AChR. These myotubes develop from the myogenic stem cells or satellite cells of the adult muscle fibers (Bischoff, 1979).

Laryngeal myotubes from *Xenopus laevis* have several distinct properties. First, like their adult counterparts (Segil *et al.*, 1983, 1986), these fibers contain androgen receptors (Fig. 2). Second, like their adult counterparts, they respond to androgens by an increase in size (Erulkar and Wetzel, 1986) (Fig. 3). Third, when cultured in the absence of neuronal elements, extrajunctional AChR are present at the membrane and these can be shown to be influenced by steroids (*vide infra*). We recognize that the properties of extrajunctional and junctional AChR differ, but the advantage of this preparation is that the results can be compared both qualitatively and quantitatively to those obtained by other techniques at junctional receptors of the neuromuscular junction of the same muscle *in vivo* (see, for example, Colquhoun and Sakmann, 1985). These quantitative comparisons are impossible for central neurons in culture and *in vivo*.

There is more known about the molecular biology, chemistry, and biophysical properties and functions of the AChR than about any other postsynaptic neurotransmitter receptor (Karlin, 1980; Changeux, 1981; Conti-Tronconi *et al.*, 1982; Peper *et al.*, 1982; Barrantes, 1983; Anholt *et al.*, 1984; Popot and Changeux, 1984; Salpeter and Loring, 1985). Finally, androgens are known to influence the numbers of AChR in steroid-sensitive muscles *in vivo* (Bleisch *et al.*, 1982, 1984; Bleisch and Harrelson, 1984). For these reasons AChR at the neuromuscular junction is a good model system for the study at a molecular level of the actions of androgens on postsynaptic neurotransmitter receptors.

E. Cell Excitability and Steroids

BACKGROUND

The morphological studies described earlier point to certain sites in the CNS whose cells concentrate steroid hormones in their milieu (e.g., Fig. 4), and these sites are obvious regions where the effects of hormones on electrophysiological activity should be and have been studied. The results have

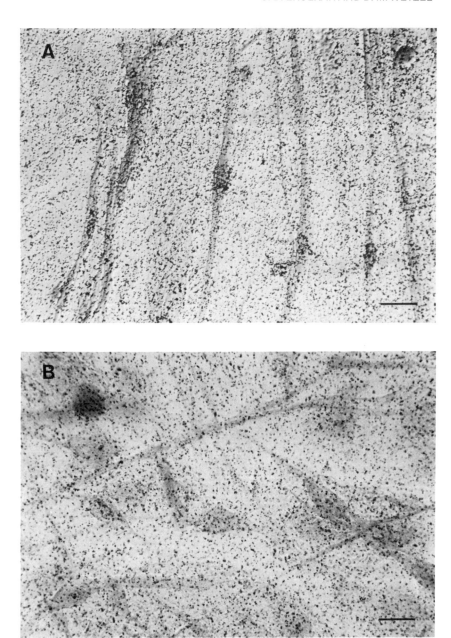

FIG. 2.   Photomicrographs of autoradiograms of *Xenopus laevis* laryngeal myotubes in culture showing accumulation of [³H]5α-DHT. (A) [³H]5α-DHT is accumulated in the nuclei of the myotubes to at least 5× background. Bar, 10 μm. (B) This accumulation is blocked by 100-fold excess untritiated 5α-DHT in control myotubes, exposed to emulsion (NTB-3) for 4 months, developed with D-19, and stained with hematoxylin and eosin Y. Hoffman modulation optics. Bar, 10 μm.

been well reviewed by Pfaff (1980) and by Moss and Dudley (1984) and we shall not expand on their excellent studies further. The end result of gonadal steroid action is reflected through some change in activity of the cell, whether it be manifested as a biophysical change at a single cell or as a behavioral change of the organism. The latter effect results from the integrated activity of populations of neurons and serves as a well-defined end point for the effects of a particular steroid. Several studies to date have been done *in vivo* and patterns of single-unit activity have been studied from anesthetized animals under different hormonal states or after addition of hormones. Studying these effects *in vivo* is complicated and has lead investigators to seek simpler systems in which to examine steroid effects on cell excitability. These preparations include *in vitro* brain stems of the rat (Pfaff, 1983; Moss and Dudley, 1984; Kow and Pfaff, 1984), hippocampus of the hamster (Teyler *et al.*, 1980; Chaia *et al.*, 1983), and isolated spinal cord of the frog (Schneider *et al.*, 1980; Erulkar *et al.*, 1981). Schneider *et al.* (1980) and Erulkar *et al.* (1981) used the isolated spinal cord from adult male *X. laevis* to examine androgen influences on motoneurons involved in the clasp reflex. The clasp response in male *X. laevis* depends upon the plasma levels of the androgens testosterone or 5α-dihydrotestosterone (5α-DHT) (Kelley and Pfaff, 1978). The clasp response provides a well-defined end point for androgen-activated behavior. This must be brought about by activation of the motoneurons leading to the two muscles mainly responsible for the motor activity, namely, the sternoradialis and the flexor carpi radialis muscles. The neurons may not be responsible for initiating the response, but they do make up the final common pathway. Subsets of motoneurons concentrate tritiated 5α-DHT in their nuclei (Erulkar *et al.*, 1981). Furthermore, there is indirect evidence that areas of the spinal cord concerned with the clasp reflex contain high levels of the enzyme testosterone 5α-reductase, which catalyzes the conversion of testosterone to 5α-DHT; the distribution of this enzyme in the spinal cord is dependent on androgen levels (Jurman *et al.*, 1982).

In electrophysiological studies it was shown that administration of androgen to the bath caused changes in patterns of activity of the motoneurons. Most notable of these changes is the increased response and synchronization of the activity of these motoneurons to the second of a pair of stimuli delivered to the dorsal roots of segment 2. These effects became prominent 7 hours following exposure of the spinal cord to androgen (Fig. 5). Cycloheximide in concentrations from 5 to 60 mg/ml prevented the increase in the response to paired dorsal root stimulation after 5α-DHT administration, but did not block the response. This suggests that the response to 5α-DHT may be mediated through genomic mechanisms. These findings are of interest because they implicate synaptic transmission as a process on which the androgen acted, although it is possible that changes may have occurred at the membrane level as well. Intracellular recordings and current

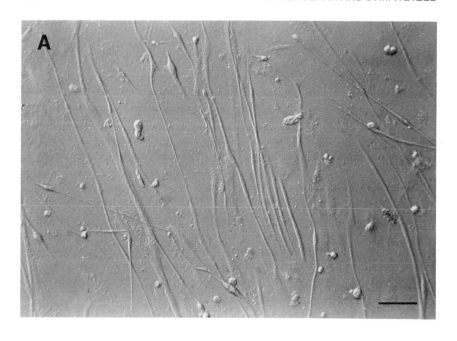

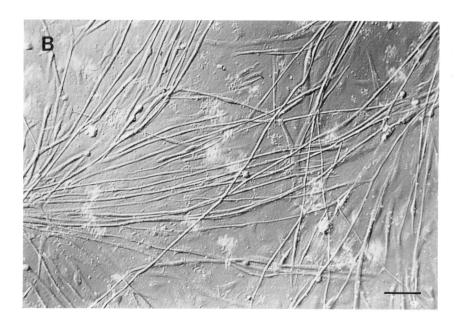

Fig.   3A and B.

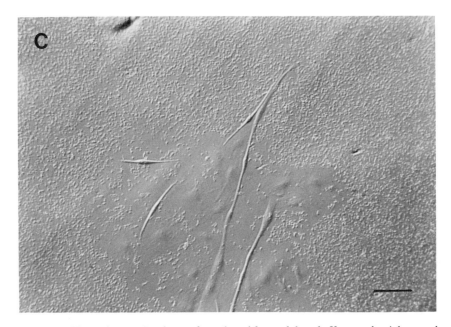

FIG. 3.   Photomicrographs of myotubes cultured from adult male *Xenopus laevis* laryngeal muscle raised in (A) control medium (50 L15/5FCS); (B) medium to which 25 nM 5α-DHT had been added 21 days prior to photography and recording; (C) medium with 25 nM 5α-DHT and 100 μM flutamide. Hoffman modulation optics. Bar, 10 μm.

injection experiments suggest that androgen levels may influence synaptic activation for spike generation at motoneurons (Fig. 6). Androgens had no significant effects on resting potentials and membrane time constants. However, it is impossible to draw firm conclusions from these experiments because it is impossible to know whether we were recording from steroid-concentrating motoneurons.

Thus it appeared that in order to obtain more conclusive results on steroid influences on synaptic transmission and membrane properties, simpler preparations were required. We have chosen to study the effects of steroids on steroid-concentrating cells in culture; the advantages and disadvantages of this approach have been outlined in Section III,C. We have found that 5α-DHT can have direct membrane effects and genomic effects on channel properties of myotube membranes.

IV.   MOLECULAR EFFECTS OF STEROIDS ON MEMBRANE PROPERTIES

A.   Comment

Although traditionally it has been thought that steroids act solely through genomic action, it is becoming increasingly clear that steroid hormones

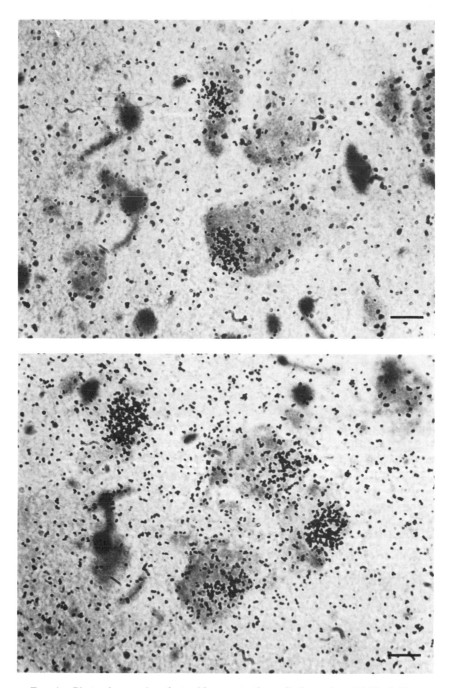

FIG. 4. Photomicrographs of steroid-concentrating cells in nucleus IX–X of *Xenopus laevis*. Exposed for 6 weeks, lightly stained with cresyl violet acetate. Bar, 20 μm. (Kelley, 1980).

can act nongenomically either through specific steroid receptors in the cytoplasm or in the membrane itself. This was first pointed out by Szego and Pietras (1981) and Baulieu (1983) and cannot be ignored any longer. Szego and Pietras (1981) make the important point that many tissues formerly thought of as not being steroid sensitive have been found to possess specific steroid-binding sites. Among these tissues can be included most skeletal muscles (Dionne *et al.*, 1979; Segil *et al.*, 1983), and as Szego and Pietras state, "It is thus likely that the true distinction between 'responsive' and 'unresponsive' cells will eventually prove to be the relative concentration of specific receptors" (cf. Lee, 1978; Pietras and Szego, 1980).

Earlier studies by Pietras and Szego (1977, 1979a,b) showed that 17β-estradiol can bind to surface membranes of isolated endometrial cells and hepatocytes of ovariectomized rats and, in 1979, the same authors, using purified plasma membrane subfractions, showed specific 17β-estradiol binding to the membranes. Furthermore, Hernandez-Pérez *et al.* (1979) showed high-affinity binding sites ($K_d = 6.6 \times 10^{-10}M$) for 17β-estradiol in crude plasma membrane fractions of human spermatozoa. They speculate that estradiol may act on human sperm cells at two organizational levels: (1) in the modification of sperm membrane structure and (2) in the interaction with specific nuclear receptors after being transported into the intracellular medium.

Farnsworth (1977) has postulated the presence of an androgen-sensitive "gatekeeper" receptor in human prostate plasma membrane and reported that 5α-DHT increases rates of phosphorylation and dephosphorylation of the cation-dependent ATPase of isolated membranes. Perhaps the best example of steroid action through membrane receptors is that of progesterone interactions with the adenylate cyclase system in *X. laevis* oocytes (Baulieu, 1983). McEwen and Parsons (1982) provided data from other authors showing extremely rapid effects of estradiol on cell firing. Whether these effects are mediated through direct action of estradiol on the membrane or through membrane receptors for the steroid is unknown.

The fact that steroids may influence the biophysical properties of membranes was pointed out by Mueller, Rudin, and colleagues (1962a,b, 1964, 1968), who pioneered the use of experimental bimolecular lipid membranes and the construction of regenerative action potentials in them. They showed that the experimental bilayers had physicochemical properties similar to the barrier structure of cell membranes and that electrical activity in these bilayers could be induced by a protein-containing cellular compound designated *excitability-inducing material* (EIM) (Mueller and Rudin, 1963). Of importance in the context of this article is that action potential phenomena and delayed rectification properties could be obtained consistently from bilayers made from a solution of 40–50% tocopherol (Vitamin E), and 3% sphingomyelin in 2:1 $CHCl_3$:CH_3OH (Mueller and Rudin, 1967, 1968). Addition of cholesterol stearate could delay deterioration

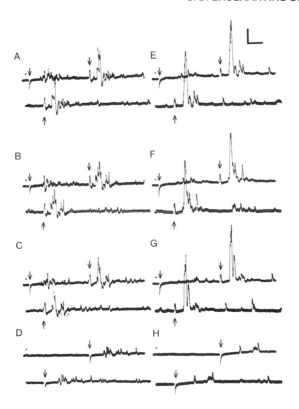

FIG. 5. Effects of 5α-DHT on spinal neuronal activity. Male *Xenopus laevis* were castrated 50 days prior to recording. (A–H) Recordings from the nerve to the sterno-radialis muscle in response to paired stimulation of the dorsal root of segment 2. In each recording, the top trace shows the responses to paired DR stimulation (arrows); the lower trace shows the same recording as the top trace but with the response to the second stimulus (at arrow) displaced to the left to show any later activity. (A–C) Consecutive recordings obtained in control Ringer's solution in (mM) 116 NaCl; 2 KCl; 1.8 CaCl$_2$; 6.2 NaHCO$_3$; 5.5 glucose; pH 7.4. (E–G) Consecutive recordings obtained 3 hours after addition of 5α-DHT ($10^{-7}\,M$) to the perfusing Ringer's solution. (D and H) Responses to single DR stimuli before and after 5α-DHT addition, respectively. Calibration point at the left of each trace indicates 50 μV; at top right, calibration is 120 μV, 20 msec. (I) Same experiment. The area under the compound action potential from the nerve to the sternoradialis muscle ($\pm$ standard error of the mean) in response to the second dorsal root stimulus is plotted against the interstimulus interval. △, Responses in control Ringer's solution; ○, responses 1 hour after addition of 5α-DHT to the perfusing medium; ▲, responses 3 hours later; ●, responses 7 hours later. (Data from Erulkar and Schneider, unpublished observations.)

of this material. EIM produces its conductance effects in membranes made from many lipid types, including oxidized cholesterol, and the rate of adsorption increases with the amount of tocopherol in the solution and is decreased by large amounts of cholesterol. Furthermore, the gating effects

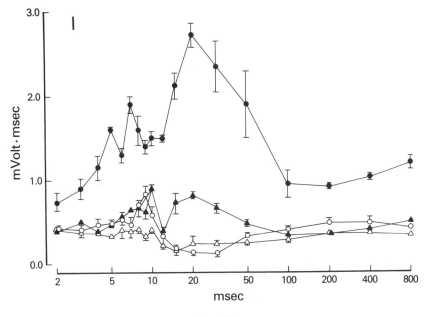

FIG. 5I.

of the EIM are strongly controlled by the lipid composition and these effects of EIM from yeast can be blocked by small quantities of cholesterol. These studies first brought to the fore the roles of steroids in the make-up and function of electrically excitable membranes. Among the direct effects of steroids on the membrane may be effects on cell-membrane fluidity (Lawrence and Gill, 1974) and local aqueous environment (Pak and Gerschfeld, 1967), which could have far-reaching effects on many membrane protein actions. These include effects on membrane conductance, channel kinetics, and the availability of neurotransmitter receptors to ligands (Heron *et al.*, 1980). In addition, Pak and Gerschfeld (1967) had shown that certain steroid hormones, including androsterone, could alter the amount of water carried by a moving monolayer of stearyl alcohol when the androsterone was added to the solution under the film. They speculated that the steroid could alter effects at biological receptor sites by acting to alter the aqueous environment.

To these effects of steroids at the membrane must be added the work of Heap *et al.* (1970), who found increases in permeability to radioactive K^+ at egg-lecithin liposomes with certain steroids, especially testosterone. Lipid permeability to K^+ was inversely related to lipid solubility of the steroid. The sequence of permeabilities for lipid solubility was the reverse of that for aqueous solubility of the steroid. Thus, the permeability effect is directly related to the aqueous solubility. In the case of steroids having polar groups that might be oriented in the hydrocarbon interior of the membrane, additional

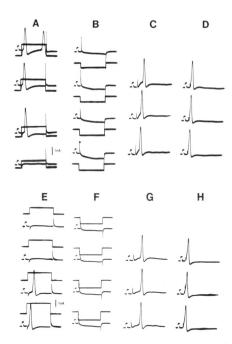

Fig. 6. Intracellularly recorded responses from sternoradialis motoneurons in isolated spinal cord of *Xenopus laevis*. (A–D) Responses from a motoneuron of an adult male injected with gonadotropin and that subsequently clasped the female. (A) Depolarizing current injected into the motoneuron. Note the development of the depolarization and second spike generation following the first action potential as current intensity is increased. (B) Hyperpolarizing currents on the same motoneuron. This motoneuron shows strong rectification properties. (C) Responses to increasing strengths (progressively from top to bottom trace) of dorsal root stimulation. Note decrease of the latency as stimulus strengths increase. (D) Responses to antidromic stimulation of the nerve to the sternoradialis muscle. All three records of constant supramaximal intensity. Calibration at beginning of each sweep, 10 mV, 5 msec. Current calibration, 1 nA. (E-H) Responses from a motoneuron of an adult male castrated 31 days prior to recording. (E) Depolarizing current injected into the motoneuron generated a single action potential even with the highest currents used. (F) Hyperpolarizing currents. Membrane tissue constants appear shorter than those for the motoneurons of the gonadotropin-treated animals. (G) Increasing strengths of dorsal root stimulation progressively from top to bottom. Note the long latency delay before a spike is generated even at the highest stimulus intensities. (H) Responses to antidromic stimulation of the sternoradialis nerve. Calibration as in A–D.

water entry into the hydrophilic region would be promoted and this would tend to increase membrane permeability.

A clue to possible mechanisms of action of steroids on membranes was provided by the studies of Strichartz *et al.* (1980), who showed that on the squid giant axon the substance phloretin, a dipolar organic compound, reversibly suppressed potassium and sodium conductances and modified their dependence on membrane potential. They concluded that the substance either acted at a receptor on the membrane or produced a diffuse dipole field in the membrane interior modifying potassium channel gating.

Heap et al. (1970) reported that early effects of steroids may be different from those after 20 hours. They speculated that an early "stabilizing" effect takes place as the steroid concentration in the lipid phase is increasing, but a later "labilizing" effect occurs when concentrations within the membrane saturate the cavities within the hydrocarbon region. These results suggest that the steroid can perhaps act at the membrane as some anesthetics are believed to (i.e., by membrane fluidization). Indeed, Selye (1941) has shown that testosterone given intraperitoneally to male and female mice can cause a long-lasting, slowly developing profound anesthesia. This is interesting because unlike some anesthetics, the steroids do not contain any ionizable groups, are highly lipophilic, and would be expected to penetrate readily into any lipid phase. Atkinson et al. (1965) described a wide range of steroids possessing structurally specific anesthetic activity. This was followed by the work of Lawrence and Gill (1974), who studied the effects of steroids on the molecular mobility and local polarity of ultrasonically dispersed vesicles of lecithin and cholesterol (liposomes) using nitroxide-labeled dipalmitoyllecithin as a molecular probe. They found that several steroids caused fluidization of the lipid bilayer at concentrations approximating those attained during anesthesia in vivo and the magnitude of the effect was linearly related to the drug concentration. Those steroids that are inactive as anesthetics produced much less disordering of the lipid bilayer. These results suggested that the lipid phase of nerve cell membranes is a site of action of general anesthetics. Indeed, a similar ability to disorder the hydrocarbon phase of the liposome bilayer that correlates very well with anesthetic potency was produced by several other anesthetics, e.g., butobarbitone, octanol, and halothane. No receptor appears to be involved.

Fluidization may alter the availability of a neurotransmitter receptor to ligand (Heron et al., 1980; Bigeon and McEwen, 1982) or coupling of a receptor to adenylate cyclase (Finidori-Lepicard et al., 1981). These latter authors found that progesterone selectively inhibits membrane-bound adenylate cyclase activity of Xenopus oocytes.

Finally, Lechleiter and Gruener (1984) have found that halothane, which increases membrane fluidity, shortens the open times of ACh-activated single channels.

Steroids may act not only on total membrane fluidization but also specifically on certain subdomains of the membrane. The lipid environment of the AChR may be an important determinant of the receptor's properties. Delipidation of the receptor leads to changes in the affinity of the receptor for agonists (Briley and Changeux, 1978; Chang and Bock, 1979), and treatment with phospholipase A_2 inhibits ion fluxes in membrane vesicles (Andreasen et al., 1979). Kilian et al. (1980) found that inclusion of

α-tocopherol in vesicles depleted of neutral lipids and into which AchR from *Torpedo californica* had been incorporated caused the generation of carbamylcholine-sensitive sodium flux. Pumplin and Bloch (1983) have shown that clusters of AchR in cultured rat myotubes are organized into two distinct membrane domains, one rich in AchR, the other poor in AchR. They found membrane structures surrounding AchR clusters were rich in cholesterol, and AchR-rich domains had more of these than did AchR-poor domains. On the other hand, Bridgman and Nakajima (1981) and Nakajima and Bridgman (1981) found that in embryonically derived myotubes from *X. laevis* most areas of the muscle plasma membrane contain sterol-specific complexes, but that these are absent from membrane areas of junctional and nonjunctional AchR.

What emerges from these studies is that species differences may define the effect of the steroid at the membrane. Indeed, Carlson *et al.* (1983) found that progesterone interacted with the phospholipid of liposomes better when cholesterol was included in the liposome. They felt that the steroid would intercalate into the phospholipid bilayers containing cholesterol and perhaps there could be some diffusion of hormone across the plasma membrane. A point made by Bloch (1982) is that although sterol-induced changes in membrane fluidity or solute permeability are viewed as meaningful indices of sterol function *in vivo*, it should be remembered that, in most methods used, bulk fluidity is measured. "There may be localized membrane regions or domains of substantially lower or higher sterol content differing in fluidity from the bulk phase" (Bloch, 1982). This fits in with the observations described above.

This raises the question whether the membrane effects of steroids are due exclusively to their intercalation into phospholipid bilayers or are mediated through membrane receptor sites for steroids, and/or are dependent upon other steroids already present in the membrane.

B. Patch-Clamp Studies

1. MEMBRANE EFFECTS

We have concentrated on how an androgenic steroid, 5α-androstan-17β-ol-3-one (5α-dihydrotestosterone) can modulate the properties of ion channels in plasma membranes of androgen-concentrating myotubes derived from adult muscle. Our results suggest that 5α-DHT may exert its effects through both membrane and genomic actions.

The morphological characterization of the myotubes in these studies has been described earlier. Suffice it to say that like their adult counterparts, these myotubes have androgen receptors. Furthermore, the AChE distribution in these myotubes is similar to that seen in denervated muscle, and thus suggests that the AChR present will have properties similar to extrajunctional AChR. This is supported by the studies described below and by those of others using embryonic myotubes in culture (Neher and Sakmann, 1976; Dreyer *et al.*, 1976).

Our first experiments were concerned with characterizing the properties of channels that were activated by acetylcholine (ACh) in myotubes raised in control culture medium (50% Leibowitz-15 medium, 5% fetal calf serum, <0.05 pM testosterone). The parameters of activation include distributions of open times and closed times, and current amplitudes, as well as current–voltage relationships.

Figure 7 shows single-channel currents recorded from an outside–out patch after the addition of 60 nM ACh to the bath. The current–voltage plot (Fig. 8) shows a linear relationship with a slope conductance of 57 pS and reversal potential close to 0 mV (see Fig. 7 legend for composition of solutions). These values are consistent with those reported by others for ACh activation of other preparations (Neher and Sakmann, 1976; Dreyer *et al.*, 1976; Clark and Adams, 1981; Hamill and Sakmann, 1981; Greenberg *et al.*, 1982; Brehm *et al.*, 1984).

The addition of 10^{-9}–$10^{-6}M$ 5α-DHT to the external solution (whether in the pipet or in the bath, depending on the patch configuration) caused a dramatic bursting of ACh-activated channel openings within a few minutes of application (Fig. 9). This effect occurred at membranes of myotubes derived from adult male laryngeal muscles and also from myotubes derived from quadriceps femoris from the same animal. The latter muscle contains low levels (3 fmol/mg protein) of androgen receptor. This fact and the fact that the effect of 5α-DHT can be obtained in excised membrane patches

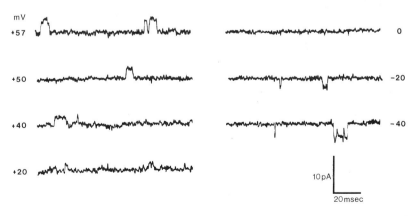

FIG. 7.   Acetylcholine (60 nM)-activated single-channel currents from laryngeal myotubes of adult male *Xenopus laevis* at different clamp potentials. Outside-out patch. Electrode contained (mM) 2 NaCl; 114 KCl; 1 EGTA; 6.2 NaHCO$_3$; 5.5 glucose, pH 7.4. Bath contained normal Ringer's solution.

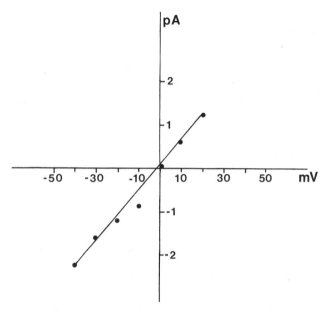

FIG. 8.   Current–voltage relationship for ACh-activated channels shown in Fig. 7. Slope conductance 57 pS.

strongly suggest that these effects of 5α-DHT are nongenomic and may be mediated by direct action on the membrane. Bursting is also elicited by 5α-DHT even in the absence of ACh (Fig. 10). It appears, therefore, that there are at least two populations of channels influenced by 5α-DHT, one that is ACh activated, the other that is not. The properties of these ACh-insensitive channels differ from those of ACh-activated channels in both kinetics and reversal potential. Thus, 5α-DHT is acting directly at the membrane to alter its physicochemical properties (as described earlier), and influences channel kinetics of these two populations and perhaps those of other populations as well.

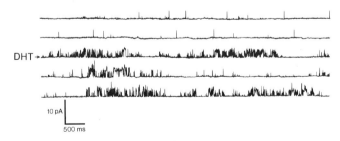

FIG. 9.   Effects of 5α-DHT on single-channel currents activated by 60 nM ACh. The top two traces show ACh-activated single-channel currents. The lower traces show the burst-like structure resulting from the addition of 10^{-6} M 5α-DHT to the medium. Similar effects have been obtained at concentrations of 10^{-9} M 5α-DHT. Outside-out patch. Membrane clamped at + 50 mV. Slope conductance 60 pS for the ACh-activated channels. It appears, however, that more than one population of channels is activated.

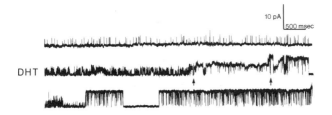

FIG. 10.   Effects of 5α-DHT on single-channel currents at laryngeal myotubes of adult male *Xenopus laevis.* Inside-out patch. The top trace shows spontaneously occurring channels present with just NFRS alone in the electrode. In the bath (i.e., at the cytoplasmic surface of the membrane) there was (mM) 114 KCl; 2 NaCl; 1 EGTA; 6.2 NaHCO₃; 5.5 glucose, pH 7.4. 5α-DHT (10^{-8} M) was added to the bath 2 minutes before recording the middle trace; the burst activity started as shown. At the first arrow a second level was recorded suggesting the activation of another channel; at the second arrow, a third level was obtained. The activity at this level was sustained. The membrane was clamped to +20 mV. There was no exogenous acetylcholine present and these records show that 5α-DHT can activate a population of channels from the internal surface. Similar results have been obtained when 5α-DHT was added at the external surface. As this burst activity can be recorded even in the presence of α-bungarotoxin, the 5α-DHT-activated channel openings cannot be those of ACh-channels.

These effects on ACh-activated channels are not similar to those of the local anesthetic QX222 as described by Neher and Steinbach (1978), or of benzocaine (Ogden *et al.,* 1981). We emphasize that these effects of 5α-DHT can be obtained from excised patches and therefore do not result from the genomic action of this steroid.

One method of analysis involves the determination of time constants for the durations of the openings and closings of the channels under different conditions. This can be achieved by the use of the *maximum likelihood* technique that allows one to obtain the best fits for the distributions of either the open or closed times as desired (Colquhoun and Hawkes, 1983). We find that the distributions of open times of channels activated by ACh fit the sums of at least three and possibly four exponentials although the contribution of each exponential to the overall distribution is different, and in some cases may be exceedingly small. Closed-time distributions also showed at least three or four time constants (Fig. 11). From the work of Neher and Steinbach (1978) and Colquhoun and Hawkes (1983) one can determine whether the substance is acting at the open channel activated by ACh. The data suggest that this is not the case and that the burst effect results from direct membrane effects (such as fluidization) and consequent disruption of the channels. Our observation that ACh-activated and voltage-dependent channel characteristics are affected by 5α-DHT independently of the presence of high levels of steroid receptors in the myotubes supports this idea. Certainly there appears to be stereospecificity in the anesthetic potencies of the steroid anesthetics in their effects on order parameters in lecithin liposomes (Lawrence and Gill, 1974). However, it appears that the orientation of the 3-OH in the 5α-DHT is critical; 5β-DHT not only fails to show any anesthetic potency, but also does not cause burst-like openings of the membrane channels

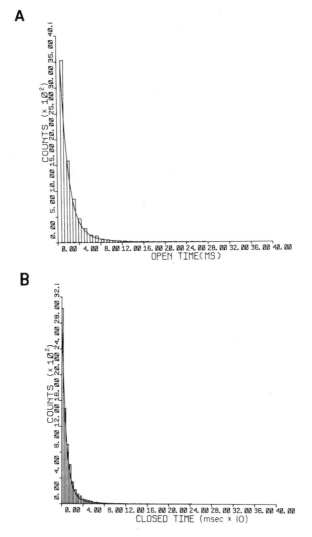

FIG. 11.   Channel kinetics in the presence of ACh. (A) Open-time histogram for 7176 channel openings. Bin width 1.05 msec; minimum event time 0.70 msec; solid line fit for sums of four exponentials; τ_1 = 0.53 msec; τ_2 = 1.45 msec; τ_3 = 3.43 msec; τ_4 = 31.0 msec with relative contributions of 10.4%, 65.8%, 23.7%, and <1%, respectively. (B) Closed-time histogram for 7176 channel closings. Bin width 5.00 msec; minimum event time at 0; solid line fit for sums of four exponentials; τ_1 = 2.5 msec; τ_2 = 8.8 msec; τ_3 = 20.2 msec; τ_4 = 118.7 msec with relative contributions of 10.5%, 62%, 27.4%, and <1%, respectively.

of laryngeal myotubes. It would appear then that hydrogen bonding is an important component of membrane-perturbing action of some drugs (Brockerhoff, 1982) and that the disordering potency may depend partly on the orientation of the hydrogen bond (see Goldstein, 1984). It is also possible

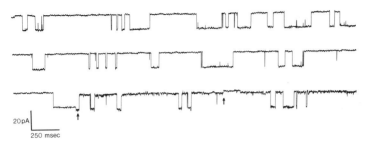

FIG. 12. Single-channel currents activated by 100 nM ACh and recorded from laryngeal myotubes of adult male *Xenopus laevis*. These myotubes had been raised in medium to which 25 nM 5α-DHT had been added 21 days prior to recording. Arrows denote opening and closing of a second channel population with lower conductance. Outside-out patch. Electrode contained (mM) 114 KCl; 2 NaCl; 1 EGTA; 6.2 NaHCO$_3$; 5.5 glucose, pH 7.4. Bath contained NFRS. Membrane clamped at -45 mV. Slope conductance 365 pS.

that ion-gating properties are altered by the production of a diffuse dipole field within the membrane. Estriol, cholesterol, and dihydroepiandrosterone do not influence channel activity.

2. GENOMIC EFFECTS

If the laryngeal myotubes are grown in a culture medium containing 25 nM 5α-DHT for a period of about 21 days (>6 days but <26 days), the channel characteristics change dramatically with respect to their amplitude and number. Figure 12 shows records in the outside-out configuration obtained at a holding potential of -45 mV from a patch of a myotube cultured under the above conditions. It can be seen that the conductance of the channels has increased greatly and that there is an increased probability of firing of at least two channels. This has been a consistent finding in our experiments—the number of AChR is increased and the conductance of the ACh-activated channels is also increased. Figure 13 shows the current–voltage relationships for these channels. It must also be pointed out that there may be a population of small conductance channels that are unaffected by the 5α-DHT treatment or (as pointed out below) are *as yet* unaffected by the 5α-DHT treatment, for it is clear that the amplitude distribution suggests that at least two populations of receptors are present (Fig. 14). We believe that these effects of 5α-DHT are genomically mediated, because (1) they are blocked by treatment with the antiandrogen flutamide (Peets *et al.*, 1974) and (2) they are *not* observed in patches from myotubes from quadriceps femoris.

If myotubes that have been grown in control medium are exposed to α-bungarotoxin (α-BTX), an irreversible cholinergic blocker, before washout and patch recording, no channel openings can be elicited by ACh

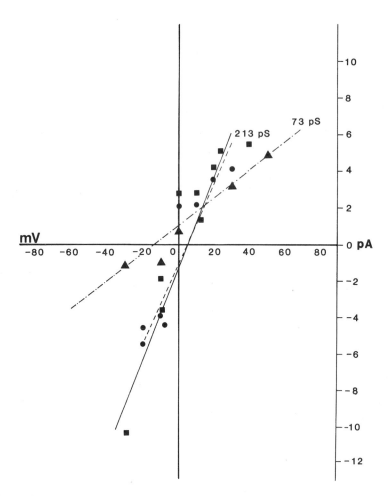

FIG. 13.   Current–voltage relations for channels recorded from myotubes grown in 25 n*M* 5α-DHT for 21 days. Outside out patch. ▲, NFRS + 60 n*M* 5α-DHT; ■, NFRS + 60 n*M* 5α-DHT + 100 n*M* ACh; ●, NFRS + 60 n*M* 5α-DHT + 500 n*M* ACh. Each point represents the mean sample of a minimum of 500 opening events with the largest number of 7000 at a single point.

administration at the external surface. If, however, the 5α-DHT (25 n*M*) is added to the culture medium 21 days prior to (1) the addition of the α-BTX, (2) washout, and (3) subsequent recordings, then small-amplitude channel openings in response to ACh can be obtained within minutes of patching (Fig. 15). In no case did we record from large-conductance channels under these conditions. This must mean that these small-conductance receptors

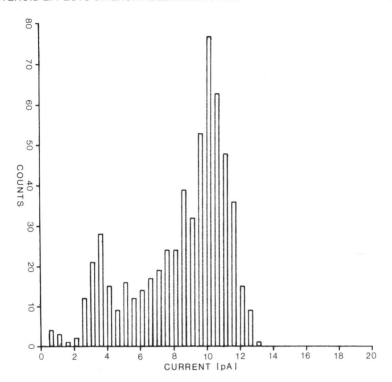

FIG. 14.   Amplitude distribution of single-channel currents activated by 60 nM ACh from laryngeal myotubes grown in 25 nM 5α-DHT for 23 days. At least two populations of channels are activated, one with lower conductance, the other with high conductance (285 pS). Holding potential, −35 mV.

had either been unmasked or had been newly inserted into the membrane, probably from the Golgi apparatus (Anderson and Blobel, 1981; Merlie *et al.*, 1983). Further evidence to support the latter concept is that we were unable to record from additional channels after having excised a patch. These were nicotinic cholinergic receptors, because both gallamine triethiodide and α-BTX blocked their activity. *d*-Tubocurarine, another cholingeric antagonist at adult muscle neuromuscular junction, caused the myotube to contract (as it does at myotubes derived from embryonic tissue) and so its action was difficult to assess. The fact that we record currents from only small-conductance channels in these preparations strongly suggests that the increase in conductance as a result of long-term 5α-DHT exposure occurs after insertion of the channel in the membrane rather than prior to its insertion.

   Our experiments suggest, therefore, that the presence of the androgen 5α-DHT can modulate the development and characteristics of the AChR of myotubes derived from male adult *X. laevis* laryngeal muscles.

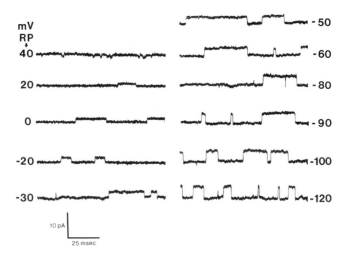

Fig. 15.   Single-channel currents at different clamp potentials of a laryngeal myotube raised in medium to which 5α-DHT was added 21 days prior to recording and which was soaked in 2 × 10^{-6} g/liter α-bungarotoxin for 3 hours prior to washout and recording. Cell attached. Electrode contained NFRS + 100 nM ACh. Small conductance ACh-activated channels can be seen.

V. MOLECULAR MECHANISMS OF STEROID HORMONE ACTION ON NONEXCITABLE CELLS

Much of our understanding of the molecular mechanisms of the action of steroids is derived from experiments on nonexcitable endocrine tissues. Because similar mechanisms may be involved in steroid action on excitable cells, we will outline briefly these mechanisms here. Most of this research has concerned the role that steroids play in gene regulation; in this regard, the use of highly steroid-sensitive cell lines and viruses that respond to steroid receptor complexes has led to rapid progress.

The two-step model of steroid hormone action described initially by Jensen et al. (1968) is still the underlying framework on which most models of steroid action are based. In this model, steroids are thought to diffuse freely through cell membranes and bind with a specific and saturable high-affinity cytoplasmic receptor. These steroid receptor complexes (SRC) are "activated" and translocated to the cell nucleus where they interact with DNA to influence its transcription, thus increasing mRNA production and eventually protein synthesis. It is thought that steroids do not turn on transcription of all genes, but rather a select group.

For most steroid responses, the fundamental elements of the model appear to be correct; that is, (1) a steroid binds to a specific receptor and (2) this receptor somehow interacts with DNA. The details of passive diffusion, cytosolic versus nuclear receptors, and how SRC may interact with DNA are not completely resolved; some current hypotheses will be discussed below. In addition, we will describe the special case of androgen action in the prostate to activate two genes that may alter processing of total cellular mRNA.

A. Steroid Diffusion

The means by which steroids enter a given cell remain poorly understood. Though it has been generally accepted that steroids diffuse freely into cells, this may not be the case. Szego and Pietras (1981) argue in an extensive review that in steroid-concentrating cells, steroids are internalized by binding to specific macromolecules on the cell membrane. However, a clear demonstration of these molecules has not yet been obtained. This is not to say that there are not membrane receptors for steroids. This was discussed in Section IV.

B. Cytoplasmic versus Nuclear Receptors

Experiments (King and Greene, 1984; Welshons *et al.*, 1984, 1985) have suggested that unoccupied steroid receptors for several steroid hormones are located primarily within the nucleus. Using monoclonal antibodies (King and Greene, 1984) and cytochalasin enucleation (Welshons *et al.*, 1984, 1985) it has been found that estrogen, glucocorticoid, and progesterone receptors when unoccupied are confined to the cell nucleus. These authors suggest that the original proposal of cytoplasmic receptors for steroids is in error and is a result of cell-fractionation procedures employed in these other earlier experiments. However, steroids still must bind to a receptor that then interacts with DNA. These more recent experiments are also supported by the work of Barrack and Coffey (1983), who found that the large majority of estradiol receptors and androgen receptors is bound to or is part of the nonextractable nuclear matrix.

C. Steroid Receptor-DNA Interactions and Gene Regulation

There has been a great deal of progress made in understanding steroid hormone regulation of gene expression. This quickly changing field has been reviewed by O'Malley *et al.* (1983), Spelsberg *et al.* (1983), Ringold (1985), and Yamamoto (1983, 1985a,b). Although steroid hormone receptor interaction with DNA is far from completely understood, some generalizations may be drawn. Steroid receptor complexes appear to bind to all DNA with low affinity; however, it is believed that it is through high-affinity binding of the SRC with specific DNA sequences that enhanced expression of only a few select genes is controlled by a given steroid. A number of such regulated genes have been described, and include (1) glucocorticoid-induced mouse mammary tumor virus (MMTV), α_{2u}-globulin, metallothionein, and lysozyme; (2) estrogen-induced ovalbumin and chick lysozyme; (3) progesterone-induced uteroglobulin; and (4) androgen-induced prostatic C3 gene (Ringold, 1985). Nitrocellulose filter binding assays, nuclease protection experiments, and DNase I foot printing experiments have shown that for the majority of the above-studied regulated genes, (1) SRC bind to "regulating" regions (or "enhancers", Yamamoto, 1985a,b) in the 5' flanking areas of these genes, often in a variable position ahead of the promoter region; (2) for some of these genes there are also binding sites in noncoding regions within the gene and in the 3' regions; and (3) that small specific chromosomal proteins may be involved in either the binding of SRC to DNA or in action after initial binding (Spelsberg et al., 1983; Ringold, 1985; Yamamoto, 1985a). Ringold (1985) outlines a number of possible mechanisms of steroid-stimulated transcription. Briefly, these are (1) SRC removes or blocks a repressor effect of the regulating region of the promoter, (2) SRC may interact directly with polymerase and stimulates polymerase function, (3) local unwinding of DNA induced by SRC binding may enhance efficiency of polymerase binding to the core promoter, (4) SRC may locally alter DNA conformation and facilitate transcriptional factors, or (5) SRC binding might alter chromatin structure in such a way as to facilitate polymerase binding. An additional model proposed by Yamamoto (1985a,b) is particularly attractive. He hypothesizes SRC as binding to enhancer regions, often in variable positions 5' of promoter regions. Binding of an SRC to an enhancer causes "release" of a transcriptional factor (probably a small chromosomal protein) which acts on the promoter. Both enhancers and promoters are able to bind multiple and different SRC or transcriptional factors, respectively, which may alter future interactions with SRC or transcriptional factors. This model is attractive in its ability to explain developmental and priming influences of steroids. (Fig. 16).

D. Posttranscriptional Control of Gene Expression

In discussing steroid regulation of gene expression, it is important to distinguish between genes that are initially activated and the role their products may play in the subsequent expression or activation of genes not directly regulated by a given steroid. As Ringold (1985) points out, this may be particularly true for sex steroids, which appear to regulate complex sets of tissue- or cell-specific genes. We can think of two examples somewhat relevant to our own work with which to explain this point.

Androgens appear to regulate the gene for ornithine decarboxylase (ODC) in mouse kidney (Janne et al., 1983; Kontula et al., 1984; Berger et

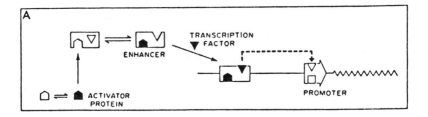

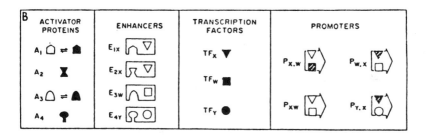

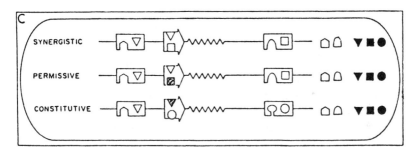

FIG. 16. Proposed model for enhancer action and multifactor gene regulation. (A) An activator protein (i.e., an activated steroid hormone receptor complex) specifically binds to an enhancer and modulates a transcription factor. This transcription factor, which may be diffusible, acts through unknown mechanisms on the promoter to initiate or facilitate initiation of transcription. (B) One assumption of the model is that different activator protein ($A_1 \ldots, A_4$): enhancer ($E_{1x}, \ldots, E_{4y}$) complexes are specifically recognized by different transcription factors (TF_x, TF_w, TF_y). These factors are "captured" by promoter sequences and with the necessary enzymes initiate RNA synthesis. The promoter sequence determines the efficiency with which different transcription factors act and are therefore rate limiting. Factors that limit the initiation rate are represented by open symbols while the factor which is the next least efficient is stippled. For example, X limits the rate of $P_{x,w}$; if this limitation is relieved, then $P_{x,w}$ is limited by W. However $P_{x,w}$ is limited by both X and W. (C) This hypothetical cell illustrates how, through different (and limited) combinations of transcription factors, enhancers and promoters, regulation may be synergistic, permissive, or constitutive. The independent action of two factors may synergistically regulate a gene (top gene). The middle gene requires two factors and thus represents permissive regulation. The lower gene is constitutively expressed since it requires an activator (A_4), which is absent, while E_{1x} provides a nonlimiting factor (TF_x). In a cell containing both A_4 and A_1 this gene would have a permissive response and in a cell without A_1 would respond to A_4 alone. (Reproduced with permission from Yamamoto, 1985a.)

al., 1984) and perhaps in some muscles (Persson and Rosengren, 1983). ODC is the rate-limiting enzyme in polyamine biosynthesis. Polyamines in turn have a wide variety of effects on cell function (Williams-Ashman and Lockwood, 1970; Bachrach, 1973); most notable here is a stabilizing effect on mRNA, decreasing its rate of degradation and stimulation of RNA polymerase activity (Caldarera *et al.*, 1968; Tabor and Tabor, 1984). Stabilizing mRNA and enhancing polymerase activity would lead to increased transcription and translation of perhaps all activated genes. Thus, the activation of this one gene for ODC would enhance production of "normal" tissue or cell-specific products.

Another demonstrated effect of androgens is enhanced transcription of rRNA genes in prostate (Liao *et al.*, 1966; Liao and Lin, 1967; Mainwaring and Derry, 1983). We can speculate that more rRNA gene transcription would result in more ribosomes and thus perhaps increased mRNA translation. As an aside, we point out that rRNA transcription is tightly coupled to polyamine synthesis (Brown and Gurdon, 1964). In addition, Jones *et al.* (1985) found that estradiol increases levels of rRNA–mRNA in rat hypothalamic neurons detected by *in situ* hybridization.

VI. CONCLUSIONS AND QUESTIONS

Steroids act at a molecular level to modulate cell function. Early exposure of a cell during development to a wide variety of steroids and other substances can influence the molecular events that occur at later stages of development. These are known as the organizational effects of the steroids. In the adult, the effects are activational rather than organizational, and it has been our aim to define some of the molecular events that underlie these activational effects. Although the mechanisms of steroid action are well understood at peripheral nonneuronal endocrine tissues, we wished to describe how steroid hormones may regulate behavior through the modulation of properties of excitable tissues. We know that steroids act at specific sites in the central nervous sytem and at muscle; we recognize that reproductive behavior is controlled by neural circuits in the nervous system. How do the steroids influence the molecular properties of the neurons that comprise the circuit? Ultrastructure, neurotransmitter levels, receptor number, and biophysical properties are all influenced by the steroids and it is the integration of these events that modifies the behavior.

We have developed a simple system as a model to study the effects of steroids on excitable cells. Some of the effects of the steroids may be directly at the membrane and not specific to steroid-sensitive cells. However, we have also shown that there is a dramatic change in ACh-activated channel properties resulting from steroid action through gene regulation.

There are now several questions that point the directions in which one can proceed to verify and extend the effects we have found.

1. Do these changes observed in cultured preparations occur in *in vivo* preparations? For instance, is AChR density altered *in vivo*? It is known (Bleisch and Harrelson, 1984) that steroids influence AChR levels in other steroid-sensitive muscles (e.g., levator ani), but these increases are correlated with increased protein and increased muscle size; the question remains whether the changes are correlated with increased receptor density at the neuromuscular junction or whether they are due to increased area of the end-plate region. Quantitative electron microscopy autoradiography and scanning electron microscopic studies of the neuromuscular junction are now being used to answer this question. Receptor binding studies indicate that the number of receptors for different neurotransmitters may be influenced by circulating physiological levels of steroids (see Table II). However, it is difficult to determine whether alterations in cell size contribute to changes in neurotransmitter receptor number in excitable cells. The neuromuscular junction may serve as an appropriate model to determine whether increased receptor number has any functional significance.

2. Does the increased conductance with long-term exposure to 5α-DHT occur *in vivo*, and, if so, what does this mean in terms of neuromuscular junction function? Do these changes occur for all neurotransmitter receptors in the central nervous system, or are they restricted only to AChR?

3. When do these influences first occur in a cell relative to its development?

4. Is the steroid effect due to second-messenger-type effects? In prostate and other tissues, androgens have been found to regulate ornithine decarboxylase transcription and thereby increase polyamine levels in the cell. Polyamines are known to influence mRNA degradation and can alter membrane properties directly. Do increased levels of polyamines play a role in increased synthesis of AChR or play a role in their conductance increase?

5. Is the increase in AChR number due to either an increase in transcription of AChR genes or a stabilizing of AChR message or some change in the processes of AChR insertion? These questions can be addressed using modern recombinant DNA techniques.

6. What is the mechanism underlying the increased ACh-activated channel conductance? Traditional biochemical techniques may be used to isolate the receptor proteins and study their composition and properties. Perhaps a more powerful approach is to use molecular biological techniques to isolate mRNA that may be responsible for the proteins involved in this conductance change. Others have found that it is possible to isolate mRNA from muscle, inject it into *X. laevis* oocytes, and obtain functional nicotinic AChR at the oocyte membrane, as revealed by patch clamp, voltage clamp,

and biochemical techniques (Miledi and Sumikawa, 1982; Miledi *et al.*, 1982; Sakmann *et al.*, 1985; White *et al.*, 1985). Does mRNA isolated from steroid-treated myotubes induce AChR with large conductances when injected into oocytes? If so, then it may be possible to use techniques of subtractive hybridization to isolate the specific messages involved in this response (Mather *et al.*, 1981; Davis *et al.*, 1984; Littman *et al.*, 1985). Clearly the resolution of the mechanisms of action of the steroids on excitable tissues will require a strongly multidisciplinary approach.

ACKNOWLEDGMENT

This work is supported by NIH Grant NS 12211.

REFERENCES

Agnati, L. F., Fuxe K., Kuonen, D., Blake, C. A., Andersson, K., Eneroth, P., Gustafsson, J. Å., Battistini, N., and Calza, L. (1981). Effects of estrogen and progesterone on central α-and β-adrenergic receptors in ovariectomized rats: Evidence for gonadal steroid receptor regulation of brain α- and β-adrenergic receptors. *In* "Steroid Hormone Regulation of the Brain" (K. Fuxe, J. Å. Gustafsson, and L. Wetterberg, eds.) pp. 237–252. Pergamon, Oxford.

Alderson, L. M., and Baum, M. J. (1981). Differential effects of gonadal steroids on dopamine metabolism in mesolimbic and nigrostriatal pathways of male rat-brain. *Brain Res.* **218**, 189–206.

Anderson, D. J., and Blobel, G. (1981). *In vitro* synthesis, glycosylation and membrane insertion of the four subunits of *Torpedo* acetylcholine receptor. *Proc. Natl. Acad. Sci. U.S.A.* **78**, 5598–5602.

Andreasen, T. J., Doerge, D. R., and McNamee, M. G. (1979). Effects of phospholipase A_2 on the binding and ion permeability control properties of the acetylcholine receptor. *Arch. Biochem. Biophys.* **194**, 468–480.

Anholt, R., Lindstrom, J., and Montal, M. (1984). The molecular basis of neurotransmission: Structure and function of the nicotine acetylcholine receptor. *In* "The Enzymes of Biological Membranes" (A. Martonosi, ed.), Vol. 3, pp. 355–401. Plenum, New York.

Arnold, A. P., and Gorski, R. A. (1984). Gonadal steroid induction of structural sex differences in the central nervous system. *Annu. Rev. Neurosci.* **7**, 413–442.

Arnold, A., Nottebohm, F., and Pfaff, D. (1976). Hormone concentrating cells in vocal control and other areas of the brain of the zebra finch (Poephila guttata). *J. Comp. Neurol.* **165**, 487–512.

Atkinson, R. M., Davis, B., Pratt, M. A., Sharpe, H. M., and Tomich, E. G. (1965). Action of some steroids on the central nervous system of the mouse. II. Pharmacology. *J. Med. Chem.* **8**, 426–432.

Bachrach, U. (1973). "Function of Naturally Occurring Polyamines." Academic Press, New York.

Barrack, E. R., and Coffey, D. S. (1983). The role of the nuclear matrix in steroid hormone action. *Biochem. Act. Horm.* **10**, 23–90.

Barrantes, F. J. (1983). Recent developments in the structure and function of the acetylcholine receptor. *Int. Rev. Neurobiol.* **24**, 259–341.

Bass, A., Gutmann, E., Hanzlikovà, V., Hàjek, I., and Syrovy, I. (1969). The effect of castration and denervation upon the contraction properties and metabolism of the levator ani muscle of the rat. *Physiol. Bohem.* **18**, 177–194.

Baulieu, E. E. (1983). Steroid membrane—adenylate cyclase interactions during *Xenopus laevis* oocyte meiosis reinitiation: A new mechanism of steroid hormone action. *Exp. Clin. Endocrinol.* **81a**,3–16.

Beach, F. A. (1948). "Hormones and Behavior." Harper, New York.

Berger, F. G., Szymanski, P., Read, E., and Watson, G. (1984). Androgen - regulated ornithine decarboxylase in RNA's of mouse kidney. *J. Biol. Chem.* **259**,7941–7946.

Biegon, A., and McEwen, B. S. (1982). Modulation by estradiol of serotonin receptors in brain. *J. Neurosci.* **2**,199–205.

Bischoff, R. (1979). Tissue culture studies on the origin of myogenic cells during muscle regeneration in the rat. *In* "Muscle Regeneration" (A. Mauro, ed.), pp. 13–29. Raven, New York.

Bleisch, W. V., and Harrelson, A. (1984). Castration affects single endplates in an androgen-sensitive muscle of the rat. *Soc. Neurosci. Abstr.* **10**, 927.

Bleisch, W. V., Harrelson, A. L., and Luine, E. N. (1982). Testosterone increases acetylcholine receptor number in the "levator ani" muscle of the rat. *J. Neurobiol.* **13**,153–161.

Bleisch, W., Luine, V. N., and Nottebohm, F. (1984). Modification of synapses in androgen sensitive muscle: I. Hormonal regulation of acetylcholine receptor number in the songbird syrinx. *J. Neurosci.* **4**,786–792.

Bloch, R. J. (1982). Sterol structure and membrane function. *CRC Crit. Rev. Biochem.* **17**,47–92.

Block, G. A., and Billiar, R. B. (1979). Effect of estradiol on the development of hypothalamic cholinergic receptors. *Biol. Reprod.* **20**: (Suppl. 1), 24A.

Breedlove, S. M. (1984). Steroid influences on the development and function of a neuromuscular system. *Prog. Brain Res.* **61**,147–170.

Brehm, P., Kullberg, R., and Moody-Corbett, F. (1984). Properties of non-junctional acetylcholine receptor channels on innervated muscle of *Xenopus laevis*. *J. Physiol. (London)* **350**,631–648.

Bridgman, P. C., and Nakajima, Y. (1981). Membrane lipid heterogeneity associated with acetylcholine receptor particle aggregates in Xenopus embryonic muscle cells. *Proc. Natl. Acad. Sci. U.S.A.* **78**,1278–1282.

Briley, M. S., and Changeux, J. P. (1978). Recovery of some functional properties of the detergent-extracted cholinergic receptor protein from *Torpedo marmorata* after reintegration into a membrane environment. *Eur. J. Biochem.* **84**,429–439.

Brockerhoff, H. (1982). Anesthetics may restructure the hydrogen belts of membranes. *Lipids* **17**,1001–1003.

Brown, D. D., and Gurdon, J. B. (1964). Absence of ribosomal RNA synthesis in the nucleate mutant of *Xenopus laevis*. *Proc. Natl. Acad. Sci. U.S.A.* **51**,139–146.

Caldarera, C. M., Moruzzi, M. S., Barbiroli, B., and Moruzzi, G. (1968). Spermine and spermidine of the prostate gland of orchiectomized rats and their effect on RNA polymerase activity. *Biochem. Biophys. Res. Commun.* **33**,266–271.

Calof, A. L., and Reichardt, L. F. (1984). Motoneurons purified by cell sorting respond to two distinct activities in myotube-conditioned medium. *Dev. Biol.* **106**,194–210.

Cardinali, D. P., and Gomez, E. (1977). Changes in hypothalamic noradrenaline, dopamine and serotonin uptake after oestradiol administration to rats. *J. Endocrinol.* **73**,181–182.

Carlson, J. C., Gruber, M. Y., and Thompson, J. E. (1983). A study of the interaction between progesterone and membrane lipids. *Endocrinology* **113**,190–194.

Chaia, N., Foy, M., and Teyler, T. J. (1983). The hamster hippocampal slice. II. Neuroendocrine modulation. *Behav. Neurosci.* **97**,839–843.

Chang, H. W., and Bok, E. (1979). Structural stabilization of isolated acetylcholine receptor: Specific interaction with phospholipids. *Biochemistry* **18**,172–179.

Changeux, J. P. (1981). The acetylcholine receptor: An "allosteric" membrane protein. *Harvey Lect.* **75**,85–254.

Clark, R., and Adams, P. (1981). ACh receptor channel populations in cultured Xenopus myocyte membranes are non-homogeneous. *Neurosci. Abstr.* **7**,838.

Clower, R. P., and DeVoogd, T. J. (1985). Anatomical correlates of laterality and endocrine state in a canary song control nucleus. *Soc. Neurosci. Abstr.* **11**,612.

Collins, W. F., III (1983). Organization of electrical coupling between frog lumbar motoneurons. *J. Neurophysiol.* **49**,730–744.

Colquhoun, D., and Hawkes, A. G. (1983). The principles of the stochastic interpretation of ion-channel mechanisms. *In* "Single Channel Recording" (B. Sakmann and E. Neher, eds.), pp. 135–175.Plenum, New York.

Colquhoun, D., and Sakmann, B. (1985). Fast events in single-channel currents activated by acetylcholine and its analogues at the frog muscle end-plate. *J. Physiol. (London)* **369**, 501–557.

Commins, D., and Yahr, P. (1982). Morphology of the gerbil preoptic area (MPOA) is sexually dimorphic and influenced by adult gonadal steroids. *Conf. Reprod. Behav. Abstr. Mich. State Univ.* p. 47.

Conti-Tronconi, B. M., and Raftery, M. A. (1982). The nicotinic acetylcholine receptor: correlation of molecular structure with functional properties. *Annu. Rev. Biochem.* **51**, 491–530.

Crowley, W. R., O'Donahue, T. L., Wachsricht, H., and Jacobowitz, D. M. (1978). Effects of estrogen and progesterone on plasma gonadotropins and on catecholamine levels and turnover in discrete regions of ovariectomized rats. *Brain Res.* **154**,345–357.

Dahlberg, E. (1982). Characterization of the cytosolic estrogen receptor in rat skeletal muscle. *Biochim. Biophys. Acta* **717**,65.

Dahlberg, E., Snochowski, M., and Gustafsson, J. A. (1980). Androgen and glucocorticoid receptors in rat skeletal muscle. *Acta Chem. Scand.(B)* **34**,141.

Davis, M. M., Cohen, D. I., Nielsen, E. A., Steinmetz, M., Paul, W. E., and Hood, L. (1984). Cell-type-specific cDNA probes and the murine I region: The localization and orientation of A_α^d. *Proc. Natl. Acad. Sci. U.S.A.* **81**,2194–2198.

Dionne, F. T., Lesage, R. L., Dubay, J. Y., and Tremblay, R. R. (1979). Estrogen binding proteins in rat skeletal and perineal muscles: *In vitro* and *in vivo* studies. *J. Steroid Biochem.* **11**,1073–1080.

Di Paolo, T., Carmichael, R., Labrii, F., and Raynaud, J. R. (1979). Effects of estrogens on the characteristics of ^{3}H spiroperidol and ^{3}H RU 24213 binding in rat anterior pituitary gland and brain. *Mol. Cell. Endocrinol.* **16**,99–104.

Di Paolo, T., Poget, P., and Labrii, F. (1981). Effects of chronic estradiol and haloperidol treatment on striatal dopamine receptors. *Eur. J. Pharmacol.* **73**,105–106.

Dohanich, G. P., Witcher, J. A., Weaver, D. E., and Clemens, L. G. (1982). Alteration of muscarinic binding in specific brain areas following estrogen treatment. *Brain Res.* **241**, 347–350.

Dreyer, E., Walther, C., and Peper, K. (1976). Junctional and extrajunctional acetylcholine receptors in normal and denervated frog muscle fibres: Noise analysis experiments with different agonists. *Pflügers Arch.* **366**,1–9.

Dubè, J. Y., Lesage, R., and Tremblay, R. R. (1976). Androgen and estrogen binding in rat skeletal and perineal muscles. *Can. J. Biochem.* **54**,50.

Dupont, A., Di Paolo, T., Cague, B., and Barden, N. (1981). Effect of chronic estrogen treatment on dopamine concentration and turnover in discrete brain nuclei of ovariectomized rats. *Neurosci. Lett.* **22**,69–74.

Eisenberg, E., and Gordan, G. S. (1950). The levator ani muscle of the rat as an index of myotrophic activity of steroidal hormones. *J. Pharmacol. Exp. Ther.* **99**,38–44.

Eisenfeld, A. J., and Axelrod, J. (1965). Selectivity of estrogen distribution in tissue. *J. Pharmacol. Exp. Ther.* **150**,469–475.

Endersby, C. A., and Wilson, C. A. (1974). The effect of ovarian steroids on the accumulation of ^{3}H-labelled monoamines by hypothalamic tissue in vitro. *Brain Res.* **73**, 321–331.

Erulkar, S. D., and Wetzel, D. M. (1986). 5-Alpha-Dihydrotestosterone alters the properties of acetylcholine-activated ion channels in androgen-concentrating myotubes. *Exp. Brain Res. Series* **14**, 220–228.

Erulkar, S. D., and Wetzel, D. M. (1987). 5α-dihydrotestosterone has both direct membrane and genomic effects on acetylcholine-activated channels. (Submitted).

Erulkar, S. D., Kelley, D. B., Jurman, M. E., Zemlan, F. P., Schneider, G. T., and Krieger, N. R. (1981). Modulation of the neural control of the clasp reflex in male *Xenopus laevis* by androgens: A multidisciplinary study. *Proc. Natl. Acad. Sci. U.S.A.* **78**, 5876–5880.

Everitt, B. J., Fuxe, K., Hokfelt, T., and Jonsson, G. (1975). Role of monoamines in the control of hormones of sexual receptivity in the female rat. *J. Comp. Physiol.* **89**, 556–572.

Farnsworth, W. E. (1977). Activities of the androgen-responsive receptor of the prostatic plasma membrane. *Invest. Urol.* **15**, 75–77.

Finidori-Lepicard, J., Schorderet-Slatkine, S., Hanoune, J., and Baulieu, E. E. (1981). Progesterone inhibits membrane-bound adenylate cyclase in *Xenopus laevis* oocytes. *Nature (London)* **292**, 255–257.

Fuxe, K., Andersson, K., Blake, C. A., Eneroth, P., Gustafsson, J. Å., and Agnati, L. F. (1981). Effects of estrogen and combined treatment with estrogen and progesterone on central dopamine, noradrenaline and adrenaline nerve terminal systems of the ovariectomized rat. Relationship of changes in amine turnover to changes in LH and prolactin secretion and in sexual behavior. *In* "Steroid Hormone Regulation of the Brain" (K. Fuxe, J. Å. Gustafsson and L. Wetterberg, eds.), pp. 73–92, Pergamon, Oxford.

Garfield, R. E., Rabideau, S., Challis, J. R. G., and Daniel, E. E. (1979). Ultrastructural basis for maintenance and termination of pregnancy. *Am. J. Obstet. Gynecol.* **133**, 308–315.

Garfield, R. E., Kannan, M. S., and Daniel, E. E. (1980). Gap junction formation in myometrium: Control by estrogens, progesterone, and prostaglandins. *Amer. J. Physiol.* **238**, C81–C89.

Goldstein, D. B. (1984). The effects of drugs on membrane fluidity. *Annu. Rev. Pharmacol. Toxicol.* **24**,43–64.

Gorski, J., Toft, D., Shyamala, G., Smith, D., and Notides, A. (1968). Hormone receptors: Studies on the interaction of estrogen with the uterus. *Recent Prog. Horm. Res.* **24**,45–80.

Greenberg, A., Nakajima, S., and Nakajima, Y. (1982). Single channel properties of newly inserted acetylcholine receptors in non-innervated embryonic muscle cells in culture. *Biophys. J.* **37**,18a.

Gurney, M. (1981). Hormonal control of cell form and number in the zebra finch song sytem. *J. Neurosci.* **1**,658–673.

Hamill, O. P., and Sakmann, B. (1981). Multiple conductance states of single acetylcholine receptor channels in embryonic muscle cells. *Nature (London)* **294**, 462–464.

Hamill, O. P., Marty, A., Neher, E., Sakmann, B., and Sigworth, F. J. (1981). Improved patch-clamp techniques for high resolution current recording from cells and cell-free membrane patches. *Pflügers Arch.* **391**,85–100.

Hamon, M., Goetz, C., Euvrard, L., Pasqualini, C., Le Dofniet, M., Kerdelhue, B., Cesselin, F., and Peillon, F. (1983). Biochemical and functional alterations of central GABA receptors during chronic estradiol treatment. *Brain Res.* **279**, 141–152.

Hanzlikova, V., and Gutmann, E. (1978). Effect of castration and testosterone adminstration on the neuromuscular junction in the levator ani muscle of the rat. *Cell Tissue Res.* **189**, 155–166.

Harder, D. R., and Coulson, B. (1979). Estrogen receptors and effects of estrogen on membrane electrical properties of coronary vascular smooth muscle. *J. Cell. Physiol.* **100**, 375–382.

Heap, R. B., Symons, A. M., and Watkins, J. C. (1970). Steroids and their interactions with phospholipids: Solubility, distribution coefficient and effect on potassium permeability of liposomes. *Biochim. Biophys. Acta* **218**,482–495.

Hernández-Pérez, O., Ballesteros, L. Ma., and Rosado, A. (1979). Binding of 17β-estradiol to the outer surface and nucleus of human spermatazoa. *Arch. Androl.* **3**, 23–29.

Heron, D. S., Shinitzky, M., Hershkowitz, M., and Samuel, D. (1980). Lipid fluidity modulates the binding of serotonin to mouse brain membranes. *Proc. Natl. Acad. Sci. U.S.A.* **77**, 7463–7467.

Hoffman, B. B., DeLean, A., Wood, C. L., Schocken, D. D., and Lefkowitz, R. J. (1979). Alpha adrenergic receptor subtypes: Quantitative assessment by ligand binding. *Life Sci.* **24**, 1739–1746.

Hruska, R. E., and Silbergeld, E. K. (1980a). Estrogen treatment enhances dopamine receptor sensitivity in the rat striatium. *Eur. J. Pharmacol.* **61**, 397–400.

Hruska, R. E., and Silbergeld, E. K. (1980b). Increased dopamine receptor sensitivity after estrogen treatment using the rat rotation model. *Science* **208**, 1466–1468.

Iverson, L. L., and Salt, P. J. (1970). Inhibition of catecholamine uptake by steroids in isolated rat heart. *Br. J. Pharmacol.* **40**, 528–530.

Janne, O. A., Kontula, K.K., Isomaa, U. U., Torkkeli, T. K., and Bardin, C. W. (1983). Androgen receptor-dependent regulation of ornithine decarboxylase gene expression in mouse kidney. *In* "Steroid Hormone Receptors: Structure and Function" (H. Eriksson and J. Å. Gustafsson, eds.). Elsevier, Amsterdam.

Jensen, E. V., and Jacobson, H. I. (1962). Basic guides to the mechanism of estrogen action. *Recent Prog. Hormo. Res.* **18**, 387–414.

Jensen, E. V., Suzuki, T., Kawashima, T., Stumpf, W. E., Jungblut, P. W., and DeSombre, E. R. (1968). A two-step mechanism for the interaction of estradiol with rat uterus. *Proc. Natl. Acad. Sci. U.S.A.* **59**, 632–638.

Johnson, A. E., Nock, B., McEwen, B. S., and Feder, H. H. (1985). Sex difference in α_1-noradrenergic receptor binding following estradiol treatment in the guinea pig assessed by quantitative autoradiography. *Neurosci. Abstr.* **11**, 772.

Jones, J. J., Chikaraishi, D. M., Harrington, C. A., McEwen, B. S., and Pfaff, D. W. (1985). Estradiol (E_2)-induced changes in rRNA levels in rat hypothalamic neurons detected by in situ hybridization. *J. Cell Biol.* **101**, 453a.

Jurman, M. E., Krieger, N. R., and Erulkar, S. D. (1982). Testosterone 5α-reductase in spinal cord of *Xenopus laevis*. *J. Neurochem.* **38**, 657–661.

Karlin, A. (1980). Molecular properties of nicotinic acetylcholine receptors. *In* "The Cell Surface and Neuronal Function" (C. W. Cotman, O. Poste, and O. L. Nicolson, eds.), pp. 191–260. Elsevier, Amsterdam.

Kelley, D. B. (1980). Auditory and vocal nuclei in the frog brain concentrate sex hormones. *Science* **207**, 553–555.

Kelley, D. B. (1981). Locations of androgen-concentrating cells in the brain of *Xenopus laevis*: Autoradiography with ^{3}H-dihydrotestosterone. *J. Comp. Neurol.* **199**, 221–231.

Kelley, D. B., and Pfaff, D. W. (1978). Generalizations from comparative studies on neuroanatomical and endocrine mechanisms of sexual behaviour. *In* "Biological Determinants of Sexual Behavior" (J. B. Hutchinson, ed.). Wiley, Chichester.

Kendall, D. A., and Narayana, K. (1978). Effect of oestradiol-17 beta on monoamine concentrations in the hypothalamus of anoestrous ewe. *J. Physiol. (London)* **282**, 44P–45P.

Kendall, D. A., and Tonge, S. R. (1977). Effects of testosterone and ethinyloestradiol on the synthesis and uptake of noradrenaline and 5-hydroxytryptamine in rat hindbrain: Evidence for a presynaptic regulation of monoamine synthesis? *Br. J. Pharmacol.* **60**, 309–310.

Kilian, P L., Dunlap, C. R., Mueller, P., Schell, M. A., Huganir, R. L. and Racker, E. (1980). Reconstitution of acetylcholine receptor from *Torpedo californica* with highly purified phospholipids: Effect of α-tocopherol, phylloquinone, and other terpenoid quinones. *Biochem. Biophys. Res. Commun.* **93**, 409–414.

King, W. J., and Greene, G. L. (1984). Monoclonal antibodies localize estrogen receptor in the nuclei of target cells. *Nature (London)* 307,745-747.

Kizer, J. S., Palkovitz, M., Zivin, J., Brownstein, M., Saavedra, J. M., and Kopin, J. J. (1974). The effect of endocrinological manipulations on tyrosine hydroxylase and dopamine β-hydroxylase activities in individual hypothalamic nuclei of the adult male rat. *Endocrinology* 95,799-812.

Koch, C., Poggio, T., and Torre, V. (1982). Retinal ganglion cells: A functional interpretation of dendritic morphology. *Philos. Trans. R. Soc. London Ser. B* 198,227-264.

Kontula, K. K., Torkkeli, T. K., Bardin, C. W., and Janne, O. A. (1984). Androgen induction of ornithine decarboxylase mRNA in mouse kidney as studied by complementary DNA. *Proc. Natl. Acad. Sci. U.S.A.* 81,731-735.

Kow, L. M., and Pfaff, D. W. (1984). Suprachiasmatic neurons in tissue slices from ovariectomized rats: Electrophysiological and neuropharmacological characterization and the effects of estrogen treatment. *Brain Res.* 16,275-286.

Krieg, M., Szalay, R., and Voigt, K. D. (1974). Binding and metabolism of testosterone and of 5α-dihydrotestosterone in bulbocavernosus/levator ani (Bcla) of male rats: In vivo and in vitro studies. *J. Steroid Biochem.* 5,453-459.

Krieg, M., Smith, K., and Elvers, B. (1980). Androgen receptor translocation from cytosol of rat heart muscle bulbocavernosus/levator ani muscle and prostate into heart muscle nuclei. *J. Steroid Biochem.* 13,577-587.

Land, B. R., Harris, W. V., Salpeter, E. E., and Salpeter, M. M. (1984). Diffusion and binding constants for acetylcholine derived from the falling phase of miniature endplate currents. *Proc. Natl. Acad. Sci. U.S.A.* 81, 1594-1598.

Lasater, E. M., and Dowling, J. E. (1985). Dopamine decreases conductance of the electrical junctions between cultured retinal horizontal cells. *Proc. Natl. Acad. Sci. U.S.A.* 82,3025-3029.

Lawrence, D. K., and Gill, E. W. (1974). Structurally specific effects of some steroid anesthetics on spin-labeled liposomes. *Mol. Pharmacol.* 11,280-286.

Lechleiter, J., and Gruener, R. (1984). Halothane shortens acetylcholine receptor channel kinetics without affecting conductance. *Proc. Natl. Acad. Sci. U.S.A.* 81,2929-2933.

Lee, S. H. (1978). Cytochemical study of estrogen receptor in human mammary cancer. *Am. J. Clin. Pathol.* 70, 197-203.

Liao, S., and Lin, A. H. (1967). Prostatic nuclear chromatin: An effect of testosterone on the synthesis of ribonucleic acid rich in citidylyl (3 ',5 ') guanosine. *Proc. Natl. Acad. Sci. U.S.A.* 57,379-386.

Liao, S., Barton, R. W., and Lin, A. H. (1966). Differential synthesis of ribonucleic acid in prostatic nuclei: Evidence for selective gene transcription induced by androgens. *Proc. Natl. Acad. Sci. U.S.A.* 55, 1953-1600.

Lin A. L., McGill H.C., Jr., and Shain, S. A. (1981). Hormone receptors of the baboon cardio-vascular system. Biochemical characterization of aortic cytoplasmic androgen receptors. *Arteriosclerosis* 1, 257-264.

Littman, D. R., Thomas, Y., Maddon, P. J., Chess, L., and Axel, R. (1985). The isolation and sequence of the gene encoding T8: A molecule defining functional classes of T lymphocytes. *Cell* 40, 237-246.

Luine, V. N., and McEwen, B. S. (1977). Effect of oestradiol on turnover of type A monoamine oxidase in brain. *J. Neurochem.* 28, 1221-1227.

Luine, V., Park, D., Joh, T., Reis, D., and McEwen, B. (1980a). Immunochemical demonstration of increased choline acetyltransferase concentration in rat preoptic area after estradiol administration. *Brain Res.* 191, 273-277.

Luine, V., Nottebohm, F., Harding, C., and McEwen, B. S. (1980b). Androgen affects cholinergic enzymes in syringeal motor neurons and muscle. *Brain Res.* 192, 89-107.

McEwen, B. S., and Parsons, B. (1982). Gonadal steroid action on the brain: Neurochemistry and neuropharmacology. *Annu. Rev. Pharmacol. Toxicol.* **22,**558-598.

McEwen, B. S., Davis, P. G., Parsons, B., and Pfaff, D. W. (1979). The brain as a target for steroid hormone action. *Annu. Rev. Neurosci.* **2,**65-112.

McEwen, B. S., Biegon, A., Rainbow, T. C., Paden, C., Snyder, L., and DeGraff, V. (1981). The interaction of estrogens with intracellular receptors and with putative neurotransmitter receptors: Implications for the mechanism of activation of sexual behavior and ovulation. *In* "Steroid Hormone Regulation of the Brain" (K. Fuxe, ed.), pp. 15-30. Pergamon, London.

McGill, H. C., Anselmo, V. C., Buchanan, J. M., and Sheridan, P. J. (1980). The heart is a target organ for androgen. *Science* **207,**775-777.

McGinnis, M. Y., Gordon, J. H., and Gorski, R. A. (1980). Time course and the localization of the effects of estrogens on glutamic acid decarboxylase activity. *J. Neurochem.* **34,** 785-792.

Mainwaring, W. I. P., and Derry, N. S. (1983). Enhanced transcription of rRNA genes by purified androgen receptor complexes *in vitro. J. Steroid Biochem.* **19,**101-108.

Marshall, J. M. (1981). Effects of ovarian steroids and pregnancy on adrenergic nerves of uterus and oviduct. *Amer. J. Physiol.* **240,**C165-C174.

Mather, E. L., Alt, F. W., Bothwell, A. L. M., Baltimore, D., and Koshland, M. E. (1981). Expression of J chain RNA in cell lines representing different stages of B lymphocyte differentiation. *Cell* **23,**369-378.

Max, S. R., Mufti, S., and Carlson, B. M. (1981). Cytosolic androgen receptor in regenerating rat levator ani muscle. *Biochem. J.* **200,**77-82.

Merlie, J. P., Sebbane, R., Gardner, S., Olson, E., and Lindstrom, J. (1983). The regulation of acetylcholine receptor expression in mammalian muscle. *Cold Spring Harbor Symp. Quant. Biol.* **48,**135-146.

Michel, G., and Baulieu, E. E. (1975). Androgen receptor in skeletal muscle. *J. Endocrinol.* **65,**31P.

Michel, G., and Baulieu, E. E. (1980). Androgen receptor in rat skeletal muscle: Characterization and physiological variations. *Endocrinology* **107,**2088-2098.

Miledi, R., and Sumikawa, K. (1982). Synthesis of cat muscle acetycholine receptors by Xenopus oocytes. *Biomed. Res.* **3,**390-399.

Miledi, R. Parker, I., and Sumikawa, K. (1982). Properties of acetylcholine receptors translated by cat muscle RNA in Xenopus oocytes. *EMBO J.* **1,**1307-1312.

Miller, M. M., Billiar, R. B., and Silver, J. (1982a). Effect of gonadal steroids on [^{125}I] α-bungarotoxin labeling of the suprachiasmatic nucleus of the rat hypothalamus. *Soc. Neurosci. Abstr.* **8,**336.

Miller, M. M., Silver, J., and Billiar, R. B. (1982b). Effects of ovariectomy on the bindings of [^{125}I]α-bungarotoxin (2.2 and 3.3) to the suprachiasmatic nucleus of the hypothalamus: An *in vivo* autoradiographic analysis. *Brain Res.* **247,**355-364.

Moawad, A. H., River, L. P., and Lin, C. C. (1985). Estrogen increases beta adrenergic binding in the preterm fetal rabbit lung. *Am. J. Obstet. Gynecol.* **151,**514-519.

Morley, B. J., Rodriguez-Sierra, J. F., and Claugh, R. W. (1983). Increase in hypothalamic nicotinic acetylcholine receptors in prepuberal female rats administered estrogen. *Brain Res.* **278,**262-265.

Morrell, J., and Pfaff, D. W. (1982). Characterization of estrogen-concentrating hypothalamic neurons by their axonal projections. *Science* **217,**1273-1276.

Morrell, J. I., Kelley, D. B., and Pfaff, D. W. (1975). Sex steroid binding in the brains of vertebrates: Studies with light microscope autoradiography. *In* "Brain Endocrine Interactions: II. The Ventricular System" (K. Knigge, D. S Scott, K. Kobayshi, and S. Ishi, eds.), pp.230-256. Karger, Basel.

Moss, R. L., and Dudley, C. A. (1984). Molecular aspects of the interaction between estrogen and the membrane excitability of hypothalamic nerve cells. *Prog. Brain Res.* **61**,3–22.

Mueller, P., and Rudin, D. O. (1963). Induced excitability in reconstituted cell membrane structure. *J. Theor. Biol.* **4**,268–280.

Mueller, P., and Rudin, D. O. (1967). Action potential phenomena in experimental bimolecular lipid membranes. *Nature (London)* **213**,603–604.

Mueller, P., and Rudin, D. O. (1968). Action potentials in bilayers. *J. Theor. Biol.* **18**, 222–258.

Mueller, P., Rudin, D. O., Tien, H. T., and Wescott, W. C. (1962a). Reconstruction of cell membrane structure *in vitro* and its transformation into an excitable system. *Nature (London)* **194**,979–980.

Mueller, P., Rudin, D. O., Tien, H. T., and Wescott, W. C. (1962b). Reconstruction of excitable cell membrane structure *in vitro*. *Circulation* **26**,1167–1170.

Mueller, P., Rudin, D. O., Tien, H. T., and Wescott, W. C. (1964). Formation and properties of bimolecular lipid membranes. *Recent Prog. Surf. Sci.* **1**,379–393.

Nakajima, Y., and Bridgman, P. (1981). Absence of filipin–sterol complexes from the membrane of active zones and acetylcholine receptor aggregates at frog neuromuscular junctions. *J. Cell Biol.* **88**,453–458.

Neher, E., and Sakmann, B. (1976). Noise analysis of drug-induced voltage clamp currents in denervated frog muscle fibres. *J. Physiol. (London)* **258**,705–729.

Neher, E., and Steinbach, J. H. (1978). Local anaesthetics transiently block currents through single acetylcholine-receptor channels. *J. Physiol. (London)* **277**,153–176.

Nicoletti, F., and Meek, J. L. (1985). Estradiol benzoate decreases nigral GABAergic activity in male rats. *Brain Res.* **332**,179–183.

O'Connor, L. H., Nock, B., and McEwen, B. S. (1985). Quantitative autoradiography of GABA receptors in rat forebrain: Receptor distribution and the effects of estradiol. *Neurosci. Abstr.* **11**,772.

Ogden, D. C., Siegelbaum, S. A., and Colquhoun, D. (1981). Block of acetylcholine-activated ion channels by an uncharged local anaesthetic. *Nature (London)* **289**,596–598.

Okun, L. M. (1981). Identification and isolation *in vitro* of neurons marked *in situ* by retrograde transport. *In* "New Approaches in Developmental Neurobiology" (D. Gottliels, organizer), pp. 109–121. Society For Neuroscience, Bethesda, Maryland.

O'Malley, B. W., Tsai, M. J., and Schrader, W. T. (1983). Structural considerations for the action of steroid hormones in eucaryotic cells. *In* "Steroid Hormone Receptors: Structure and Function" (H. Erikkson and J. A. Gustafsson, eds.), pp.307–327. Elsevier, Amsterdam.

Owman, C., and Sjoberg, N. O. (1973). Effect of pregnancy and sex hormones on the transmitter level in uterine short adrenergic neurons. *In* "Frontiers in Catecholamine Research" (E. Usdin, and S. H. Snyder, eds.), pp. 795–801. Pergamon, Oxford.

Pak, C. Y. C. and Gershfeld, N. L. (1967). Steroid hormones and monolayers *Nature (London)* **214**,888–889.

Paul, S. M., Axelrod, J., Saavedra, J. M., and Skolnick, P. (1979). Estrogen-induced efflux of endogenous catecholamines from the hypothalamus *in vitro*. *Brain Res.* **178**,499–505.

Peets, E. A., Henson, M. F., and Neri, R. (1974). On the mechanism of the antiandrogenic action of flutamide (α,α,α-trifluoro-2-methyl-4 '-nitro-m-propionotoluidide) in the rat. *Endocrinology* **94**,532–540.

Peper, K., Bradley, R. J., and Dreyer, I. (1982). The acetylcholine receptor at the neuromuscular junction. *Physiol. Rev.* **62**,1271–1340.

Persson, L., and Rosengren, E. (1983). Polyamine metabolism in muscles of mice and rats. *Acta Physiol. Scand.* **117**,457–460.

Pfaff, D. W. (1980). "Estrogens and Brain Function: Neural Analysis of a Hormone-controlled Mammalian Reproductive Behavior." Springer-Verlag, New York.

Pfaff, D. W. (1983). Impact of estrogens on hypothalamic nerve cells: Ultrastructural, chemical, and electrical effects. *Recent Prog. Horm. Res.* **39**,127–179.

Pfaff, D. W., and Keiner, M. (1973). Atlas of estradiol-concentrating cells in the central nervous system of the female rat. *J. Comp. Neurol.* **151**,121–158.

Pfaff, D. W., and McEwen, B. S. (1983). Actions of estrogens and progestins on nerve cells. *Science* **219**,808–814.

Pfaff, D., and Modianos, D. (1985). Neural mechanisms of female reproductive behavior. *Handb. Behav. Neurobiol.* **7**,423–493.

Pietras, R. J., and Szego, C. M. (1977). Specific binding-sites for oestrogen at the outer surfaces of isolated endometrial cells. *Nature (London)* **265**,69–72.

Pietras, R. J., and Szego, C. M. (1979a). Estrogen receptors in uterine plasma membrane. *J. Steroid Biochem.* **11**,1471–1483.

Pietras, R. J., and Szego, C. M. (1979b). Metabolic and proliferative responses to estrogen by hepatocytes selected for plasma membrane binding-sites specific for estradiol-17β. *J. Cell. Physiol.* **91**,145–160.

Pietras, R. J., and Szego, C. M. (1980). Partial purification and characterization of oestrogen receptors in subfractions of hepatocyte plasma membranes. *Biochem. J.* **191**,743–760.

Popot, J. L., and Changeux, J. P. (1984). Nicotinic receptor of acetylcholine: Structure of an oligomeric integral membrane protein. *Physiol. Rev.* **64**,1162–1239.

Portaleone, P., Genazzani, E., Pagnini, G., Crispino, A., and DiCarlo, F. (1980). Interaction of estradiol and 2-hydroxy-estradiol with histamine receptors at the hypothalamic level. *Brain Res.* **187**,216–220.

Pumplin, D. W., and Bloch, R. J. (1983). Lipid domains of acetylcholine receptor clusters detected with saponin and filipin. *J. Cell Biol.* **97**, 1043–1054.

Rainbow, T. C., DeGraff, V., Luine, V. N., and McEwen, B. S. (1980). Estradiol-β increases the number of muscarinic receptors in hypothalamic nuclei. *Brain Res.* **198**,239–243.

Rall, W. (1977). Core conductor theory and cable properties of neurons. Handb. Physiol. p. 39–97.

Ringold, G. M. (1985). Steroid hormone regulation of gene expression. *Annu. Rev. Pharmacol. Toxicol.* **25**,529–566.

Roberts, J. M., Insel, P. A., Goldfien, R. D., and Goldfien, A. (1977). Alpha Adrenoreceptors but not β-adrenoreceptors increase in rabbit uterus with oestrogen. *Nature (London)* **270**,624–625.

Rubinstein, N. A., Erulkar, S. D., and Schneider, G. T. (1983). Sexual dimorphism in the fibers of a "clasp" muscle of *Xenopus laevis. Exp. Neurol.* **82**,424–431.

Saartok, T., Dahlberg, E., and Gustafsson, J. Å. (1984). Relative binding affinity of anabolic androgenic steroids: Comparison of the binding to the androgen receptors in skeletal muscle and in prostate as well as sex-hormone binding globulin. *Endocrinology* **114**,2100–2106.

Sakmann, B., Methfessel, C., Mishina, M., Takahashi, T., Takai, T., Kurasaki, M., Fukuda, K., and Numa, S. (1985). Role of acetylcholine receptor subunit in gating of the channel. *Nature (London)* **318**,538–543.

Salpeter, M. M., and Loring, R. H. (1985). Nicotine acetylcholine receptors in vertebrate muscle: Properties, distribution and neural control. *Prog. Neurobiol.* **25**,297–325.

Sar, M., and Stumpf, W. E. (1977). Distribution of androgen target cells in rat forebrain and pituitary after ^{3}H-dihydrotestosterone administration. *J. Steroid Biochem* **8**,1131–1135.

Schneider, G., Rubinstein, N., Zemlan, F. P., and Erulkar, S. D. (1980). Hormonal modulation of central and peripheral components of "fast" and "slow" skeletal nerve–muscle systems in *Xenopus laevis. Neurosci. Abstr.* **6**,560.

Segil, N., Silverman, L., Kelley, D. B., and Rainbow, T. C. (1983). Androgen binding in the laryngeal muscle of *Xenopus laevis*: Sex differences and hormonal regulation. *Soc. Neurosci. Abstr.* **9**,1093.

Segil, N., Silverman, L., and Kelley, D. B. (1987). Androgen receptor recognition in a sexually dimorphic muscle. *Gen. Comp. Endocrinol.* (in press).

Selye, H. (1941). Anesthetic effect of steroid hormones. *Proc. Soc. Exp. Biol.* **46**,116–121.

Simpkins, J. W., Kalra, S. P., and Kalra, P. S. (1983). Variable effects of testosterone on dopamine activity in several microdissected regions in the preoptic area and medial basal hypothalamus. *Endocrinology* **112**,665–669.

Snochowski, M., Dahlberg, E., and Gustafssen, J. Å. (1980). Characterization and quantification of the androgen and glucocorticoid receptors in cytosol from rat skeletal muscle. *Eur. J. Biochem.* **111**,603–616.

Sotelo, L., and Grofova, I. (1976). Ultrastructural features of the spinal cord. *In* "Frog Neurobiology" (R. Llinas and W. Precht, eds.). Springer-Verlag, Berlin.

Souccar, C., Lapa, A. J., and Ribiro do Valle, J. (1982). The influence of testosterone on neuromuscular transmission in hormone sensitive mammalian skeletal muscles. *Muscle Nerve* **5**,232–237.

Spelsberg, T. C., Littlefield, B. A., Seelke, R., Dani, G. M., Toyoda, H., Boyd-Leinen, P., Thrall, C., and Kon, O. L. (1983). Role of specific chromosomal proteins and DNA sequences in the nuclear binding sites for steroid receptors. *Recent Prog. Horm. Res.* **39**,463–517.

Strichartz, G. R., Oxford, G. S., and Ramon, F. (1980). Effects of the dipolar form of phloretin on potassium conductance in squid giant axons. *Biophys. J.* **31**,229–246.

Stumpf, W. E., Sar, M., and Aumüler, G. (1977). The heart: A target organ for estradiol. *Science* **196**,319–321.

Szego, C. M., and Pietras, R. J. (1981). Membrane recognition and effector sites in steroid hormone action. *Biochem. Act. Horm.* **VIII**,307–463.

Tabor, C. W., and Tabor, H. (1984). Polyamines. *Annu. Rev. Biochem.* **53**,749–790.

Teyler, T. J., Vardaris, R. M., Lewis, D., and Rawitch, A. B. (1980). Gonadal steroids: Effects on excitability of hippocampal pyramidal cells. *Science* **209**,1017–1019.

Thibert, P. (1986). Androgen sensitivity of skeletal muscle: nondependence of the motor nerve in the frog forearm. *Exp. Neurol.* **91**, 559–570.

Thorbert, G. (1979). Regional changes in structure and function of adrenergic nerves in guinea-pig uterus during pregnancy. *Acta Abstr. Gynecol. Suppl. 2* **79**,5–32.

Tobias, M., and Kelley, D. B. (1985). Physiological characterization of the sexually dimorphic larynx in *Xenopus laevis. Neurosci. Abstr.* **11**,496.

Toran-Allerand, C. D. (1980). Sex steroids and the development of the newborn mouse hypothalamus and preoptic area *in vitro.* Morphological correlates and hormonal specificity. *Brain Res.* **189**,413–427.

Tremblay, R. R., Dubè, J. Y., Ho-Kim, M. A., and Lesage, R. (1977). Determination of rat muscles androgen-receptor complexes with methyltrienclone. *Steroids* **29**,185–196.

Tucek, S., Kostirova, D., and Gutmann, E. (1976). Effects of castration, testosterone and immobilization on the activities of choline acetyltransferase and cholinesterase in rat limb muscles. *J. Neurol. Sci.* **27**,363–372.

Vacas, M. I., and Cardinali, D. P. (1980). Effect of estradiol on α- and β-adrenoceptor density in medial basal hypothalamus cerebral cortex and pineal gland of ovariectomized rats. *Neurosci. Lett.* **17**,73–77.

Van de Kar, L., Levine, J., and VanOrden, L. S. (1978). Serotonin in hypothalamic nuclei: Increased content after castration of male rats. *Neuroendocrinology* **27**,186–192.

Venable, J. H. (1966a). Constant cell populations in normal testosterone-deprived, and testosterone-stimulated levator ani muscles. *Am. J. Anat.* **119**,263–270.

Venable, J. H. (1966b). Morphology of the cells of normal, testosterone-deprived and testosterone-stimulated levator ani muscles. *Am. J. Anat.* **119**,271–302.

Vyskocil, F., and Gutmann, E. (1977). Electrophysiological and contractile properties of the levator ani muscle after castration and testosterone administration. *Pflügers Arch.* **368**,105–109.

Wagner, H. R., Crutcher, K. A., and Davis, J. N. (1979). Chronic estrogen treatment decreases β-adrenergic responses in rat cerebral cortex. *Brain Res.* **171**,147–151.

Wainman, P., and Shipounoff, G. L. (1941). The effects of castration and testosterone propionate on the striated perineal musculature in the rat. *Endocrinology* **29**,975–978.

Wallis, C. J., and Suttge, W. G. (1980). Influence of oestrogen and progesterone on glutamic acid decarboxylase activity in discrete regions of rat brain. *J. Neurochem.* **34**,609–613.

Welshons, W. V., Lieberman, M. E., and Gorski, J. (1984). Nuclear localization of unoccupied oestrogen receptors. *Nature (London)* **307**,747–749.

Welshons, W. V., Krummel, B. M., and Gorski, J. (1985). Nuclear localization of unoccupied receptors for glucocorticoids, estrogens, and progesterone in GH_3 cells. *Endocrinology* **117**,2140–2147.

Wetzel, D. M., Haerter, U. L., and Kelley, D. B. (1986). A proposed neural pathway for vocalization in South African clawed Frogs, *Xenopus laevis*. *J. comp. Physiol. A* **157**,749–761.

White, M., Mayne, K. M., Lester, H. A., and Davidson, N. (1985). Mouse–torpedo hybrid acetylcholine receptors: Functional homology does not equal sequence homology. *Proc. Natl. Acad. Sci. U.S.A.* **82**,4852–4856.

Wiesel, F. A., Fuxe, K., Hokfelt, T., and Agnati, L. E. (1978). Studies on dopamine turnover in ovariectomized or hypophysectomized female rats. Effects of 17β-estradiol benzoate, ethynodiol diacetate and ovine prolactin. *Brain. Res.* **148**,399–411.

Wilkinson, M., Herdon, H., Pearce, M., and Wilson, C. (1979). Radioligand binding studies on hypothalamic noradrenergic receptors during the estrous cycle or after steroid injection in ovariectomized rats. *Brain Res.* **168**,652–655.

Wilkinson, M., Herdon, H., and Wilson, C. A. (1981). Gonadal steroid modification of adrenergic and opiate receptor binding in the central nervous system. *In* "Steroid Hormone Regulation of the Brain" (K. Fuxe, J. Å. Gustafssen, and W. Wetterburg, ed.), pp.253–263. Permagon, Oxford.

Williams-Ashman, H. G., and Lockwood, D. H. (1970). Role of polyamines in reproductive physiology and sex hormone action. *Ann. N.Y. Acad. Sci.* **171**,882–894.

Wirz-Justice, A., Hackmann, E., and Lichtsteiner, M. (1974). The effect of oestradiol dipropionate on monoamine uptake in rat brain. *J. Neurochem.* **22**,187–189.

Yamamoto, K. R. (1983). On steroid receptor regulation of gene expression and the evolution of hormone-controlled gene networks. *In* "Steroid Hormone Receptors: Structure and Function. Nobel Synposium 57" (H. Eriksson, J. Å. Gustafsson, and B. Hogberg, eds.). Elsevier, Amsterdam.

Yamamoto, K. R. (1985a). Steroid receptor regulated transcription of specific genes and gene networks. *Annu. Rev. Gene.* **19**,209–252.

Yamamoto, K. R. (1985b). Hormone-dependent transcriptional enhancement and its implications for mechanisms of multifactor gene regulation. *In* "Molecular Developmental Biology: Expressing Foreign Genes" (L. Bogorad and G. Adelman, eds.). Liss, New York (in press).

Young, W. L. (1961). The hormones and mating behavior. *In* "Sex and Internal Secretions" (W. L. Young, ed.), pp.1173–1239. Williams & Wilkins, Baltimore.

Zigmond, R. E. (1975). Binding, metabolism, and action of steroid hormones in the central nervous system *In Handbook Psychopharmacol.* (L. L. Iversen, S. D. Iversen, and S. H. Snyder, eds.). pp. 239–328. Plenum, New York.

Estradiol-Regulated Neuronal Plasticity

CHARLES V. MOBBS AND DONALD W. PFAFF

The Rockefeller University
New York, New York 10021

I. Introduction
II. Lordosis
III. Sexual Dimorphisms
IV. Reproductive Senescence
V. E2-Induced Proteins
References

Neuronal functions regulated by estradiol provide an excellent model system for studying the regulation of functional and anatomical connectivity over three temporal domains: (1) efficacy of neuronal connections (estradiol-dependent lordosis in adults), (2) formation of neuronal connections [estradiol-dependent sexual dimorphism in the central nervous system (CNS) in neonates], and (3) degeneration of neuronal connections (estradiol-dependent reproductive senescence in old females). The advantages of this model system include easy experimental access to several levels of neuronal plasticity, ease of manipulation of the system, ease of measurement of functional outputs, and extensive information on the cellular, biochemical, and molecular effects of estradiol. Since much is now known of the anatomy and physiology of these estradiol-dependent processes, the biochemistry of these processes can be addressed. For example we have discovered an estradiol-induced protein synthesized in and transported to brain areas that control lordosis. By studying the induction of specific proteins by estradiol, we hope to examine the biochemical mechanisms by which estradiol alters neuronal connectivity. These systems offer the possibility of discovering common biochemical mechansisms that regulate synaptic facilitation, development, and aging.

191

I. INTRODUCTION

The nervous system may be viewed as functioning by selecting and regulating connections between neurons (Edelman and Mountcastle, 1978). For example, the modification of behavior by experience may involve strengthening weak synaptic connections, a form of functional plasticity. An extreme form of this notion is that all potential memories are coded in neuronal circuits regardless of experience, and that experience serves only to select the appropriate neuronal associations by selectively strengthening (or suppressing) synaptic connections. Since at any given moment an organism can exhibit only a very small component of its potential behavioral repertoire, the efferents of circuits that give rise to competing behaviors must be functionally weakened and/or the efferents from the dominating behavior must be strengthened. Development of the nervous system also involves selecting appropriate anatomical connections and eliminating inappropriate connections (Mark, 1974), and so involves the regulation of anatomical plasticity. The regulation of neuronal connectivity is also of clinical importance, because impaired connectivity may contribute to disorders of the nervous system and aging (Cotman, 1986).

What governs functional and anatomical connectivity within the nervous system? We are addressing this problem using as a model system estradiol-regulated neuronal functions, in particular the female rodent mating behavior lordosis. A main attractive feature of this model system is that at least three level of neuronal plasticity are experimentally accessible. First, estradiol can make functional an otherwise nonfunctional neuronal connection; in the presence of estradiol, flank and perineal stimulation elicit lordosis, but in the absence of estradiol this response is completely absent (Pfaff, 1980). Second, estradiol regulates the formation of particular anatomical connections during development; transient neonatal exposure to estradiol causes permanent sexual dimorphisms in certain neuronal circuits and robust sex differences in the regulation of a number of neuroendocrine functions, including lordosis (Gorski, 1986). Third, estradiol causes morphological changes in the brain and functional impairments associated with reproductive senescence (Finch et al., 1984); estradiol may therefore cause selective loss of functional connectivity. We know of no other model of the nervous system in which these levels of neuronal plasticity can be manipulated by a single agent. An appeal of this model is therefore that explanations of each level of plasticity may inform explanations of other levels. It may even be possible to discover common mechanisms regulating the different levels of plasticity.

Estradiol-regulated neuronal functions as model systems have other advantages as well. First, estradiol levels are easily manipulated over several orders of magnitude, from well below to well above physiological levels, without adverse side effects. Although estradiol modifies neuronal function, it is not produced in the nervous sytem, but instead is normally produced by the ovaries and arrives in the brain via plasma (although in some cases neurons can produce estradiol by metabolizing blood-borne testosterone, discussed below). Ovariectomy is a simple procedure and estradiol is easily replaced by injection or implant. Because the reproductive system is relatively insulated from other systems necessary for homeostatic maintenance, these manipulations have little impact on the health of the animal or on neuronal functions not directly affected by estradiol. The ease and simplicity of these manipulations contrast with the difficulty and complex physiological results of manipulating the levels of other neurotransmitters and hormones that modify neuronal functions.

Another advantage of the model is that many effects of estradiol are qualitative and extremely easy to measure. Lordosis is almost completely absent in females without estradiol and in males with or without estradiol (Pfaff, 1980). In the presence of estradiol, the lordotic response to flank stimulation is robust, is quantitatively correlated with dose and duration of exposure to estradiol, and can be measured reliably, repeatedly, and quickly (about 5 seconds/measurement). Estradiol also causes qualitative changes in courting and maternal behavior. Biochemical (e.g., progesterone-receptor induction) and hormonal (e.g., secretion of luteinizing hormone) effects of estradiol are similarly robust, although not as easy to measure. Uterine weight and vaginal epithelial cornification are easy to measure and are highly correlated with plasma estradiol levels (Mobbs et al., 1985b).

The utility of estradiol-regulated functions as models for other biological problems stems in part from the relatively detailed understanding of the cellular, biochemical, and molecular mechanisms by which estradiol acts. Estradiol generally acts by increasing the transcription of specific genes, leading to the synthesis of specific protein products (O'Malley and Means, 1974). This increased transcription is mediated by a receptor, which has recently been cloned (Walter et al., 1985), whose binding to chromatin is altered by estradiol (Jensen et al., 1982). The regulation of ovalbumin and vitellogenin by estradiol has become a powerful model for studying the regulation of eukaryotic transcription. For example, estradiol induces DNase hypersensitive sites on chromatin and DNA demethylation when it activates vitellogenin (Burch and Weintraub, 1983) and ovalbumin (Kaye et al., 1986) transcription. Because estradiol appears to act through similar mechanisms in many different tissues, these data should prove useful in understanding how estradiol regulates neuronal functions.

Below we will describe in more detail the regulation by estradiol of lordosis, sexual dimorphism in the brain, and reproductive senescence. We will then discuss how the estradiol-induced proteins in the brain may be used to study the biochemical regulation of neuronal connectivity.

II. LORDOSIS

The lordosis response in female rats is directly elicited by flank and perineal stimulation, but prior exposure to estradiol is necessary before the stimulus and the response are coupled. The behavior consists of a stereotypic dorsiflexion of the vertebral column, accompanied by elevation of the head and extension of the hind legs (Pfaff and Lewis, 1974), normally elicited by the male rat during mating or proestrus. The response can also be elicited experimentally by manual stroking and pressure of the same regions, and can be quantified using either males or manual stroking as the stimulus (Pfaff, 1980). The degree of lordosis response is quantitatively related to dose and duration of exposure to estradiol (Pfaff, 1980), and to exposure to other pharmacological manipulations (Harlan *et al.*, 1984). In ovariectomized rats, plasma estradiol must be elevated by injection or implants at least 12–24 hours before the lordosis response can be demonstrated (Parsons *et al.*, 1982). The minimum exposure to estradiol necessary to elicit lordosis involves two 1-hour exposures to estradiol implants spaced 4–13 hours apart, followed by an injection of progesterone (Parsons *et al.*, 1982).

The anatomy of the minimal neural circuitry regulating lordosis suggests that estradiol acts directly on neurons in the ventral medial hypothalamus (VMH), whose facilitatory effects are registered in the midbrain central gray (MCG). Systemic injection of tritiated estradiol followed by autoradiographic analysis demonstrates specific uptake in neuronal groups associated with the reproductive system (Pfaff and Keiner, 1973); these regions contain high-affinity estradiol receptors, as demonstrated by biochemical competitive binding studies of microdissected brain regions (McEwen *et al.*, 1972). One major locus for estradiol uptake is in and around the lateral portion of the ventral medial nucleus of the hypothalamus (VMH) (Pfaff and Keiner, 1973). Electrolytic lesions of the VMH, and knife cuts isolating the VMH from the midbrain, abolish the lordosis response (Pfaff, 1980). Lesions of other regions that heavily concentrate estradiol have little effect, or, in the preoptic area, actually facilitate lordosis. Lesions of the central midbrain gray (MCG), the midbrain reticular formation, the medullary reticular formation, and isolation of the spinal cord from the brain, all block lordosis (Pfaff, 1980); these regions have relatively few cells that concentrate estradiol (Pfaff and Keiner, 1973). Combining autoradiography with horseradish peroxidase (HRP) retrograde transport techniques shows that about 30% of the neurons of the VMH that concentrate estradiol project to the MCG (Morrell and Pfaff, 1982). Tracing techniques have demonstrated major projections from the MCG to the medullary reticular formation, and to the MCG from the medullary reticular formation and the spinal cord (Morrell and Pfaff, 1983). The reticular formation receives projections from the MCG and spinal cord, and projects to the MCG and spinal cord (Morrell and Pfaff, 1983).

Studies that utilize stimulation or blockage of electrical activity suggest that the VMH neurons modulate MCG function but are not directly in the circuit mediating stimulus and response, whereas synapses in the MCG appear to be directly involved in the circuit. Blocking electrical activity in the MCG with tetrodotoxin (TTX) immediately blocks lordosis responsiveness (Rothfeld *et al.*, 1986), and electrical stimulation of the MCG causes immediate enhancement of lordosis (Sakuma and Pfaff, 1979). In contrast, blocking electrical activity in the VMH with TTX decreases lordosis responsiveness, but only after a delay of 40 minutes (Harlan *et al.*, 1983b), and stimulation of the VMH increases lordosis with a similar delay (Pfaff and Sakuma, 1979). Therefore, activity of the estradiol-concentrating VMH neurons is not immediately necessary for the lordosis response, but activity of (MCG) neurons that receive VMH projections is necessary. This suggests that although estradiol acts directly on cell bodies in the VMH, facilitatory consequences of estradiol on neuronal connectivity are registered in the MCG.

Electrophysiological studies suggest that estradiol acts through VMH neurons to alter the electrical excitability of MCG neurons. Estradiol has little effect on the spontaneous electrical activity of many VMH neurons, but increases the number of very slowly firing cells (Bueno and Pfaff, 1976). Estradiol and electrical stimulation of VMH cells significantly increased spike potentials in MCG somatodendritic complexes in response to antidromic stimulation of the medullary reticular formation (Sakuma and Pfaff, 1980). Furthermore, lesions of the VMH reduce responsiveness of the MCG to antidromic stimulation (Sakuma and Pfaff, 1980).

The effects of estradiol on lordosis may involve synthesis of specific proteins in the VMH and transport of these proteins to the MCG (Pfaff, 1983). As mentioned above, in all systems studied in detail, the effects of estradiol involve increased gene transcription, leading to increased synthesis of specific proteins (O'Malley and Means, 1974). Ultrastructural studies suggest a profound effect of estradiol on the protein synthetic apparatus of neurons in the VMH. These effects include increased rough endoplasmic reticulum stacking, increased numbers of dense core vesicles, and increased nucleolus-associated chromatin (Cohen and Pfaff, 1981; Cohen *et al.*, 1984; Meisel and Pfaff, 1985a; Jones *et al.*, 1985a). The magnitude of these ultrastructural effects of estradiol correlates with the onset of lordosis (Meisel and Pfaff, 1985a). In the VMH, intracranial implants of anisomycin, which blocks protein synthesis, reversibly inhibit lordosis, but anisomycin in the MCG has no effect on lordosis (Meisel and Pfaff, 1985b). Infusing colchicine into the VMH, which blocks axonal transport, also inhibits lordosis (Harlan *et al.*, 1982).

These data suggest that estradiol induces the synthesis of protein products which are transported from the VMH and, when released, increase the efficacy of synapses within the MCG. In the absence of estradiol these synapses are weak or nonfunctional. After exposure to estradiol, projections from VMH neurons cause these synapses to become functional, thus completing the circuit allowing lordosis to occur in response to flank stimulation. Although other levels of the neuroaxis are certainly involved in carrying out the response, the bulk of the evidence suggests that the most likely area at which synaptic facilitation occurs is the MCG. Identification of the protein products induced by estradiol in the VMH, and examination of how these products alter the electrical excitability of synapses in the MCG, should provide a powerful tool in studying the regulation of synaptic efficacy.

III. SEXUAL DIMORPHISMS

Responses to certain stimuli differ greatly between males and females; in rodents, these differences appear to be mediated by differences in the nervous system caused by exposure to sex steroids, particularly estradiol. As mentioned above, male rats almost never exhibit lordosis, regardless of hormone environment or stimulation, whereas in females this response is extremely robust in the presence of estradiol (Pfaff, 1980). Courting behavior, parental behavior, territorial marking, and aggressive behavior also all exhibit qualitative differences between the sexes (Gorski, 1986; Goy and McEwen, 1980). In addition, regulation of reproductive hormones such as luteinizing hormone (LH) and prolactin differ between males and females. For example estradiol is more effective in suppressing LH after gonadectomy in females than in males (Gee *et al.*, 1984). These sex differences, for example in the regulation of lordosis, persist even when plasma hormone levels are the same between the sexes after gonadectomy and steroid replacement (Barraclough and Gorski, 1962; Harris and Levine, 1965). Therefore, sexually different responses appear to be due to sex differences in the neuronal connectivity of circuits regulating these behaviors and pituitary secretions.

Morphological, functional, and biochemical differences between the nervous systems of males and females clearly exist, although the functional correlates of these differences are not always clear. Sex differences have been reported in the morphology of nuclei in the VMH (Dorner and Staudt, 1968) and in the pattern of synaptic terminals in the preoptic area (Raisman and Field, 1973), the arcuate nucleus (Matsumoto and Arai, 1976a), and the amygdaloid nucleus (Nihizuka and Arai, 1981). A much more dramatic difference exists in the sexually dimorphic nucleus (SDN) of the preoptic area of rats (Gorski *et al.*, 1980). The sexual dimorphism of this nucleus is easily detected by visual inspection with a low-power microscope; the volume of this nucleus is three to five times greater in males than in females (Gorski *et al.*, 1980). However, the functions of this nucleus are unknown. Electrolytic lesions of the SDN do not disrupt male copulatory behavior (Arendash and Gorski, 1983a), although grafting the SDN from neonatal males into neonatal females greatly increased both masculine and feminine sexual behavior of these females in adulthood (Arendash and Gorski, 1982). Electrical connectivity between the VMH and the MCG appears to differ between males and females, based on antidromic studies (Sakuma and Pfaff, 1981). The influence of estradiol on progesterone receptors (Rainbow *et al.*, 1982), and choline acetyltransferase (Luine and McEwen, 1983) is reported to be greater in females than in males.

These sex differences in behavior and morphology are due largely to sex differences in hormone levels during a critical period that, in rats, lasts from birth until about 5 days after birth. If males are gonadectomized on the day of birth, in adulthood they will show feminine lordotic responses in the presence of estradiol (Gorski, 1971). If neonatally gonadectomized rats of either sex are given a single injection of testosterone the day after birth, they will not show lordotic responses as adults (Gorski, 1971). Ovariectomy of neonatal females has little or no effect on lordotic response in adulthood (Gorski, 1971). All of the morphological and behavioral sex differences discussed above exhibit similar dependency on the presence of testosterone in the first few days after birth (Dorner and Staudt, 1968; Dohler *et al.*, 1982a,b). Testosterone is increasingly ineffective in masculinizing the nervous system after day 5, and after day 11 has almost no permanent effect, even when implanted intracerebrally (Hayashi and Gorski, 1974). These data suggest that the female nervous system conforms to a hormone-neutral pattern, and that the presence of sex steroids can convert that pattern to a male pattern.

However, estradiol, not testosterone, appears to be the proximal agent for inducing sexual differentiation. Testosterone is converted to estradiol in the brain by aromatase (Lieberberg *et al.*, 1977). If this conversion is blocked by neonatal injection of aromatase inhibitors, intact males will develop female behaviors, including lordosis responsiveness, and neural morphologies (Gorski, 1985). Furthermore, neonatal injection of estradiol is about 10 times more effective than testosterone in inducing male patterns in females (Mobbs *et al.*, 1985b). Finally, 2 days after birth, cells of the SDN concentrate and retain moxestrol, a synthetic estrogen that does not bind to α-fetoprotein, but do not retain methyltrienolone, a synthetic androgen that does not aromatize to estrogen (Shoonmaker *et al.*, 1983). Females do not usually develop male patterns because estradiol in neonates is usually bound by α-fetoprotein and therefore is not accessible to the nervous system; testosterone is not so bound, and thus has free access to the nervous system, where it is converted to estradiol.

The mechanism by which estradiol induces permanent changes in neuronal connectivity may involve the induction of specific growth factors (Gorski, 1986). Estradiol induces dramatic neurite outgrowth in explant culture of neonatal mouse hypothalamus (Toran-Allerand, 1976), and specifically stimulates dendritic growth (Toran-Allerand et al., 1983). In vivo, estradiol appears to stimulate synaptogenesis (Matsumoto and Arai, 1976b). Furthermore, grafts of preoptic tissue transplanted into neonatal females and examined in adulthood were five-fold larger if, after transplantation, the females were given injections of testosterone proprionate (presumably acting through conversion to estradiol) (Arendash and Gorski, 1983b). In adults with hypothalamic deafferentation, estradiol stimulated synaptogenesis (Arai et al., 1978).

These studies suggest that estradiol may induce growth in neurons in which it acts, which allows these neurons to make contact more readily with the appropriate target neurons. In the uterus estradiol may induce factors in the endometrium that are secreted into the lumen and regulate the dramatic growth of the uterus in response to estradiol (Kuivanen and DeSombre, 1985). It seems likely that, in addition to growth factors, other chemoaffinity factors are induced by estradiol and lead to connective specifity. The identity of these growth factors or other proteins induced by estradiol in neonates would provide a powerful tool for determining how connectivity and neuronal phenotype are permanently determined during development.

IV. REPRODUCTIVE SENESCENCE

Reproductive senescence in female rodents is due in part to neuroendocrine impairments caused by exposure to estradiol. Regular 4- or 5-day estrous cycles in rodents begin at puberty, 1–3 months after birth (Nelson et al., 1982). By 7–10 months after birth, these cycles become longer and less regular (Nelson et al., 1982). By 13–15 months of age, most females have ceased cycling altogether, and by 20 months of age essentially no females exhibit estrous cycles (Nelson et al., 1982; Aschheim, 1983). Most rodents live for more than 30 months, and at the time reproductive cycles cease are in good health (Nelson et al., 1982; Mobbs et al., 1984b). Lengthening and loss of reproductive cycles are due to ovarian and neuroendocrine impairments, although the degree to which each impairment contributes to loss of cycles varies between species, strains, and even individuals within a strain. The number of ovarian follicles decreases exponentially with age in rodents (Gosden et al., 1983); however, reproductive cycles generally cease before the follicles are completely depleted (Gosden et al., 1983). Middle-aged rodents given ovarian grafts from young rodents exhibit fewer estrous cycles than young rodents given similar grafts from young rodents (Mobbs *et al.*, 1984b). In young ovariectomized female rodents, a preovulatory-like surge of LH can be induced by an appropriate schedule of exogenous estradiol; in middle-aged rodents, this estradiol-induced LH surge is greatly diminshed (Mobbs *et al.*, 1984b; Peng and Huang, 1972), and this diminution is correlated with the loss of ability to support estrous cycles. Under a different schedule, estradiol in young ovariectomized rodents also suppresses LH secretion and increases prolactin; middle-aged rodents show altered sensitivity to estradiol under these conditions (Mobbs *et al.*, 1985a). These effects of estradiol in young rodents are correlated with changes in norepinephrine turnover (Honma and Wuttke, 1980), estradiol-receptor binding, progesterone-receptor induction, and metabolic activity in preoptic and hypothalamic nuclei (Wise, 1986); all of these estradiol-induced neurochemical changes are diminished in middle-aged female rats (Wise, 1983, 1986; Wise and Camp, 1984).

Some morphological changes in the brain during aging correlate with reproductive senescence. In the arcuate nucleus, middle-aged female rodents exhibit much greater glial proliferation and astrocytic reactivity, indicative of neuronal degeneration, than young females (Schipper *et al.*, 1981). Similarly, the arcuate nucleus of middle-aged female hamsters exhibits less binding to estradiol, as determined by steroid autoradiography, than the arcuate nucleus of young hamsters (Blaha and Lamperti, 1983).

Many of these functional impairments and morphological changes associated with reproductive senescence are caused by exposure to estradiol. Middle-aged female rodents, if ovariectomized when young, have only minor demonstrable impairments in the ability to produce an LH surge in response to estradiol or to support estrous cycles given young ovarian grafts (Aschheim, 1983; Mobbs *et al.*, 1984a), in contrast to the impairments in mice ovariectomized when middle-aged. Similarly, age-correlated changes in the response of LH and prolactin to constant levels of estradiol are prevented when mice are ovariectomized when young (Mobbs *et al.*, 1985b). Furthermore, age-correlated changes in arcuate nucleus morphology are prevented in rats and mice by ovariectomy when young (Schipper *et al.*, 1981). Conversely, in young mice high levels of estradiol impair the ability to produce an LH surge and to support estrous cycles with young ovaries (Mobbs *et al.*, 1984b), and also induce morphological changes in the arcuate nucleus similar to those correlated with aging (Brawer *et al.*, 1978). Physiological but constant levels of estradiol can also induce impairments similar to those associated with reproductive senescence (Mobbs and Finch, unpublished). These studies suggest that exposure to estradiol during reproductive cycles causes impairments in neuronal connections sensitive to estradiol, leading to a loss in the ability to support further reproductive cycles (Finch *et al.*, 1984).

The mechanism by which estradiol impairs neuronal connections during aging may be related to the mechanism by which estradiol regulates development. As discussed previously, cumulative exposure to estradiol during aging appears to decrease the degree to which LH is suppressed by estradiol (Mobbs *et al.*, 1985a). At puberty, the ability of estradiol to suppress LH is substantially decreased (Ramaley, 1979), and the reduction in this sensitivity is regulated by estradiol (Docke and Dorner, 1974). This reduced suppression of LH is thought to help trigger the onset of puberty (Ramaley, 1979). Therefore, reduction in sensitivity to estradiol during aging may simply reflect the continuation of a process that is necessary for the development of puberty. This idea is consistent with the "pleiotropic" theory of aging (Williams, 1957), which holds that senescence is a reflection of developmental processes that are advantageous early in adulthood but become lethal later. These later lethal effects are not selected against because they are expressed after offspring have been born.

V. E2-INDUCED PROTEINS

We have suggested that identification of proteins regulated by E2 in the nervous system might prove useful in studying neuronal plasticity. One strategy we have adopted to discover such proteins involves labeling proteins with radioactive amino acids in neural tissue, and separating these proteins by two-dimensional gel electrophoresis. We then examine for differences in the protein patterns from rats treated with E2 compared with ovariectomized rats with no E2. We and others using different labeling and detection methods have found such E2-regulated protein spots on two-dimensional gels (Scouten *et al.*, 1985; Jones *et al.*, 1986; Mobbs *et al.*, 1987). Such spots give the molecular weight and isoelectric point of E2-regulated proteins; these physical properties will suggest or rule out potential E2-regulated proteins by comparison with proteins whose functions and properties have been published. Antibodies can also be generated against such proteins; such antibodies can be useful even if the function of the antigen remains unknown. Finally, sequence data can be generated from such proteins; such data can be useful in generating antisera, and can also be used to screen cDNA libraries. We will discuss these strategies in the context of an E2-regulated protein that may be correlated with lordosis, and thus may be related to synaptic facilitation. This same approach may be useful in studying E2-regulated development and senescence.

Using *in vivo* labeling followed by two-dimensional gel elctrophoresis, we discovered a protein that appeared to be induced by E2 in the VMH and transported to the MCG (Mobbs *et al.*, 1987). [^{35}S]Methionine and -cysteine were microinfused into the VMH of ovariectomized rats; eight rats were given E2 implants and eight were given sham implants at the time of microinfusion. Fourteen hours after the end of microinfusion, the rats were sacrificed and several brain regions, including the VMH and MCG, were obtained by microdissection. Diffusion of the label was minimal because radioactivity incorporated into protein was 10-fold-higher in the VMH than in regions more than 1 mm lateral to the VMH. Homogenates from VMH and MCG tissue were loaded onto polyacrylamide tube gels for isoelectric focusing. These tube gels were then placed on top of sodium dodecylsulfate (SDS) discontinous polyacrylamide gels for separation by molecular size. The SDS gels were treated for autoradiography, exposed 2–10 weeks to X-ray film, and developed.

Typical protein spot patterns obtained in such studies are shown in Fig. 1. In general, over 250 spots are reproducibly detectable. The only spot that appeared to be consistently present in E2-treated rats but absent in control rats was a spot corresponding to an M_r of 70, and pI of about 5.9 (Fig. 1). A spot of the same M_r but pI 5.8 appeared to be inversely correlated with the induced spot. This E2-induced spot appears both in the VMH (Fig. 1) and in the MCG (Fig. 2). Spots detected in the MCG are probably synthesized in the VMH and transported to the MCG, because the radioactivity in the MCG is reduced by over 80% when the VMH is infused with colchicine. Therefore, this protein appears to be induced in the VMH by E2 and transported to the MCG. These features of the 70k protein are consistent with the properties expected of a protein involved in the regulation of lordosis.

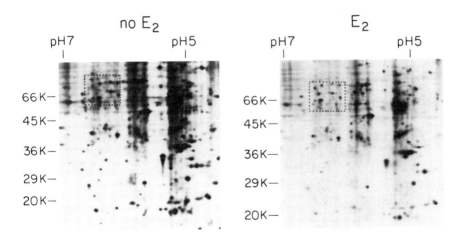

FIG. 1. Representative two-dimensional gel patterns from the ventromedial hypothalamus of individual ovariectomized rats, either given no estradiol replacement (no E2) or given estradiol-containing capsules (E2) 14 hours before sacrifice. [^{35}S]Methionine (600 μCi) and [^{35}S]cysteine (300 μCi) were infused into the VMH 14 hours before sacrifice and microdissection. The VMH was microdissected and 500,000 trichloroacetic acid-precipitatable counts were separated by two-dimensional gel electrophoresis (O'Farrell, 1975), prepared for autoradiography, and exposed to X-ray film at -70 °C for 1 week. The boxed area is displayed in more detail in Fig. 2.

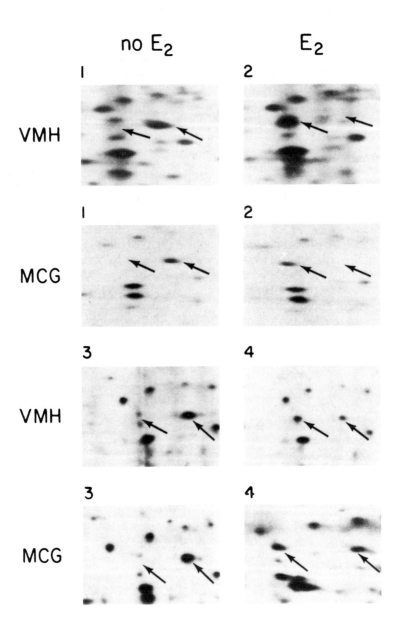

FIG. 2.   Estradiol-regulated protein spots from the VMH and MCG of individual rats. Rats were treated and tissue was prepared and analyzed as described in Fig. 1. The area shown corresponds to the boxed area in Fig. 1. The number in the upper left of each box indicates the individual animal number. Arrows indicate spots regulated by E2.

Although in this study only one protein was found to be consistently and qualitatively changed, other studies have reported quantitative effects of E2, as well as sex differences, on individual protein spots from brain tissues (Scouten *et al.*, 1985; Gold *et al.*, 1983; Jones *et al.*, 1986). In addition, a previous study from this laboratory detected an E2-regulated protein transported to the MCG using a similar labeling paradigm, followed by analysis with high-performance liquid chromatography (HPLC) (Pfaff *et al.*, 1984); the molecular weight of this protein was similar to the protein we detected with two-dimensional gel analysis. We have substantiated the qualitative induction of the 70K protein by quantitative analysis (Fig. 3). Furthermore, we have detected qualitative effects of E2 on protein synthesis in pituitary (Fig. 4). Although we have not extensively studied uterine proteins, we can resolve over 800 proteins from uterus (Fig. 5) and others have reported E2-regulated effects using gels with less resolution (Korach *et al.*, 1981). These studies show that it is feasible to discover E2-induced proteins using radioactive labeling followed by two-dimensional gel electrophoresis.

How can the identification of these proteins help elucidate the mechanism by which E2 alters neuronal connectivity? By itself, the discovery of a

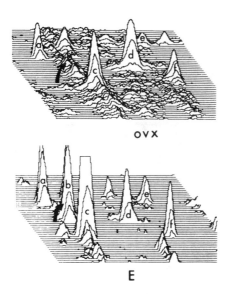

OVX

E

FIG. 3.   Graphic representation of computerized densitometric analysis of one estradiol-regulated protein spot. Rats were treated and tissue was prepared and subjected to two-dimensional gel analysis as described in Figs. 1 and 2. Autoradiograms were quantitated using the Biomed 2D flatbed laser densitometer and associated software on an Apple IIe microcomputer. Height on the graph corresponds to the density of each pixel. Arrow indicates spot regulated by E2.

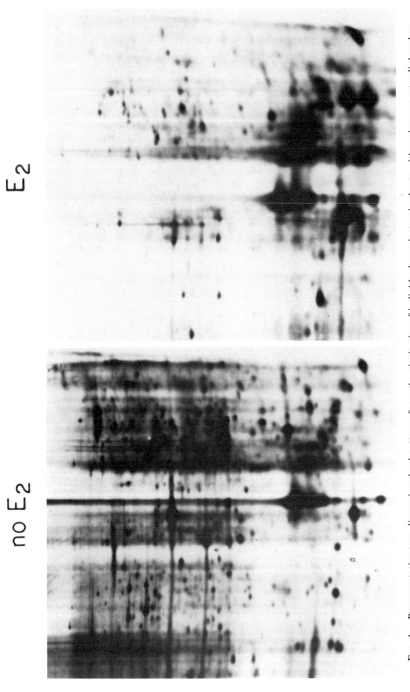

no E₂   E₂

FIG. 4.   Representative two-dimensional gel patterns from the pituitaries of individual ovariectomized rats, either given no estradiol replacement (no E2) or given estradiol-containing capsules (E2) 1 week before sacrifice. Pituitaries were bisected and incubated in 200 μl minimal essential medium for 30 minutes at 37°C in a 5% carbon dioxide atmosphere. Then [³⁵S]methionine (300 μCi) and [³⁵S]cysteine (150 μCi) in 25 μl were added to the well and the tissue was incubated for an additional 4 hours. The tissue was homogenized and 500,000 trichloroacetic acid-precipitable counts were subjected to separation by two-dimensional gel electrophoresis (O'Farrell, 1975), prepared for autoradiography, and

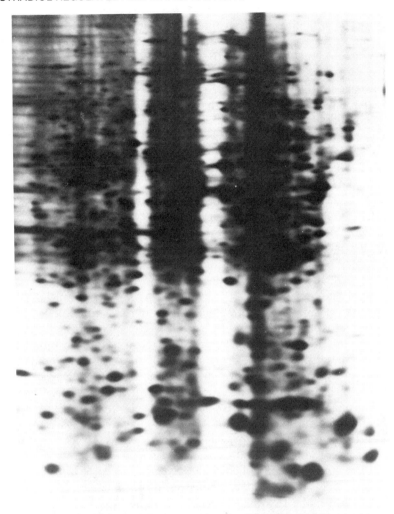

Fig. 5. Representative two-dimensional gel pattern from uterine tissue of an individual ovariectomized rat. The uterus was removed, a small section was diced with fine scissors, and the tissue was incubated, prepared, and analyzed as in Fig. 4.

specific protein spot induced by E2 is of little direct value, other than adding support to the already well-established notion that E2 acts by inducing the synthesis of specific proteins. The physiological function of the protein may be very difficult to ascertain, although, as we will discuss, the function of the protein may not be necessary or even useful information to exploit other knowledge of the protein. The most direct uses of two-dimensional gels involve determination of molecular weight and isoelectric point and the generation of antibodies. Of possible use would be the determination of some amino acid sequences. To exemplify these strategies we will discuss how we may use the discovery of the E2-induced protein spot from VMH tissue to further study E2-regulated neuronal connectivity.

Knowledge of M_r and pI of an E2-regulated protein suggests its function based on comparison with previously characterized proteins. For example, the E2-induced protein spot in Figs. 1 and 2 corresponds to an M_r of about 70,000 and pI of about 5.9. There are at present no comprehensive databases that allow a listing of all characterized proteins by M_r and/or pI. A pair of tables exists which, between them, list most proteins for which both pI and M_r were published by 1979 (Rhigetti and Tudor, 1981; Rhigetti and Caravaggio, 1976). Another table, largely redundant, lists pI values for about 400 proteins (Malamud and Drysdale, 1978); we know of no comprehensive table that lists known proteins by M_r. These tables list proteins by alphabetical order, making comparison based on M_r and pI cumbersome but possible. The values on such tables, especially for pI, must be considered relative and dependent on the system in which they were measured. As a general rule we consider a 10% difference in M_r and a unit difference in pI as within the bounds of variability, so for our E2-induced protein we consider as possible candidates proteins with M_r ranging from 63,000 to 77,000, and pI from 5.5 to 7.5. Use of the Righetti tables in conjunction with these critera suggested the proteins in Table I as possible candidates.

Of over a thousand entries given by these tables, only 20 proteins were within even the extremely liberal criteria we set. Of these 20, only 3 had M_r and pI reasonably similar to the E2-induced protein we found. Two are of questionable relevance: albumin and DNA-binding protein. The remaining protein is the thyronine receptor. An argument against this identity is that the 70K protein found by two-dimensional gels represents about 0.01% of the total protein on the gel, whereas the thyronine receptor in brain probably represents less than 0.0001%; under our protocol it is unlikely that we could even detect the thyronine receptor. Another argument against the identity is that it seems unlikely that the receptor would be transported to the MCG, whereas the 70K protein does seem to be transported. Nevertheless, we hope to identify with antibodies and ligand binding exactly where on our gels the thyronine receptor is likely to migrate and if it would colocalize exactly with the 70K protein.

Less systematically, identification can be suggested based on molecular weight (much more commonly reported and more reliable than pI) and an interesting functional property. For example, we broadly consider as a candidate for the 70K protein any protein that may have properties related to neuronal plasticity. One postsynaptic theory of synaptic facilitation suggests a role for proteases in activating glutamate receptors (Lynch and

TABLE I
POSSIBLE CANDIDATES FOR E2-INDUCED PROTEIN

Protein	Species	Tissue	Subunit M_r	pI
5'-AMP aminohydrolase	Human	Submandibular	70,000	5.5
Amylase	Human	Saliva	64,000	6.5
Thyronine receptor	Dog	Kidney	70,000	5.7
DNA-binding protein	Rat	Brain	68,000	5.9
Esterase I	Rat	Liver	70,000	6.3
Estrogen receptor	Rat	Uterus	68,000	6.4
α-Fetoprotein	Mouse	Plasma	70,000	5.4
Glucouronidase	Rat	Preputial gland	72,000	6.15
Glutamate decarboxylase	Human	Brain	67,000	5.4
Oxaloacetate decarbylase	Fish	Muscle	63,000	6.6
Proteinase		*Chromobacterium livium*	72,000	7.15
Proteinase		*Chromobacterium livium*	67,000	6.15
Protein kinase, cGMP	Bovine	Lung	74,000	5.4
Protein S, vitamin K	Human	Plasma	69,000	5.5
Albumin	Human	Plasma	69,000	5.8
Carboxylesterase E2	Rat	Liver	70,000	6.6
Glycerol dehydrogenase	Rabbit	Fat	66,000	6.3
Malic enzyme	*Ascaris*	Muscle	66,400	6.6
Phosphoglucomutase	Rabbit	Muscle	64,900	6.8
Ribosyltransferase	Human	Erythrocyte	68,000	6.5

Baudry, 1984). Furthermore, E2 induces proteolytic activity in the uterus (Hendrickson and Dickerman, 1983), and one protease, plasminogen activator, has an M_r of about 70K (Kneifel *et al.*, 1982) and may be involved in neuronal development (Kalderon, 1985). We are therefore pursuing the possibility that the 70K protein may be related to plasminogen activator and may have proteolytic activity. One presynaptic theory suggests that synaptic facilitation involves increased release of neurotransmitter (Kandel and Schwartz, 1982); one mechanism by which this might occur would be increased synthesis of proteins involved in synaptic release. A candidate for the E2-induced 70K protein might therefore be chromogranin A, a protein with a reported M_r of 68K which is distributed throughout the diffuse neuroendocrine system and is associated with secretory vesicles (Angeletti and Hickey, 1985). Such hypotheses can be tested by immunoblot (Western) analysis of a two-dimensional gel or by immunocytochemistry in VMH and MCG.

Even without identification, the discovery of an E2-induced spot may be useful as a marker, especially if used to generate in antibody. Protein spots from two-dimension gels may be used directly as immunogens. Even if this is not feasible, knowledge of the M_r and pI makes purification of the protein straightforward by using gel-exclusion chromatography followed by chromatofocusing. Purification can be monitored by two-dimensional gels and the purified protein can be used to produce polyclonal antibodies. The antibody can be used to label the subset of neurons presumably most relevant to the regulation of connectivity by E2, which then allows a more specific analysis of the effects of E2. For example, E2 may facilitate synapses in the MCG by increasing the number of spines or postsynaptic densities in target neurons, as has been suggested for long-term potentiation (Siekovitz, 1985). Presently, this hypothesis can be addressed only by analyzing all synapses in the MCG, where an effect of E2 may be relatively minor in the context of the more general integrative activities of the MCG. More likely to be informative would be to analyze only those neurons postsynaptic to neurons containing, in this case, the E2-induced 70K protein synthesized in the VMH and transported to the MCG. Labeling by an appropriate antibody, followed by electron microscopic analyis, would facilitate such a study.

Finally, proteins separated on two-dimensional gels may be electroeluted and sequenced directly (Aebersold et al., 1986). As little as 50 ng of a 50K protein can give sequences of up to 20 residues. Normally 100- to 1000-fold more protein is necessary to generate antibodies (Vaitukaitis, 1981). Even partial sequence information may be used to generate oligonucleotides for screening cDNA libraries using a λgt10 library (Walter et al., 1985). cDNA probes could be used to examine regulation using in situ hybridization (McCabe et al., 1986; Shivers et al., 1986). Using this method, E2 has been shown to increase ribosomal RNA (Jones et al., 1986b) and metenkephalin mRNA (Romano et al., 1986). Sequence information may also be used to synthesize peptides, which may in turn allow the generation of a battery of antibodies. These antibodies may in turn be used directly, as discussed in the preceding paragraph, or may be used to screen a cDNA library constructed with the λgt11 vector (Young and Davis, 1983). Complete sequence analysis, generated most easily from the cDNA libraries, facilitates comparison by sequence homology with other characterized proteins. Such comparisons often suggest functional domains on a previously uncharacterized protein, even if in other domains there is little sequence homology. Sequence analysis may also suggest the production of peptides by cleavage of a precursor at basic residues. These peptides could be synthesized, and antibodies against them produced, to test their effects on behavior, followed by, for example, infusion into the MCG.

We have argued that a general problem in neurobiology concerns the regulation of neuronal connectivity, and that E2 regulates neuronal connectivity at the functional and anatomical level over three functional domains. Much data suggests that the effects of E2 are mediated by the synthesis of specific proteins. We have tried to demonstrate the plausibility of discovering these proteins using metabolic labeling and two-dimensional gel electrophoresis. We have also tried to outline how the discovery of such proteins could lead to explaining how these proteins mediate the effects of E2. Understanding how E2 regulates neuronal connectivity may provide a general mechanism common to the regulation of development, efficacy of neuronal connections, and aging.

NOTE ADDED IN PROOF

We have recently proposed that the E2-induced 70K protein is a member of the 70K heat-shock/uncoating ATPase protein family.

ACKNOWLEDGMENTS

We appreciate the contributions of the many colleagues cited in this papter, and conversations with Dr. Zita Wenzel and Dr. Joel Rothfeld have been very helpful. This work has been supported by NIG grants AGO5326 and HD05751.

REFERENCES

Aebersold, R. H., Teplow, D. B., Hood, L. E., and Kent, S. B. H. (1986). Electroblotting onto activated glass: High efficiency preparation of proteins from analytical SDS-polyacrylamide gels for direct sequence analysis. *J. Biol. Chem.* **261**, 4229–4238.

Angeletti, R. H., and Hickey, W. F. (1985). A neuroendocrine marker in tissues of the immune ystem. *Science* **230**, 89–90.

Arai, Y., Matsumoto, A., and Nishizuka, M. (1978). Synaptogenic action of estrogen on the hypothalamic arcuate nucleus (ARCN) of the developing brain and of the deafferented adult brain in female rats. *In* "Hormones and Brain Development" (G. Dorner and M. Kawakami, eds.), Elsevier/North Holland Biomedical Press, Amsterdam.

Arendash, G. W., and Gorski, R. A. (1982). Enhancement of sexual behavior in female rats by neonatal transplantation of brain tissue from males. *Science* **217**, 1276–1278.

Arendash, G. W., and Gorski, R. A. (1983a). Effects of discrete lesions of the sexually dimorphic nucleus of the preoptic area or other medical preoptic regions on the sexual behavior of male rats. *Brain Res. Bull.* **10**, 147–154.

Arendash, G. W., and Gorski, R. A. (1983b). Testosterone-induced enhancement of male medial preoptic tissue transplant volumes in female recipients: A "neuronotrophic" action of testosterone. *Soc. Neurosci. Abstr.* **9**, 307.

Aschheim, P. (1983). Relation of neuroendocrine system to reproductive decline in female rats. *In* "Neuroendocrinology of Aging" (J. Meites, ed.), p. 73. Plenum Press, New York.

Barraclough, C. A., and Gorski, R. A. (1962). Studies on mating behavior in the androgen-sterilized rat and their relation to the hypothalamic regulation of sexual behavior in the female rat. *J. Endocrinol.* **25**, 175–182.

Blaha, G. C., and Lamperti, A. A. (1983). Estradiol target neurons in the hypothalamic arcuate nucleus and lateral ventromedial nucleus of young adult, reproductively senescent and monosodium-glutamate lesioned female golden hamsters. *J. Gerontol* **38**, 335–343.

Bleier, R., Byne, W., and Siggelkow, I. (1982). Cytoarchitectonic sexual dimorphisms of the medial preoptic and anterior hypothalamic areas in guinea pig, rat, hamster, and mouse. *J. Comp. Neurol.* **212**, 118–130.

Brawer, J. R., and Finch, C. E. (1983). Normal and experimentally altered aging processes in the rodent hypothalamus. *In* "Experimental and Clinical Interventions in Aging" (R.F Walker and R. L. Cooper, eds.), p. 45. Dekker, New York.

Brawer, J. R., Naftolin, F., Martin, J., and Sonnenschein, C. (1978). Effects of a single injection of estradiol valerate on the hypothalamic arcuate nucleus and on reproductive function in the female rat. *Endocrinology* **103**, 510.

Brawer, J. E., Ruf, K. I., and Naftolin, F. (1980). Effects of estradiol-induced lesions of the arcuate nucleus on gonadotropin release in response to preoptic stimulation in the rat. *Neuroendocrinology* **30**, 144.

Brawer, J. R., Schipper, H., and Naftolin, F. (1980). Ovary-dependent degeneration in the hypothalamic arcuate nucleus. *Endocrinology* **107**, 274.

Brawer, J., Schipper, H., and Robaire, B. (1983). Effects of long term androgens and estradiol exposure on the hypothalamus. *Endocrinology* **112**, 194.

Bueno, J., and Pfaff, D. W. (1976). Single unit recording in hypothalamus and preoptic area of estrogen-treated and untreated ovariectomized female rats. *Brain Res.* **101**, 67–78.

Burch, J. B. E., and Weintraub, H. (1983). Temporal order of chromatin structural changes associated with activation of the major chicken vitellogenin gene. *Cell* **33**, 65–76.

Carrer, H. F., and Aoki, A. (1982). Ultrastructural changes in the hypothalamic ventro-medical nucleus of ovariectomized rats after estrogen treatment. *Brain Res.* **240**, 221–233.

Casanueva, F., Cocchi, D., Locatelli, V., Flauto, C., Zambotti, F., Bestetti, G., Rossi, G. L., and Muller, E. (1982). Defective central nervous system dopaminergic function in rats with estrogen-induced pituitary tumors, as assessed by plasma prolactin concentrations. *Endocrinology* **110**, 590.

Cohen, R. S., and Pfaff, D. W. (1981). Ultrastructure of neurons in the ventromedial nucleus of the hypothalamus in ovariectomized rats with or without estrogen treatment. *Cell Tissues Res.* **217**, 451–470.

Cohen, R. S., Chung, S. K., and Pfaff, D. W. (1984). Alteration by estrogen of the nucleoli in nerve cells of the rat hypothalamus. *Cell Tissues Res.* **235**, 485–489.

Cotman, C. W., ed. (1986). "Synaptic Plasticity." Guilford Press, New York.

Docke, F., and Dorner, G. (1974). Oestrogen and the control of gonadotrophin secretion in the immature rat. *J. Endocrinol.* **63**, 285.

Dohler, D. D., Coquelin, A., Davis, F., Hines, M., Shryne, J. E., and Gorski, R. A. (1982a). Differentiation of the sexually dimorphic nucleus in the preoptic area of the rat brain is determined by the perinatal hormone environment. *Neurosci. Lett.* **33**, 295–298.

Dohler, K. D., Hines, M., Coquelin, A., Davis, F., Shryne, J. E., and Gorski, R. A. (1982b). Pre- and postnatal influence of diethylstilboestrol on differentiation of the sexually dimorphic nucleus in the preoptic area of the female rat brain. *Neuroendocrinol. Lett.* **4**, 361–365.

Dorner, G., and Staudt, J. (1968). Structural changes in the preoptic anterior hypothalamic area of the male rat, following neonatal castration and androgen substitution. *Neuroendocrinology* **3**, 136–140.

Edelman, G. M., and Mountcastle, V. B. (1978). "The Mindful Brain." MIT Press, Cambridge, Massachusetts.

Edwards, D. A., and Pfeifle, J. K. (1981). Hypothalamic and midbrain control of sexual receptivity in the female rat. *Physiol. Behav.* **26**, 1061–1067.

Finch, C. E., Felicio, L. S., Mobbs, C. V., and Nelson, J. F. (1984). Ovarian and steroidal influences on neuroendocrine aging processes in female rodents. *Endocrine Rev.* **5**, 467–497.

Gee, D. M., Flurkey, K. Mobbs, C. V., Sinha, Y. N., and Finch, C. E. (1984). The regulation of luteinizing hormone and prolactin in C57BL/6J mice: Effects of estradiol implant size, duration of ovariectomy, and aging. *Endocrinology* **114**, 685–693.

Gold, M. A., Heydorn, W. E., Creed, J. G., and Jacobowitz, D. M. (1983). Sex differences in specific proteins in the preoptic medial nucleus of the rat hypothalamus. *Neuroendocrinology* **37**, 470–472.

Gorski, R. A. (1971). Gonadal hormones and the perinatal development of neuroendocrine function. *In* "Frontiers in Neuroendocrinology, 1971" (L. Martini and W. F. Ganong, eds.), pp. 237–290. Oxford University Press, New York.

Gorski, R. A. (1986). Gonadal hormones as putative neurotrophic substances. *In* "Synaptic Plasticity" (C. W. Cotman, ed.). Guilford Press.

Gorski, R. A. (1987). Comparative aspects of sexual differentiation of the brain. *In* "The Control of the Onset of Puberty. II" (M. M. Grumbach, P. C. Sizonenko, and M. L. Aubert eds.), Academic Press, Orlando, Florida, in press.

Gorski, R. A., Gordon, J. H., Shryne, J. E., and Southam, A. M. (1978). Evidence for a morphological sex difference within the medial preoptic area of the rat brain. *Brain Res.* **148**, 333-346.

Gorski, R. A., Harlan, R. E., Jacobson, C. D., Shryne, J. E., and Southam, A. M. (1980). Evidence for the existence of a sexually dimorphic nucleus in the preoptic area of the rat. *J. Comp. Neurol.* **193**, 529-539.

Gosden, R. G., Laing, S. C., Felicio, L. S., Nelson, J. F., and Finch, C. E., (1983). Imminent oocyte exhaustion and reduced follicular recruitment mark the transition to acyclicity in aging C57BL/6J mice. *Biol. Reprod.* **28**, 255.

Goy, R. W., and McEwen, B. S. (1980). "Sexual differentiation of the Brain." MIT Press, Cambridge, Massachusetts.

Greenough, W. T., Carter, C. S., Steerman, C., and DeVoogd, T. J. (1977). Sex differences in dendritic patterns in hamster preoptic area. *Brain Res.* **126**, 63-72.

Grey, G. D., Tennent, B., Smith, E. R., and Davidson, J. M. (1980). Luteinizing hormone regulation and sexual behavior in middle-aged female rats. *Endocrinology* **107**, 187.

Harlan, R. E., Gordon, J. H., and Gorski, R. A. (1979). Sexual differentiation of the brain: Implications for neuroscience. *In* "Reviews of Neuroscience" (D. Schneider, ed.), pp. 31-71. Raven, New York.

Harlan, R. E., Shivers, B. D., Moss, R. L., Shryne, J. E., and Gorski, R. A. (1980). Sexual performance as a function of time of day in male and female rats. *Biol. Reprod.* **23**, 64-71.

Harlan, R. E., Shivers, B. D., Kow, L.-M., and Pfaff, D. W. (1982). Intrahypothalamic colchicine infusions disrupt lordotic responsiveness in estrogen-treated female rats. *Brain. res.* **238**, 153-167.

Harlan, R. E., Shivers, B. D., and Pfaff, D. W. (1983a). Midbrain microinfusions of prolactin increase the estrogen-dependent behavior, lordosis. *Science* **219**, 1451-1453.

Harlan, R. E., Shivers, B. D., Kow, L.-M. and Pfaff, D. W. (1983b). Estrogenic maintenance of lordotic responsiveness: Requirement for hypothalamic action potentials. *Brain Res.* **268**, 67-78.

Harlan, R. E., Shivers, B. D., and Pfaff, D. W. (1984). Lordosis as a sexually dimorphic neural function. *In* "Progress in Brain Research" (G. J. de Vries, ed.), Vol. 61. Elsevier, Amsterdam.

Harris, G. W., and Levine, S. (1965). Sexual differentiation of the brain and its experimental control. *J. Physiol. (London)* **181**, 379-400.

Hayashi, S., and Gorski, R. A. (1974). Critical exposure time for androgenization by intracranial crystals of testosterone propionate in neonatal female rats. *Endocrinology* **94**, 1161.

Henrikson, K. P., and Dickerman, H. W. (1983). An estrogen-stimulated, calcium-dependent tosylarginine methyl ester (TAME) hydrolase in immature rat uterus. *Mol. Cell. Endocrinol.* **32**, 143-156.

Honma, K., and Wuttke, W. (1980). Norepinephrine and dopamine turnover rates in the medial preoptic area and the medial basal hypothalamus of the rat brain after various endocrinological manipulations. *Endocrinology* **106**, 1848-1853.

Huynh, T. V., Young, R. A., and Davis, R. W. (1985). *In* "DNA Cloning: A Practical Approach" (D. M. Glover, ed.), Vol. 1, pp. 98-121. IRL, Oxford.

Jensen, E. V., Greene, G. L., Closs, L. E., DeSombre, E. R., and Nadji, M. (1982). Receptors reconsidered: a 20-year perspective. *Recent Prog. Horm. Res.* **38**, 1-40.

Jones, K. J., Pfaff, D. W., and McEwen, B. S. (1985a) Early estrogen-induced nuclear changes in rat hypothalamic ventromedial neurons: An ultrastructural and morphometric analysis. *J. Comp. Neurol.* **239**, 255-266.

Jones, K.J., Chikarishi, D. M., Harrington, C. A., McEwen, B. S., and Pfaff, D. W. (1985b). Effects of estradiol on ribosomal RNA in the ventral medial hypothalamus using in situ hybridization. *Mol. Brain. Res.* **1**, 45-152.

Jones, K. J., McEwen, B. S., and Pfaff, D. W. (1986). Effects of estradiol (E2) on protein synthesis in vitro, in the ventromedial hypothalamic nucleus (VMN) and preoptic area (POA) of the female rat. *Soc. Neurosci. Abstr.* **11**, 738.

Kalderon, N. (1985). Extracellular proteolysis in CNS histogenesis: Expression of plasminogen activator activity by the differentiating astrocytes. *Soc. Neurosci. Abstr.* **11**, 1148.

Kandel, E. R., and Schwartz, J. H. (1982). Molecular biology of learning: Modulation of transmitter release. *Science* **218**, 433-443.

Kaye, J. S., Pratt-Kaye, S., Bellard, M., Dretzen, G., Bellard, F., and Chambon, P. (1986). Steroid hormone dependence of four DNase I-hypersensitive regions located within the 7000-bp 5' -flanking segment of the ovalbumin gene. *EMBO J.* **5**, 277-285.

Kneifel, M., Leytus, S. P., Fletcher, E., Weber, T., Mangel, W. F., and Katzenel, B. S. (1982). Uterine plasminogen activator activity—Modulation by steroid hormones. *Endocrinology* **111**, 493-499.

Korach, K. S., Harris, S. E., and Carter, D. B. (1981). Uterine proteins influenced by estrogen exposure—Analysis by 2-dimensional gel electrophoresis. *Mol. Cell. Endocrinal.* **21**, 243-254.

Krieger, M. S., Conrad, L. C. A., and Pfaff, D. W. (1979). An autoradiographic study of the efferent connections of the ventromedial nucleus of the hypothalamus. *J. Comp. Neurol.* **183**, 785-816.

Kuivanen, P. C., and DeSombre, E. R. (1985). The effects of sequential administration of 17B estradiol on the synthesis and secretion of specific proteins in the immature rat uterus. *J. Steroid Biochem.* **22**, 439-451.

Lieberberg, I., and McEwen, B. S. (1977). Brain cell nuclear retention of testosterone metabolites, 5-alpha-dihydrotestosterone and estradiol-17-beta in adult rats. *Endocrinology* **100**, 588-597.

Lu, J. H. K., Damassa, D. A., Gilman, D. P., Judd, H. L., and Sawyer, C. H. (1980). Differential patterns of gonadotrophin responses to ovarian steroids and to LH releasing hormone between constant estrous and pseudopregnant states in aging rats. *Biol. Reprod.* **23**, 345.

Luine, V. N., and McEwen, B. S. (1983). Sex differences in cholinergic enzymes of diagonal band nuclei in the rat preoptic area. *Neuroendocrinology* **36**, 475-482.

Luine, V. N., and Rhodes, J. C. (1983). Gonadal hormone regulation of MAO and other enzymes in hypothalamic areas. *Neuroendocrinology* **36**, 235-241.

Luine, V. N., Kylechevskaya, R. I., and McEwen, B. S. (1974). Oestrogen effects on brain and pituitary enzyme activities. *J. Neurochem.* **23**, 925-934.

Luine, V. N., Khylchevskaya, R. I., and McEwen, B. S. (1975). Effects of gonadal hormones on enzyme activities in brain and pituitary of male and female rats. *Brain Res.* **86**, 283-292.

Lynch, G., and Baudry, M. (1984). The biochemistry of memory: A new and specific hypothesis. *Science* **224**, 1057-1063.

McCabe, J. T., Morrell, J. I., Ivell, R., Schmale, H., Richter, D., and Pfaff, D. W. (1986). In situ hybridization technique to localize rRNA and mRNA in mammalian neurons. *J. Histochem. Cytochem.* **34**, 45-50.

McEwen, B. S., Davis, P. G., Parsons, B., and Pfaff, D. W. (1979). The brain as a target for steroid hormone action. *Annu. Rev. Neurosci.* **2**, 65-112.

McEwen, B., Plapinger, L., Wallach, G., and Magnus, C. (1972). Properties of cell nuclei isolated from various regions of rat brain: Divergent characteristics of cerebellar cell nuclei. *J. Neurochem.* **19,** 1159–1170.

Malamud, D., and Drysdale, J. W. (1978). Isoelectric points of proteins: A table. *Anal. Biochem.* **86,** 620–647.

Mark, R. (1974). "Memory and Nerve Cell Connections." Oxford University Press, London.

Matsumoto, A., and Arai, Y. (1976a). Developmental changes in synaptic formation in the hypothalamic arcuate nucleus of female rats. *Cell Tissue Res.* **169,** 143–156.

Matsumo, A., and Arai, Y. (1976b). Effect of estrogen on early postnatal development of synaptic formation in the hypothalamic arcuate nucleus of female rats. *Neurosci. Lett.* **2,** 79–82.

Meisel, R. L., and Pfaff, D. W. (1985a). Brain region specificity in estradiol effects on neuronal ultrastructure in rats. *Mol. Cell. Endocrinol.* **40,** 159–166.

Meisel, R. L., and Pfaff, D. W. (1985b). Specificity and neural sites of action of anisomycin in the reduction or facilitation of female sexual behavior in rats. *Hormones Behav.* **19,** 237–251.

Mobbs, C. V., Kannegieter, L. S., and Finch, C. E. (1985a) Delayed anovulatory syndrome induced by estradiol in female C57BL/6J mice: Age-like neuroendocrine, but not ovarian, impairments. *Biol. Reprod.* **32,** 1010–1017.

Mobbs, C. V., Cheyney, D., Sinha, Y. N., and Finch, C. E. (1985b). Age-correlated and ovary-dependent changes in relationships between plasma estradiol and luteinizing hormone, prolactin, and growth hormone in female C57BL/6J mice. *Endocrinology* **116,** 813–820.

Mobbs, C. V., Flurkey, K., Gee, D. M., Yamamoto, K., Sinha, Y. N., and Finch, C. E. (1984a). Estradiol-induced adult anovulatory syndrome in female C57BL/6J mice: Age-like neuroendocrine, but not ovarian impairments. *Biol. Reprod.* **30,** 556–563.

Mobbs, C. V., Gee, D. M., and Finch, C. E. (1984b). Reproductive senescence in female C57BL/6J mice: Ovarian impairments and neuroendocrine impairments that are partially reversible and delayable by ovariectomy. *Endocrinology* **115,** 1653–1662.

Mobbs, C. V., Harlan, R. H., and Pfaff, D. W. (1987). An estradiol-induced protein synthesized in the ventral medial hypothalamus and transported to the midbrain central gray. *J. Neurosci.,* in press.

Morell, J. I., and Pfaff, D. W. (1982). Characterization of estrogen-concentrating hypothalamic neurons by their axonal projections. *Science* **217,** 1273–1275.

Morrell, J. I., and Pfaff, D. W. (1983). Retrograde HRP identification of neurons in the rhombencephalon and spinal cord of the rat that project to the dorsal midbrain. *Am. J. Anat.* **167,** 229–240.

Morrell, J. I., Greenberger, L. M., and Pfaff, D. W. (1981). Hypothalamic, other diencephalic, and telencephalic neurons that project to the dorsal midbrain. *J. Comp. Neurol.* **201,** 589–620.

Morrell, J. I., McGinty, J. F., and Pfaff, D. W. (1985). A subset of β-endorphin- or dynorphin-containing neurons in the medial basal hypothalamus accumulates estradiol. *Neuroendocrinology* **41,** 417–426.

Nelson, J. F., Felicio, L. S., Randall, P. K., Sims, C., and Finch, C. E. (1982). A longitudinal study of estrous cyclicity in aging C57BL/6J mice. I. Cycle frequency, length, and vaginal cytology. *Biol. Reprod.* **27,** 327.

Nishizuka, M., and Arai, Y. (1981). Organizational action of estrogen on synaptic pattern in the amygdala—Implications for sexual differentiation of the brain. *Brain Res.* **213,** 422–426.

O'Malley, B. W., and Means, A. R. (1974). Female steroid hormones and target cell nuclei. *Science* **183,** 610–620.

Parsons, B., McEwen, B. S., and Pfaff, D. W. (1982). A discontinuous schedule of estradiol treatment is sufficient to activate progesterone-facilitated feminine sexual behavior and to increase cytosol receptors for progestins in the hypothalamus of the rat. *Endocrinology* **110**, 613–619.

Peng, M. T., and Huang, H.-H. (1972). Aging of hypothalamic-pituitary-ovarian function in the rat. *Fertil. Steril.* **23**, 535.

Pfaff, D. W. (1968). Autoradiographic localization of radioactivity in the rat brain after injection of tritiated sex hormones. *Science* **161**, 1355–1356.

Pfaff, D. W. (1970). Nature of sex hormone effects on rat sex behavior: Specificity of effects and individual patterns of response. *J. Comp. Physiol. Psychol.* **73**, 349–358.

Pfaff, D. W. (1973). Luteinizing hormone-releasing factor potentiates lordosis behavior in hypophysectomized ovariectomized female rats. *Science* **182**, 1148–1149.

Pfaff, D. W. (1980). "Estrogens and Brain Function." Springer-Verlag, New York.

Pfaff, D. W., and Keiner, M. (1973). Atlas of estradiol-concentrating cells in the central nervous system of the female rat. *J. Comp. Neurol.* **151**, 121–158.

Pfaff, D. W., and Lewis, C. (1974). Film analyses of lordosis in female rats. *Hormones Behav.* **5**, 317–335.

Pfaff, D. W., and Sakuma, Y. (1979). Faciliation of the lordosis reflex of female rats from the ventromedial nucleus of the hypothalamus. *J. Physiol.* **288**, 189–202.

Pfaff, D. W., Rosello, L., and Blackburn, P. (1984). Proteins synethesized in medial hypothalamus and transported to midbrain in estrogen-treated female rats. *Exp. Brain. Res.* **57**, 204–207.

Rainbow, T. C., Parsons, B., and McEwen, B. S. (1982). Sex differences in rat brain oestrogen and progestin receptors. *Nature (London)* **300**, 648–650.

Raisman, G., and Field, P. (1973). Sexual dimorphism in the neuropil of the preoptic area of the rat and its dependence on neonatal androgen. *Brain Res.* **54**, 1.

Ramaley, J. A. (1979). Development of gonadotropin regulation in the prepubertal mammal. *Biol. Reprod.* **20**, 1.

Rhighetti, P. G., and Caravaggio, T., (1976). Isoelectric points and molecular weights of proteins. A table. *J. Chromotogr.* **127**, 1–28.

Rhigetti, P. G., and Tudor, G. (1981). Isoelectric points and molecular weights of proteins. A new table. *J. Chromatogr.* **220**, 115–194.

Romano, G. J., Harlan, R. E., Shivers, B. D., Howells, R. D., and Pfaff, D. W. (1986). Estrogen increases proenkephalin mRNA levels in the mediobasal hypothalamus of the rat. *Soc. Neuroscience Abstr.* **12**, 692.

Rothfeld, J. M., Harlan, R. E., Shivers, B. D., and Pfaff, D. W. (1986). Reversible disruption of lordosis via midbrain infusions of procaine and tetrodotoxin. *Pharmacol. Biochem. Behav.* **25**, 857–863.

Sakuma, Y., and Pfaff, D. W. (1979). Mesencephalic mechanisms for integration of female reproductive behavior in the rat. *Am. J. Physiol.* **237**, R285–R290.

Sakuma, Y., and Pfaff, D. W. (1980). Excitability of female rat central gray cells with meduallary projections: Changes produced by hypothalamic stimulation and estrogen treatment. *J. Neurophysiol.* **44**, 1012–1023.

Sakuma, Y., and Pfaff, D. W. (1981). Electrophysiological determination of projections from ventromedial hypothalamus to midbrain central gray: differences between female and male rats. *Brain Res.* **225**, 184–188.

Sakuma, Y., and Pfaff, D. W. (1983). Modulation of the lordosis reflex of female rats by LHRH, its antiserum and analogs in the mesencephalic central gray. *Neuroendocrinology* **36**, 218–224.

Schipper, H., Brawer, J. R., Nelson, J. F., Felicio, L. S., and Finch, C. E. (1981). The role of gonads in the histologic aging of the hypothalamic arcuate nucleus. *Biol. Reprod.* **25**, 413.

Schoonmaker, J. M., Breedlove, S. M., Arnold, A. P., and Gorski, R. A. (1983). Accumulation of steroid in the sexually-dimorphic nucleus of the preoptic area in the neonatal rat hypothalamus. *Soc. Neurosci. Abstr.* **9**, 1094.

Scouten, C. W., Heydorn, W. E., Creed, J. G., Malsbury, C. W., and Jacobowitz, D. M., (1985). Proteins regulated by gonadal steroids in the medial preoptic and ventromedial hypothalamic nuclei of male and female rats. *Neuroendocrinology* **41**, 237–245.

Shivers, B. D., Harlan, R. E., Pfaff, D. W., and Schacter, B. S. (1986). Combination of immunocytochemistry and in situ hybridization in the same tissue section of rat pituitary. *J. Histochem. Cytochem.* **34**, 39–43.

Siekevitz, P. (1985). The postsynaptic density: A possible role in long-lasting effects in the central nervous system. *Proc. Natl. Acad. Sci.* **82**, 3494–3498.

Toran-Allerand, C. D. (1976). Sex steroids and the development of the newborn mouse hypothalamus and preoptic area in vitro: Implications for sexual differentiation. *Brain Res.* **106**, 407–412.

Toran-Allerand, C. D. (1981). Gonadal steroids and brain development. *Trends Neurosci.* **4**, 118–121.

Toran-Allerand, C. D., Gerlach, J. L., and McEwen, B. S. (1980). Autoradiographic localization of 3H-estradiol related to steroid responsiveness in cultures of the hypothalamus and preoptic area. *Brain Res.* **184**, 517–522.

Toran-Allerand, C. D., Hashimoto, K., Greenough, W. T., and Saltarelli, M. (1983). Sex steroids and the development of the newborn mouse hypothalamus and preoptic area in vitro: III. Effects of estrogen on dendritic differentiation. *Dev. Brain Res.* **7**, 97–101.

Vaitukaitis, J. L. (1981). Production of antisera with small doses of immunogen- Multiple intradermal injections. *Methods Enzymol.* **73**, 46–52.

Walter, P., Green, S., Greene, G., Krust, A., Bornert, J.-M., Jeltsch, J-M., Staub, A., Jensen, E., Scrace, G., Waterfield, M., and Chambon, P. (1985). Cloning of the human estrogen receptor cDNA. *Proc. Soc. Natl. Acad. Sci. U.S.A.* **82**, 7889–7893.

Williams, G. C. (1957). Pleiotropy, natural selection, and the evoluation of senescence. *Evolution* **11**, 398.

Wise, P. M. (1983). Aging of the female reproductive system. *Rev. Biol. Res. Aging* **1**, 195.

Wise, P. M., and Camp, P. (1984). Changes in concentration of estradiol nuclear receptors in the preoptic area, medial basal hypothalamus, amygdala and pituitary gland of middle-aged and old cycling rats. *Endocrinology* **114**, 92.

Wise, P. M., Ratner, A., and Penke, G. T. (1976). Effect of ovariectomy on serum prolactin concentrations in old and young rats. *J. Reprod. Fertil.* **47**, 363.

Wise, P. M., London, E. D., Cohen-Becker, I. R., and Weiland, N. G. (1985). Circadian rhythm of local cerebral glucose utilization (LCGU) in estradiol-treated ovariectomized rats: Effect of age. *Soc. Neurosci. Abstr.* **11**, 951.

Young, R. A., and Davis, R. W. (1983). Efficient isolation of genes using antibody probes. *Proc. Natl. Acad. Sci. U.S.A.* **80**, 1194–1198.

CURRENT TOPICS IN MEMBRANES AND TRANSPORT, VOLUME 31

Steroid Hormone Influences on Cyclic AMP-Generating Systems

ALLAN HARRELSON[1] AND BRUCE MCEWEN

The Rockefeller University
New York, New York 10021

I. INTRODUCTION

Steroid hormones regulate a variety of physiological systems and processes, including intermediary metabolism, function of the immune system, ionic transport in intestinal and renal epithelia, hypothalamic and pituitary hormone synthesis and secretion, and certain aspects of behavior. Indeed, the range and variety of steroid actions are so immense that it is difficult to perceive any common cellular mechanism whereby steroids exert their effects, other than the fact that in most cases steroids have their primary action on gene transcription (Yamamoto, 1985). What genes are regulated by steroids? Are there some common biochemical pathways and processes

---

[1] Present address: Department of Biological Sciences, Stanford University, Stanford, California 94305.

(e.g., glycolysis, methylation reactions, ATPase activity) that are focal points for steroid action? One possibility for such a common process is the metabolism of cyclic nucleotide second messengers, in particular the synthesis and degradation of cyclic adenosine $3',5'$-monophosphate (cAMP) by adenylate cyclase and phosphodiesterase. In this article we shall review the evidence that steroids influence the activity of the cyclic AMP-generating system, after first describing the structure and function of adenylate cyclase. The examples will be presented according to organ or tissue, starting with the nervous system. Then various mechanisms will be discussed by which steroids may influence cAMP generation and breakdown. This article will be limited to discussing cAMP and adenylate cyclase. Although there is evidence for steroid influence on cyclic guanosine $3',5'$-monophosphate (cGMP) levels (Morton and Truman, 1985; Amechi et al., 1985), little is known about the biochemistry and physiology of cGMP synthesis.

Before beginning the article there are two important qualifications. First, steroids may often alter cellular biochemistry without having any detectable effect on cAMP generation. For example, although glucocorticoids increase the synthesis of the hepatic enzyme tyrosine aminotransferase (TAT) both in vivo and in vitro, Granner et al. (1968) showed that, in hepatoma cells cultured in vitro, adrenocorticosteroids stimulated a 10- to 15-fold increase in TAT synthesis without having any effect on cAMP levels. Second, where steroid hormones do alter cellular cAMP levels, they may have separate, and even seemingly opposite, effects within the same cAMP-linked biochemical system or pathway. This is the case in the liver and the heart. In these tissues, norepinephrine affects cAMP content as well as the rates of gluconeogenesis and glycogen depletion. An increase in the level of cAMP is correlated with the activation of both gluconeogenesis and glycogen depletion. This response to norepinephrine by these tissues is controlled by glucocorticoids, but in a way that is almost paradoxical; i.e., while glucocorticoids decrease the stimulation of adenylate cyclase by norepinephrine, they strongly enhance the potency of cAMP to elicit both gluconeogenesis and glycogenolysis (Schaeffer et al., 1969; Miller et al., 1971; Exton et al., 1972; Chan et al., 1979; Davies et al., 1981). This seemingly contradictory action of the steroid hormone suggests that, although steroids may alter the function of adenylate cyclase in one way, they may affect cAMP-dependent enzymes in the opposite direction. Thus, a complete understanding of the biochemical mechanism of a steroid action requires knowing far more than just how adenylate cyclase is affected.

II. ADENYLATE CYCLASE

Adenylate cyclase consists of a group of proteins that stimulate the synthesis of cAMP from adenosine triphosphate (ATP). A detailed overview of adenylate cyclase is beyond the scope of this article, but can be found in a number of reviews (Smigel et al., 1984; Birnbaumer et al., 1985; Gilman, 1985). The primary component of the adenylate cyclase complex is the so-called catalytic subunit, a large protein that resides in the inner face of the plasmalemma and is responsible for the actual hydrolysis of ATP (Fig. 1). Several groups have reported partial purification of the catalytic subunit (Coussen et al., 1985; Pfeuffer et al., 1985; Yeager et al., 1985); in all cases the catalytic subunit was found to have an approximate molecular weight of 135,000–150,000. The genes coding for the catalytic subunit from Escherichia coli and from yeast have been cloned and the primary structure of the protein deduced (Aiba et al., 1984; Casperson et al., 1985; Katoaka et al., 1985). When purified from brain tissue, the protein has been found to contain a calmodulin-binding site, which may impart calcium sensitivity (Coussen et al., 1985; Pfeuffer et al., 1985; Yeager et al., 1985).

The true engine of the adenylate cyclase complex is two heterotrimeric proteins known as the guanyl nucleotide-binding stimulatory coupling protein (Ns) and the guanyl nucleotide-binding inhibitory coupling protein (Ni) (Fig. 1) (Birnbaumer et al., 1985). The basal adenylate cyclase activity of the catalytic subunit is fairly low, and it is Ns that is responsible for stimulating that activity to significant levels. The magnitude and efficiency of that stimulation can in turn be inhibited by Ni. Both Ns and Ni consist of α-, β-, and γ-subunits; the β- and γ-subunits seem to be identical between Ns and Ni, whereas the α-subunits seem to differ, and hence to endow Ns and Ni with their distinctive properties (Northrup et al., 1980; Codina et al., 1983; Bokoch et al., 1983; Hildebrandt et al., 1984).

The action of the guanyl nucleotide-coupling proteins requires both the binding of magnesium and the binding and subsequent hydrolysis of guanosine triphosphate (GTP) (Sigel et al., 1984; Gilman, 1985; Birnbaumer et al., 1985). Activation seems to involve binding of GTP and Mg^{2+}, followed by dissociation of the α-subunit from the β- and γ-subunits; it is the α-subunit that interacts with the catalytic subunit (Northrup et al., 1980; Codina et al., 1983; Bokoch et al., 1983). Hydrolysis of GTP and release of the α-subunit from the catalytic subunit terminate the activation cycle, at which point the α-, β-, and γ-subunits reunite. A similar inhibition cycle occurs for Ni. Thus, in the unstimulated state, adenylate cyclase activity reflects the steady state between the activation cycle of Ns and the inhibition cycle of Ni (reviewed in Birnbaumer et al., 1985). It has become clear that Ns and Ni may have other actions than modification of the adenylate cyclase catalytic subunit and may be subject themselves to modification by other types of receptor mechanisms than those that activate or inhibit adenylate cyclase. This will be discussed further in Section IV (Nakamura and Ui, 1985; Gilman, 1985).

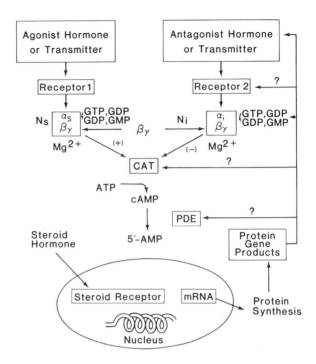

FIG. 1. Possible sites of steroid effects' on cyclic AMP generation. This schematic diagram shows how adenylate cyclase is activated by agonist hormones acting through Ns, or inhibited via antagonist hormones acting through Ni. The cAMP generated is broken down by phosphodiesterase (PDE). CAT, Catalytic subunit of adenylate cyclase. The components of the stimulatory and inhibitory coupling proteins, Ns and Ni, are shown, as is the likelihood that they share two subunits, β and γ. (+), Stimulates; (−), inhibits. Steroid hormones, acting via the intracellular receptors that affect genomic activity, induce or repress gene products. These products may either be part of the adenylate cyclase system outlined in the figure or they may regulate it. In addition, a few direct effects of steroids have been identified (e.g., progesterone in the *Xenopus* oocyte) but the exact site of action is not known.

Activation or inhibition of adenylate cyclase by hormones acting via cell surface receptors accelerates either the activation cycle of the inhibition cycle, depending upon the particular type of receptor activated (Fig. 2). The hormone binds to its receptor, which causes a conformational change in the receptor that endows it with the ability to interact with the appropriate GTP-binding protein. It is not yet known what features of the receptor molecule allow it to interact with its guanyl nucleotide-binding protein.

III. STEROID MODULATION OF cAMP METABOLISM

There are many instances in which changes in steroid hormone exposure bring about changes in hormone-stimulated cAMP activity. For the most part, these are not direct effects of the steroid on adenylate cyclase itself. A summary of these effects is presented in Table I, according to the major steroid classes. These examples indicate that steroid regulation of adenylate cyclase is highly tissue or brain region specific. There is little indication of a common steroid effect occurring in all tissues. Rather, a steroid may increase the cAMP response of one tissue to an agonist while diminishing the response of another tissue to the same agonist. This type of specificity is described in the text below, where some of the examples from Table I are grouped according to tissue or organ.

A. Nervous System

The tissue-specific nature of steroid regulation of the cAMP system is particularly well illustrated by studies on the nervous system. Given the extreme interconnectedness of the brain and the mutual interactions that exist between nerve cells, it may be somewhat surprising that modulation of adenylate cyclase by steroids differs greatly between brain regions.

1. Hypothalamus

The first studies of steroid effects on cyclic nucleotide synthesis in the central nervous system (CNS) were conducted on the hypothalamus. Acute administration of estradiol to intact female rats increases basal hypothalamic cAMP levels (Gunaga *et al.*, 1974). Subsequent studies of estrogen action on hypothalamus incubated *in vitro* also revealed a stimulation of cAMP accumulation, but these effects occurred at such high concentrations of estrogen as to be nonphysiological (Gunaga and Menon, 1973; Weissman *et al.*, 1975; Weissman and Skolnick, 1975). Later studies, using *in vivo* administration of steroids to female rats, showed that both estradiol and progesterone have significant effects on norepinephrine-stimulated cAMP accumulation in hypothalamic slices, estradiol being generally stimulatory and progesterone inhibitory (Etgen and Petitti, 1986).

Dopamine-stimulated adenylate cyclase is significantly elevated in female rat medial hypothalamus during proestrus, when estrogen levels are highest (Barr *et al.*, 1983). The same authors reported that castration enhanced catecholamine-stimulated adenylate cyclase in medial hypothalamus (Barr *et al.*, 1983). Although analysis of these apparent sex differences is not complete, these studies suggest that the sexual differentiation of the hypothalamus may

TABLE I
STEROID MODULATION OF CYCLIC AMP METABOLISM[a]

Steroid	Tissue	Agonist	Effect	Reference
A. Adrenal steroids				
Dex	Embryonic rat calvaria	PTH, PGE₂, NaF, GppNHp	Increased cAMP accumulation	Chen and Feldman (1978, 1979)
	Calvaria-derived cell lines	PTH	Increased cAMP accumulation	Ng et al. (1979)
	Osteosarcoma-derived cell lines	PTH, Isop	Enhance AC activity	Rodan et al. (1984)
TA	Osteosarcoma-derived cell lines	PTH, forskolin	Increased cAMP accumulation	Catherwood (1985)
Adx	Rat myocardial cells	Isop	Increased EC_{50} for stimulation	Davies et al. (1981)
Dex, Cort	3T3-L1 adipocyte cell line	Isop	Increased cAMP accumulation	Lai et al. (1982)
Dex	Hepatoma cells	PGE₂, catecholamines	Increased cAMP accumulation	Manganiello and Vaughan (1972); Manganiello et al. (1972); Ross et al. (1977)
Cort, Dex	Transformed fibroblasts	Isop, cholera toxin GTP	Increased cAMP and AC	Johnson and Jaworski (1983)
Cort	Rat liver	Isop, epinephrine	Decreased cAMP levels and AC activity	Bitensky et al. (1970); Exton et al. (1972); Wolfe et al. (1976)
Adx	Rat fat pads	Epinephrine	Increased cAMP accumulation	Exton et al. (1972)
Adx	Rat fat cells	Isop, GppNHp	Decreased AC activity	Thotakura et al. (1982)

Dex	Rat adipocyte membranes	N-6-Phenylisopropyl-adenosine	Decreased AC activity	Mazancourt et al. (1982)
Dex	Rat kidney medullary membranes	Vasopressin	Increased maximal AC activity	Rajerison et al. (1974)
F	Human lymphocytes		Increase cAMP response	Marone et al. (1980); Lee and Reed (1977)
Cort, Dex	Rat anterior pituitary cells	VIP, CRF	Decrease cAMP accumulation	Rotsztejn et al. (1981); Bilezikjian and Vale (1983); de Souza et al. (1985)
Dex	Rat anterior pituitary cells	CRF	Prevents desensitization of AC	Ceda and Hoffman (1986)
Cort	Rat frontal cortex and limbic forebrain	NE	Decrease cAMP accumulation	Mobley and Sulser (1980a,b); Mobley et al. (1983)
Cort, Dex	Rat hippocampus	NE, Isop, VIP	Decrease cAMP accumulation	Roberts et al. (1984,); Harrelson et al. (1983, 1987); Harrelson and McEwen (1985)
Dex	Rat hippocampus and amygdala	Histamine	Increase cAMP accumulation	Harrelson et al. (1987)
Dex	Rat amygdala and septum	VIP	Decrease cAMP accumulation	Harrelson et al. (1987)
Dex, F	Human astrocytoma cell line, C6 glioma cells, human neuro-blastoma	Isop, GppNHp, NaF, PGE_2	Increase cAMP accumulation and AC activity	Foster and Harden (1980); Brostrom et al. (1974); Forray and Richelson (1985)
Aldo	Toad bladder epithelium	Vasopressin	Increase cAMP accumulation	Stoff et al. (1972)
Dex, Cort, DOC	Rat anterior pituitary	PGE_2	Decrease basal cAMP and cAMP accumulation	Tornello et al. (1984)

(continued)

TABLE I (continued)

Steroid	Tissue	Agonist	Effect	Reference
B. Vitamin D				
D3	Bone cells	PTH, PGE$_2$	Inhibition of cAMP accumulation	Crowell et al. (1981); Kent et al. (1980)
D3	Osteosarcoma cell line	PTH, PGE$_2$, calcitonin, forskolin, Isop	Inhibit stimulation of AC	Kubata et al. (1985)
D3	Monocyte-like cell line	Isop, forskolin	Inhibition of cAMP accumulation	Rubin and Catherwood (1984)
D3	Osteosarcoma cell line	PTH, forskolin	Blocks glucocorticoid enhancement of PTH- and forskolin-stimulated AC	Catherwood (1985)
C. Gonadal Steroids				
E	Female rat hypothalamus		Increase basal cAMP levels	Gunaga and Menon (1974); Gunaga et al. (1974); Weissman et al. (1975); Weissman and Skolnik (1975)
E	Female rat hypothalamus and preoptic area	NE	Increase cAMP accumulation	Etgen and Petitti (1985)
P	E-primed female rat hypothalamus and preoptic area	NE	Decrease cAMP accumulation	Etgen and Petitti (1985)
E	Medial hypothalamus of proestrous female rat	Dopamine	Increase stimulation of AC	Barr et al. (1983)

Cast	Medial hypothalamus of male rat	Catecholamines	Enhances stimulation of AC	Barr et al. (1983)
Ovx	Female rat striatum and nucleus accumbens	Dopamine	Decreased stimulation of AC at 4 days and increased at 4–6 weeks	Kumakura et al. (1979)
E	Female rat cerebral cortex	Isop	Decreased stimulation of AC	Wagner et al. (1979)
Cast	Male rat hippocampus	Histamine	Decreased cAMP accumulation	Harrelson and McEwen (unpublished)
E + P	Female rat hippocampus	Isop, histamine	Decreased cAMP accumulation	Harrelson and McEwen (unpublished)
E	Rat ovarian granulosa cells	FSH	Enhance cAMP accumulation	Richards et al. (1976)
E	Rabbit corpus luteum	LH, Isop	Inhibit cAMP accumulation	Knecht et al. (1984)
E	Human endometrium membranes	NaF	Increases basal AC activity	Bergamini et al. (1985)
Estrone	Rat uterus	NaF	Increase basal and NaF-stimulated AC	Bekiari et al. (1984)
P	Xenopus oocyte	None	Inhibits AC	Finidori-Lepicard et al. (1981)
P	Female rat liver	Epinephrine	Decrease stimulation of AC	Bitensky et al. (1970)
P	Osteosarcoma cell line	PTH	Inhibition of AC	Rodan et al. (1984)

[a] Examples are grouped according to the steroid or endocrine manipulation which was utilized, as indicated on the left. Dex, Dexamethasone; Adx, adrenalectomy; Cort, corticosterone; Aldo, aldosterone; DOC, deoxycorticosterone; D3, 1,25-dihydroxyvitamin D; TA, triamcinolone acetonide; F, cortisol; Cast, castration; Ovx, ovariectomy; E, estradiol; P, progesterone; Isop, isoproterenol; PTH, parathyroid hormone; PGE, prostaglandin E; NE, norepinephrine; FSH, follicle-stimulating hormone; AC, adenylate cyclase; LH, luteinizing hormone; NaF, sodium fluoride.

be mirrored by sexual differentiation in steroid regulation of adenylate cyclase. They may also indicate that where a brain region has disparate roles in feedback regulation of the pituitary–gonadal axis, there may be highly specific regulation of adenylate cyclase by different steroids.

2. STRIATUM

In the corpus striatum and nucleus accumbens, the response of dopamine-stimulated adenylate cyclase to gonadal steroids differs from that in the hypothalamus. Ovariectomy has a biphasic effect on dopamine-stimulated adenylate cyclase activity in striatum and nucleus accumbens, with ovariectomized rats having increased cAMP formation in response to dopamine 4 days after surgery but lesser basal and dopamine-stimulated adenylate cyclase after 4 and 6 weeks (Kumakura *et al.*, 1979). Thus, changes in adenylate cyclase after a change in steroid status may be time dependent.

The existence of estrogen effects is surprising, since the striatum–accumbens region has very few if any specific receptors for female gonadal steroids (Pfaff and Keiner, 1973; Parsons *et al.*, 1982). That changes are observed in this area suggests that the proximate effects of the steroids may in fact occur in other brain regions that influence adenylate cyclase in the striatum via neural connections.

3. CEREBRAL CORTEX

Cerebral cortex is another area where steroid modulation of adenylate cyclase takes place. Chronic estradiol treatment of female rats has been shown to decrease β-adrenergic-stimulated cAMP accumulation in slices of cerebral cortex (Wagner *et al.*, 1979). The cerebral cortex has relatively few estradiol receptors, suggesting that the primary action of estradiol might be on another brain region that projects to the cortex. This interpretation gains support from the observation that chronic exposure to estradiol increases turnover of norepinephrine in the cortex of female rats and increases release of catecholamines from neural tissue *in vitro* (Paul *et al.*, 1979). Thus, Wagner *et al.* (1979) interpreted the decrease in cAMP response as resulting from the desensitization of adenylate cyclase subsequent to an increase in cortical norepinephrine turnover induced by estradiol.

The adrenergic activation of cortical adenylate cyclase activity is also modified by adrenocortical hormones. Mobley and Sulser (1980) reported that adrenalectomy increases norepinephrine-stimulated accumulation of cAMP in slices from rat frontal cortex. They later found that this was due to the decrease in serum glucocorticoids following adrenalectomy (Mobley and Sulser, 1980a,b; Mobley *et al.*, 1983). Adrenalectomy increased norepinephrine-stimulated cAMP accumulation by approximately 70% in slices of frontal cortex (Mobley and Sulser, 1980a). Basal cAMP levels were not altered. The change in cAMP generation reflected an increase in maximal responsiveness of the adenylate cyclase system, and not a decrease in the EC_{50} (i.e., half-maximally effective agonist concentration) for norepinephrine stimulation. The period following adrenalectomy was 14 days, and corticosterone (Cort) replacement effects occurred over 3–5 days, indicating that Cort does not directly inhibit adenylate cyclase. Rather, the lag period may involve gene transcription. Replacement with Cort was effective in hypophysectomized animals, suggesting that the steroid action was directly on the brain, and not via regulating adrenocorticotrophic hormone secretion from pituitary (Mobley et al., 1983).

These experiments do not pinpoint, however, what components of the cortical cAMP response mechanism are altered by glucocorticoids. The effects of adrenalectomy and Cort were obtained in the presence of phosphodiesterase inhibitor, suggesting that the changes resulted from alterations in a component of adenylate cyclase, and not phosphodiesterase. This was confirmed by direct measurement of phosphodiesterase activity (Mobley et al., 1983). Nor was there any alteration in basal or NaF-stimulated adenylate cyclase in membranes prepared from frontal cortex in adrenalectomized animals, suggesting that changes in the catalytic subunit of adenylate cyclase were not responsive for the steroid effect. The effects of adrenalectomy and Cort replacement were not observed when the slices were stimulated with the β-adrenergic agonist isoproterenol, which suggests that α-adrenergic receptors might be involved. However, there was no change in the number or affinity of β- or α-adrenergic receptors after adrenalectomy, implying that changes in receptor numbers were not responsible for the change. At present, the exact mechanism remains obscure, but it may involve steroid regulation of adenylate cyclase-coupling factors such as Ns and Ni.

4. Hippocampus and Other Limbic Brain Regions

Regulation of adrenergic-stimulated adenylate cyclase by glucocorticoids also occurs in parts of the limbic system, where the highest levels of glucocorticoid receptors are found (McEwen et al., 1986). Mobley and Sulser (1980a,b) reported that removal of the adrenal glands from rats caused norepinephrine-stimulated accumulation of cAMP to increase in limbic forebrain (composed of olfactory tubercles, rostral limbic nuclei, nucleus accumbens, septalnuclei, and parts of the anterior amygdaloid nuclei). This increase in adrenergic-stimulated adenylate cyclase after adrenalectomy is similar to that found in the cortex and is supported by the results of Roberts et al. (1984), who found that adrenalectomy caused a small (approximately 20%) increase in norepinephrine-stimulated cAMP levels in slices from rat hippocampus. No change in basal cAMP levels was observed. The effect of adrenalectomy was reversed by Cort replacement in vivo. Adrenalectomy effects could be mimicked by in vivo administration of metopirone, an antagonist of glucocorticoid synthesis. The experiments were run in the presence of phosphodiesterase inhibitors and hence the results presumably represent changes in adenylate cyclase and not phospho-diesterase. Because there was no alteration in the number or properties of hippocampal β-adrenergic receptors following adrenalectomy (Roberts *et al.*, 1984), the effects on cAMP formation may have resulted from hormonal influences on Ns, Ni, or adenylate cyclase.

Our own work supports these previous investigators and indicates that glucocorticoids and gonadal steroids modulate cyclic nucleotide synthesis in the hippocampus and in the amygdala in response to β-adrenergic agonists and histamine as well as the neuropeptide vasoactive intestinal peptide (VIP) (Harrelson *et al.*, 1983, 1987; Harrelson and McEwen, 1985). In the hippocampus, treatments with the synthetic glucocorticoid dexamethasone (Dex) or Cort suppress VIP-stimulated cAMP accumulation, while at the same time they enhance histamine-stimulated cAMP accumulation. The ability of adrenalectomy to alter adenylate cyclase is specific for each brain region, in that not all the limbic brain areas respond the same way to adrenalectomy or glucocorticoid replacement. For example, β-adrenergic stimulation of cAMP accumulation in the amygdala and septum was not affected by adrenalectomy.

Suppressive effects of glucocorticoid administration on VIP-stimulated cAMP generation were seen in two other regions of the limbic system, the amygdala and the septum, but not in other regions such as the frontal cortex and the olfactory bulbs. Thus, glucocorticoid regulation of VIP-stimulated cAMP accumulation is region specific and occurs in brain areas such as the hippocampus and amygdala that have high levels of glucocorticoid receptors (McEwen *et al.*, 1986) and are involved in neuroendocrine regulation of the hypothalamic–pituitary–adrenal axis. The effects of adrenalectomy and *in vivo* Dex replacement were apparent only after 2 or more days, and Dex replacement *in vitro* was ineffective. This suggests, once again, that an indirect, transcriptionally mediated mechanism might be operative. Interestingly, the ability of Dex to regulate the hippocampal response to VIP seemed to depend partially upon the presence of adrenocorticotrophic hormone (Harrelson and McEwen, 1985, 1987a).

Gonadal steroids also regulate the cAMP system in the hippocampus (Harrelson and McEwen, 1987b). Histamine-stimulated cAMP generation was diminished in hippocampal slices 6 days after castration of male rats, but there was no effect on VIP or norepinephrine stimulation of cAMP formation. Treatment of ovariectomized female rats with estradiol and progesterone had no effect on VIP-stimulated cAMP accumulation in hippocampal slices, but did suppress cAMP formation in response to both histamine and isoproterenol. Although there was no effect of gonadal steroid treatment on the VIP-stimulated cAMP response, Dex replacement of adrenalectomized female rats induced an increase in the response to VIP. This result is opposite to that in male rats, in which Dex treatment diminished VIP-stimulated adenylate cyclase. Thus, in the hippocampus we observe regulation by gonadal and adrenal steroids in a manner that is highly region specific and may be sexually differentiated.

In summary, despite the great connectedness and mutual interdependence of brain cells, the CNS shows a great deal of regional specificity in steroid regulation of the cAMP response. These regional differences may reflect the great heterogeneity of neurons. There is also the possibility that a steroid action on any particular brain nucleus may be manifested at synaptic terminals some distance away, and thus an effect directly on one small brain area might influence a wide variety of other brain regions.

B. Pituitary

Like the brain, the pituitary gland is an important site for the integration of hormonal signals involved in many physiological processes. However, owing to the lesser diversity of its cell populations and the relative ease whereby pituitary cells can be cultured *in vitro*, it furnishes a more tractable system than the brain for analysis of steroid effects. As a result, we are learning more about the cellular mechanisms involved in steroid regulation of the cAMP system, and it is also evident that steroid modulation of cAMP levels is important in the feedback control of pituitary hormone secretion.

For example, glucocorticoids strongly inhibit adrenocorticotrophic hormone release from pituitary corticotrophs, and they have been found to modulate cAMP levels both *in vivo* and in dispersed pituitary cells. Both Cort and Dex decrease cAMP accumulation in the anterior pituitary gland elicited by corticotropin-releasing factor (CRF) (Bilezikjian and Vale, 1983; de Souza *et al.*, 1985). Because CRF-stimulated adrenocorticotrophic hormone release is dependent upon increased cAMP synthesis, it seems likely that glucocorticoid modulation of the corticotroph cAMP system is a key component of the feedback inhibition of adrenocorticotrophic hormone synthesis and release.

Changes in the CRF receptors appear to be an important component of the mechanism of adrenal steroid action, because adrenalectomy leads to a decrease in the number of pituitary receptors for CRF and a concomitant shift to the right in the dose–response curve for CRF-stimulated adenylate cyclase (de Souza *et al.*, 1985). Both these phenomena may reflect the result of agonist-induced down-regulation caused by the dramatic increase in CRF release into the portal circulation after adrenalectomy.

Yet the pituitary corticotroph may be capable of a negative response to glucocorticoids independently of continual exposure to CRF. Bilezikjian and Vale (1983) reported that treatment of cultured rat anterior pituitary cells with 20 nM Dex for 18 hours, in the absence of CRF, blocked CRF-stimulated adrenocorticotrophic hormone release and attenuated cAMP production. Accumulation of cAMP was reduced by 50–60%, due both to decreased maximal response and a shift in the CRF dose–response curve to higher doses. The effect was specific for glucocorticoids and occurred over a Dex concentration range of 1–100 nM; Cort was effective at higher doses, beginning at 250 nM. Although the mechanism of the Dex effect was not investigated directly, it did not seem to involve phosphodiesterase, because it was also seen in the presence of a phosphodiesterase inhibitor, 3-isobutyl-1-methylxanthine.

Another peptide involved in glucocorticoid feedback on the pituitary is VIP. Although the exact role of VIP in the modulation of hormone secretion by the anterior pituitary is not completely clear, it seems that VIP may have a synergistic effect upon CRF-stimulated adrenocorticotrophic hormone release and may affect growth hormone release and also prolactin secretion (Tilders *et al.*, 1984; Rostene, 1984). Like CRF, VIP stimulation of the pituitary causes a large increase in cAMP accumulation. However, regulation of the anterior pituitary adenylate cyclase response to VIP by glucocorticoids differs somewhat from that seen to CRF. Rotsztejn *et al.* (1981) found that in primary cultures of rat pituitary cells, Dex inhibited the maximal cAMP response to VIP without detectably altering the EC_{50} for VIP stimulation. There were no alterations in basal cAMP levels caused by Dex, and the effect did not seem to be due to changes in phosphodiesterase activity. This regulatory effect, which is restricted to steroids that are glucocorticoid agonists, is notable for the rapid time course of onset—later than 5 minutes after steroid addition but as early as 15 minutes—and the extremely low doses of Dex that are effective—in the range of 0.01 to 1 pM. Notwithstanding this rapid time course, Dex action is blocked by cycloheximide. It is not clear which components of the adenylate cyclase system might be affected in such a rapid manner and yet involve changes in protein synthesis (Rotsztejn *et al.*, 1981).

C. Gonads, Gametes, and Reproductive Tract

Another prominent steroid hormone target tissue is the reproductive system, and here also there is evidence that steroids exert their effects in part by altering cAMP levels. In the rat ovarian granulosa cell, estradiol enhances the stimulation of cAMP levels by follicle-stimulating hormone, while in the rabbit corpus luteum it attenuates the adenylate cyclase response to luteinizing hormone and the β-adrenergic agonist isoproterenol (Richards *et al.*, 1976; Knecht *et al.*, 1984). As in the brain and pituitary, the mechanism of the steroid effect may involve changes in the properties or levels of Ns or Ni, as well as alterations in the number of receptors. Moreover, the time course of the estrogen effect indicated that it was not a direct activation of the adenylate cyclase enzyme complex itself, but rather involved changes in gene transcription (Richards *et al.*, 1976).

The reproductive system also provides an example in which a steroid appears to modulate cAMP levels by a direct interaction with the adenylate cyclase enzyme complex, namely, the *Xenopus* oocyte, in which progesterone stimulates meiotic maturation by inhibiting cAMP production. Apparently progesterone binds to a specific membrane receptor site, which subsequently affects adenylate cyclase activity (Finidori-Lepicard *et al.*, 1981; Jordana *et al.*, 1981). The inhibition occurs at physiological progesterone concentrations and involves an interaction with a GTP-binding protein different from Ni (Olate *et al.*, 1984; Goodhardt *et al.*, 1984; Sadler *et al.*, 1984). The exact relationship between the membrane receptor site for progesterone and this GTP-binding protein is not yet clear, although it seems fairly certain that they are not the same protein (Blondeau and Baulieu, 1984). Thus, although the mechanism is different and does not involve an effect on gene transcription, the presumed site of the steroid effect on cyclase—coupling proteins—is the same as is postulated for the effect described above, which appears to involve genomic activation and protein synthesis.

D. Liver and Fat Cells

Steroid hormones, especially glucocorticoids, have profound effects on the biochemistry of liver and fat tissues, and steroid regulation of cAMP metabolism in cells derived from liver and adipose tissues has been known for some time. In the liver, adrenalectomy leads to a large increase in β-adrenergic receptor-stimulated adenylate cyclase activity, which is reversed by restoration of glucocorticoids. Exton *et al.* (1972) found that 5–7 days after adrenalectomy, the cAMP response of rat liver to perfusion with epinephrine is elevated by about 100%, with no change in basal liver cAMP levels. Wolfe *et al.* (1976) reported that adrenalectomy caused a three- to fivefold increase in rat liver β-adrenergic-stimulated adenylate cyclase, concomitant with an increase in β-adrenergic receptors, and no changes in basal adenylate cyclase activity or in NaF-stimulated adenylate cyclase.

Adrenalectomy also makes the liver refractory to cAMP-induced glycogenolysis and gluconeogenesis, whereas glucocorticoid replacement allows intracellular cAMP to once again stimulate glucose metabolism (Bitensky *et al.*, 1970; Exton *et al.*, 1972). In other words, glucocorticoids diminish the response of the cAMP-generating system to epinephrine, while they make the cells more sensitive to the cAMP which is produced. As a result of this latter action, glucocorticoids are said to have a "permissive" effect on epinephrine-stimulated glucose metabolism in rat liver, by permitting cAMP to promote glycogen breakdown and gluconeogenesis (Bitenskey *et al.*, 1970; Exton *et al.*, 1972).

Although regulation of cAMP levels by glucocorticoids in adipocytes differs in some ways from that in the liver, the general story is very much the same, namely, effects at both the level of cAMP and in post-adenylate cyclase, cAMP-dependent biochemical phenomena (Exton et al., 1972). However, steroid regulation differs somewhat between the tissues, in that adrenalectomy has a biphasic effect on adenylate cyclase in adipose tissue, with an initial increase followed by a long-term decrease. Adrenalectomy for 5–7 days enhances epinephrine-stimulated cAMP formation in fat pads from rats (Exton et al., 1972). From 1 to 2 weeks following adrenalectomy, isoproterenol-stimulated adenylate cyclase is decreased by 30–40% relative to sham-operated controls in isolated adipocytes (Thotakura et al., 1982). The delayed decrease in response to adrenalectomy seemed to involve both differences in receptors and other components of the adenylate cyclase system, because there was a concomitant 50% decrease in β-receptor number and a reduction in the maximal response of adenylate cyclase to activation by guanylyl imidodiphosphate (GppNHp) (Thotakura et al., 1982).

Interactions with other receptors plays a role in this steroid regulation in adipocytes. When fat cells from adrenalectomized animals were incubated with adenosine deaminase, to remove the effects of adenosine on cAMP generation, isoproterenol-stimulated adenylate cyclase was increased to normal values, implying that the stimulatory regulatory ability of glucocorticoids depends in part upon adenosine (Thotakura et al., 1982). A complex role for adenosine in regulation of adenylate cyclase by steroids was implied by Mazancourt et al. (1982), who reported that the adenosine receptor agonist N-6-phenylisopropyladenosine (PIA) inhibits adenylate cyclase in adipocyte membranes from intact rats, while adipocyte membranes from adrenalectomized rats show a stimulation of adenylate cyclase by PIA at concentrations below 10 nM but on inhibition at PIA concentrations greater than 100 nM. The stimulatory effect of adrenalectomy on adenosine-stimulated adenylate cyclase in adipocytes was blocked by Dex replacement for 7 days (Mazancourt et al., 1982). The authors suggested that perhaps adrenalectomy leads to the appearance of new and different adenosine receptors or produces changes in receptor–adenylate cyclase coupling.

In this regard, there is evidence that glucocorticoids may have differential effects on receptor subtypes. Lai et al. (1982) reported that treatment of 3T3-L1 preadipocytes and adipocytes with Dex caused a loss of β(1)-type adrenergic receptors and an increase in the number of β(2)-adrenoreceptors with a concomitant increase in isoproterenol-stimulated adenylate cyclase. The effect of steroid treatment is first evident at 6 hours and reaches a maximum at 48 hours, indicating a possible genomic action.

E. Bone and Connective Tissue

There is substantial evidence that steroids regulate cyclic nucleotide levels in cells of bone and connective tissue and that this regulation may be important in the hormonal control of calcium metabolism and skeletal growth. Several different hormones are involved in controlling calcium metabolism from the skeleton, and they interact with one another in complex ways and act on subpopulations of cells in bone. Not only does one steroid alter the cAMP response to several agonists, but this steroid regulation is in turn itself modulated by another steroid. In particular, glucocorticoids and vitamin D_3 have been shown to modulate the effectiveness of a variety of agents in stimulating the accumulation of cAMP in bone and bone-derived cells. For example, although vitamin D_3 by itself has no direct effect on cAMP levels in bone (Wong et al., 1977), treatment of isolated bone cells with vitamin D_3 inhibits the stimulation of adenylate cyclase by parathyroid hormone (PTH) within 6 hours with a maximal response at 24 hours (Crowell et al., 1981). Stimulation of cAMP accumulation by prostaglandin E (PGE) is also attenuated by vitamin D_3 (Kent et al., 1980).

Similarly, in the osteosarcoma cell line UMR 106-06, vitamin D_3 decreases the response of adenylate cyclase to PTH and PGE, as well as to calcitonin, forskolin, and the β-adrenergic agonist isoproterenol, without altering basal levels of cAMP (Kubata et al., 1985). For PTH, calcitonin, and PGE, vitamin D_3 decreases the maximum stimulation of adenylate cyclase; in the case of PTH, the steroid also increases the half-maximal PTH concentration from 0.3 to 5 nM. As with isolated bone cells, the steroid action does not occur immediately upon addition of vitamin D_3 to the incubation medium, suggesting once again that it does not involve a significant direct interaction of the steroid with any of the components of adenylate cyclase. Instead, the effectiveness of PTH and isoproterenol were diminished within 48 hours, whereas the responses to calcitonin, PGE, and forskolin were only altered after 96 hours, and thus may involve genomic regulation of components of the adenylate cyclase system.

Vitamin D_3 treatment modifies the properties of the catalytic component of adenylate cyclase, because basal Ns activity remained unchanged and the vitamin D_3 effect was not altered by inhibition of Ni with islet-activating protein (Kubata et al., 1985). There were also changes in number of calcitonin receptors. (Receptors for the other agonists were not measured.) The diminished effectiveness of forskolin might reflect changes in the catalytic subunit itself, or in some undefined component acting closely with it. Kubata et al. (1985) suggest that perhaps vitamin D_3 alters the intracellular concentration or distribution of Mg^{2+}, thus reducing the cyclase activation rate.

Results very similar to these have been reported for a bone-derived monocyte-like cell line. Rubin and Catherwood (1984) found that exposure to vitamin D_3 diminishes the accumulation of cAMP elicited by isoproterenol. The effect was detectable by 8 hours, following steroid addition, and was maximal at 48 hours, at which time the response to isoproterenol was reduced by 55% relative to controls not exposed to vitamin D_3. The response to forskolin was also decreased by vitamin D_3.

Compared to vitamin D_3, glucocorticoids tend to cause opposite changes in the responsiveness of bone cells. Dex administration greatly enhances parathyroid hormone-stimulated cAMP accumulation in embryonic rat calvaria, calvaria-derived cells, and osteosarcoma-derived cell lines (Chen and Feldman, 1978, 1979; Ng et al., 1979; Rodan et al., 1984). In the latter system, β-adrenergic activation of adenylate cyclase also increases after Dex treatment (Rodan et al., 1984). The action of glucocorticoids seems broadly similar in all three cases. In bone cells from fetal rat calvaria, Dex potentiates the PTH-induced cAMP response by one- to twofold in the presence of theophylline, suggesting that adenylate cyclase activity is affected (Chen and Feldman, 1978, 1979). Although basal cAMP levels are not altered by Dex, there is a reduction of phosphodiesterase activity by approximately 40%, which suggests that increases in cAMP levels might result from decreases in phosphodiesterase activity as well as from enhanced activity of adenylate cyclase. The effect on adenylate cyclase becomes evident 24 hours after exposure to Dex; interestingly, Dex does not have to be present during the entire incubation period for it to be effective (Chen and Feldman, 1978). This strongly implies that the steroid is acting by regulating gene transcription.

When bone-derived cells were separated into osteoblast-enriched and osteoclast-enriched subpopulations, it was discovered that, although both cell groups possess roughly the same number of glucocorticoid receptors and although both are inhibited in their growth by glucocorticoids, glucocorticoids enhanced the cAMP response to PTH primarily in the osteoblast fraction (Chen and Feldman, 1979). The complexity of the mechanism became further evident when the authors found that the increase in adenylate cyclase activity reflected both greater sensitivity to PTH as well as increased maximal response. Moreover, in the osteoblast-enriched population, Dex caused an elevation in NaF- and GppNHp-stimulated adenylate cyclase. Yet both cell populations showed increased PGE-stimulated cAMP accumulation when treated with Dex, suggesting that many of the differences in glucocorticoid regulation of adenylate cyclase between the two populations were fairly subtle in nature. Curiously, although both cell types responded to calcitonin with increased cAMP, this response was not altered by Dex in either cell population (Chen and Feldman, 1979).

The effects of glucocorticoids in cultured newborn rat calvaria are similar. Ng et al. (1979) found that after 24 hours in culture with a μM cortisol there was an approximately 100% increase in PTH-stimulated cAMP levels. This effect was enhanced by the phosphodiesterase inhibitor β-isobutyl-1-methylxanthine, suggesting that regulation of phosphodiesterase was not the reason for the enhancement of cAMP levels. The Dex action did not seem to be a direct competitive inhibition, since one-half hour incubation with cortisol had no effect. In the osteosarcoma-derived cell line ROS 17/2.8, Rodan et al. (1984) found that Dex increases the maximal stimulation of cAMP levels by PTH and also makes the cells more sensitive to PTH. Since Dex had no effect on basal cAMP levels or on the activities of either high- or low-affinity phosphodiesterase, it seems likely that the effect is specific for adenylate cyclase. The particular component of adenylate cyclase that is affected is unknown, however, because steroid treatment did not alter the ability of GppNHp or NaF to stimulate adenylate cyclase and forskolin-stimulated cAMP accumulation was not affected. An indirect mechanism of Dex action was implied by the authors' finding that the enhancement of PTH-stimulated cAMP accumulation was first detectable 12 hours after addition of the steroid and increased with time after this point. Interestingly, Rodan et al. (1984) also found that incubation of these cells for 3 days with progesterone (1 μM), an antiglucocorticoid, caused a significant inhibition of PTH-stimulated adenylate cyclase.

Not surprisingly, there is evidence that the glucocorticoid action on bone tissues overlaps with regulation by vitamin D_3 at some level. Vitamin D_3 blocks glucocorticoid potentiation of PTH-stimulated cAMP synthesis in the 17/2.8 osteosarcoma cell line (Catherwood, 1985). Treatment with vitamin D for 24–48 hours was required, suggesting an indirect mechanism of action for the steroid. Similar results were obtained with forskolin stimulation of adenylate cyclase, suggesting that the level of regulation of adenylate cyclase might be distal to the PTH receptor. Conversely, in cultures of osteoblast-like cells derived from rat calvaria, glucocorticoids potentiate several vitamin D-dependent biochemical responses (Chen et al., 1986). This augmentation comes about due to a glucocorticoid-induced increase in the number of specific receptors for vitamin D. Thus, regulation of adenylate cyclase in the presence of both steroids may involve dynamic changes at the genomic level.

Glucocorticoids have been found to modulate cAMP accumulation in connective tissue-derived cells, in a manner somewhat similar to that seen in bone cells. In cultured fibroblasts, glucocorticoids increase the cAMP response to isoproterenol and cholera toxin (Johnson and Jaworski, 1983). Two days of treatment with glucocorticoids is required, suggesting once more that the action is not directly on the adenylate cyclase system, but instead indirect, probably via changes in gene transcription. The changes seemed to result from alterations in adenylate cyclase, because phosphodiesterase activity was unaffected by glucocorticoids. However, it was not clear from the results which component of adenylate cyclase is altered by the glucocorticoids. Basal adenylate cyclase activity in fibroblast membranes was unchanged, but GTP-stimulated advenylate cyclase was elevated, whereas NaF-stimulated adenylate cyclase showed a decrease. The number of β-adrenergic receptors was unchanged, but agonist affinity was elevated threefold in glucocorticoid-treated cultures. The authors expressed doubt that this increased affinity could explain the results, since the steroid effect was seen at high, saturating concentrations of isoproterenol (Johnson and Jaworski, 1983).

IV. MECHANISMS

Based on the examples listed in Table I and discussed above, one can make several generalizations and speculations about how steroids modulate cAMP accumulation.

A. Steroid Effects Directly on Adenylate Cyclase

In a few systems steroids have been reported to affect adenylate cyclase directly, without the apparent mediation of gene transcription. As described above, the evidence for such a mechanism is most persuasive in the case of the *Xenopus* oocyte. Progesterone stimulates meiotic division by causing an inhibition of oocyte adenylate cyclase activity and a subsequent drop in cytoplasmic cAMP levels. Progesterone acts via a membrane-bound receptor that is stereospecific, affected by guanyl nucleotides, and operates at progesterone concentrations around 1 μM (Sadler *et al.*, 1981; Finidori-Lepicard *et al.*, 1981; Jordana *et al.*, 1981). Although the precise molecular changes involved are not known, it is thought that they do not involve alterations in the activities of Ns or Ni (Olate *et al.*, 1984; Goodhart *et al.*, 1984; Sadler *et al.*, 1984). Rather, the response to progesterone includes decreased phosphorylation of a 48-kDa oocyte membrane protein (Blondeau and Baulieu, 1985). The steroid action may also involve calcium effects on adenylate cyclase, as progesterone causes a redistribution of membrane Ca^{2+} concomitant with its inhibition of adenylate cyclase (reviewed in Morrill *et al.*, 1981).

There have been a few other reports of direct steroid effects on adenylate cyclase, such as that by Bergamini *et al.* (1985), showing that *in vitro* addition of estradiol to membranes from human endometrium causes a three- to fourfold increase in adenylate cyclase activity. Taken together, these results establish that in a limited number of cases steroids alter adenylate cyclase directly at the membrane level, without an intermediate step of steroid-induced gene transcription.

B. Gene Transcription

In almost all other cases the initial action of the steroid hormone takes time to develop and may occur at the level of gene transcription. The time lag is usually on the order of several hours, from which it is reasonable to suppose that steroids first cause alterations in levels of gene products, which subsequently lead to changes in adenylate cyclase activity. This conclusion is supported by the fact that, except for those cases noted in Section II, steroids are inactive on the adenylate cyclase system when they are tested in membrane or broken cell preparations.

A further indication that changes in gene products may be involved is the observation that in some cases the effects of steroids on adenylate cyclase are blocked by inhibitors of protein synthesis. For example, Rotsztejn et al. (1981) found that the rapid effect of glucocorticoids on pituitary adenylate cyclase was eliminated in the presence of protein-synthesis inhibitors. Forray and Richelson (1985) found that the enhancement of PGE-stimulated adenylate cyclase by Dex in cultured neuroblastoma was blocked by inhibitors of protein and RNA synthesis.

What genes are affected? Steroids may produce some of their influence by altering transcription of the genes that code for the various molecular components of the adenylate cyclase system, i.e., the receptors, the stimulatory and inhibitory coupling proteins, and even, the catalytic subunit. Direct evidence is lacking. However, in several cases (e.g., Wolfe et al., 1976; de Souza et al., 1985) the steroid-induced changes in adenylate cyclase are associated with corresponding changes in receptor numbers. Whether these changed receptor numbers are due to their genes being transcribed differently, or to other processes such as decreased degradation, or agonist-related up- or down-regulation (see below) remains unknown.

However, the notion that steroids control transcription of only a few genes of the adenylate cyclase system may be oversimplified. The major reason is that the directions of steroid effects on adenylate cyclase are tissue specific; i.e., the same effect is rarely seen throughout the body. Consider, for example, the effects of glucocorticoids on β-adrenergic-stimulated adenylate cyclase. Glucocorticoids inhibit β-stimulated adenylate cyclase in the hippocampus, the cerebral cortex, the liver, and fat cells; and in cultured fibroblasts they enhance β-stimulation. Therefore, tissue-specific factors must somehow modify the influence of glucocorticoids on β-receptors and the adenylate cyclase system. It thus seems unlikely that glucocorticoids affect transcription of the genes coding for β-adrenergic receptors or other adenylate cyclase components the same way in each of these tissues.

The cloning of genes coding for adenylate cyclase component proteins in several species (Aiba *et al.*, 1984; Casperson *et al.*, 1985; Kataoka *et al.*, 1985), and the modest steps that have been achieved in the discovery of tissue-specific transcriptional regulation, give hope that in the not-too-distant future such interactions at the genomic level may be detected.

C. Alterations in Adenylate Cyclase Due to Other Second Messengers and Agonists

It is possible that some of the effects of steroids on the adenylate cyclase system may not involve changes in the amounts of the constituent proteins of the adenylate cyclase system, but instead involve modification of their activities, for example, by interactions with other agonists or second messenger systems. In such instances, the possible involvement of gene expression and the nature of the gene products involved remain open questions. For example, the hormone action might be at the level of the coupling protein. In cultured rat astrocytes somatostatin inhibits β-adrenergic receptors–Ns coupling, and in dog and rat myocardium, muscarinic agonists modulate catecholamine receptor–Ns coupling (Watanabe *et al.*, 1978; Yamada *et al.*, 1980; Rougon *et al.*, 1983; Niehoff and Mudge, 1985).

A second avenue of steroid regulation is via other agonists that activate the adenylate cyclase system. Londos *et al.* (1985) have shown that endogenous adenosine has a powerful influence on basal and hormone-stimulated adenylate cyclase activity, and they have further suggested that adenosine may have a subtle influence on the effectiveness of hormones that are adenylate cyclase agonists as well as those hormones that do not utilize cAMP as a second messenger. Most of the studies on steroid regulation of adenylate cyclase discussed above did not address the issue of whether some of the changes seen may be due to modifications of the actions of adenosine on the adenylate cyclase system.

A third possibility is that steroid actions first affect second-messenger systems other than adenylate cyclase, and that these messengers subsequently alter the ability to generate cAMP, e.g., by phosphorylation of receptors, coupling proteins, or adenylate cyclase itself. Numerous reports suggest that two second-messenger systems—adenylate cyclase and the hormone-stimulated phosphotidylinositol (PI) cycle—may interact, even to the point of sharing as a component the inhibitory guanyl nucleotide-binding protein Ni (Gilman, 1985). For example, in mast cells, Ca^{2+}-mediated histamine release and the stimulation of inositol triphosphate breakdown are blocked by islet-activating protein, a specific modifier of Ni (Nakamura and Ui, 1985). Conversely, Bell *et al.* (1985) reported that in S49 lymphoma cells, phorbol esters—activators of C kinase—increase β-adrenergic-stimulated adenylate cyclase by enhancing the interaction between the α-subunits of Ns and the catalytic subunit of adenylate cyclase. This suggests that some of the steroid effects on adenylate cyclase discussed above may actually derive from steroid action on parts of the PI cycle, whose products subsequently exert their effects on adenylate cyclase. It is known, for example, that glucocorticoids can modify the responsiveness of C kinase and receptors coupled to PI hydrolysis in the pituitary, and that this action might indirectly affect the responsiveness of adenylate cyclase (Abou-Samra *et al.*, 1986).

Such a mechanism might explain some of the paradoxical results found by investigators looking solely at adenylate cyclase. For example, Mobley *et al.* (1983) found that glucocorticoids modify adrenergic-stimulated cAMP synthesis in slices of frontal cortex; yet this effect was blocked not by antagonists of the β-adrenergic receptor—the receptor subtype that is directly coupled to adenylate cyclase—but instead by antagonists of α-adrenergic receptors, which themselves have little or no effect on adenylate cyclase. It is conceivable, in this case, that glucocorticoids regulate an α-stimulated activity, which in turn modifies β-stimulated adenylate cyclase. Recent data support this interpretation (Stone *et al.*, 1986).

D. Agonist-Related Desensitization and Up-Regulation

Another factor to consider when speculating about mechanisms of steroid modulation of the cAMP system is the possibility that the steroid effects may be on the production and release of agonists that activate adenylate cyclase. Such action might also involve genomic regulation and take time to appear. For example, early studies of cAMP accumulation in the hypothalamus, noted above, indicated that estradiol treatment had a stimulatory effect, which was probably indirect; i.e., the initial action of estradiol was on presynaptic cells, where the steroid inhibited uptake or promoted release of catecholamines (Paul *et al.*, 1979). The higher ambient levels of these agonists lead to a chronic occupation of postsynaptic adrenergic receptors and subsequent increase in cAMP levels in these cells. When increased levels of agonist lead eventually to a diminished cAMP response to stimulation, we refer to this as agonist-induced desensitization. Many of the observed instances where steroids negatively affect adenylate cyclase may be the result of agonist-induced desensitization. Conversely, up-regulation of receptors following a decrease in agonist stimulation might also occur as a result of a steroid inhibition of agonist release and/or production.

Most or all of the neurotransmitter receptors implicated in steroid modulation of adenylate cyclase are known to exhibit agonist-induced desensitization under some physiological circumstances. For example, VIP-stimulated adenylate cyclase in the pineal shows both agonist-induced desensitization and heterologous desensitization to norepinephrine, and these effects can be mimicked by alterations in the circadian rhythm of the light–dark cycle (Yuwiler, 1983; Kaku et al., 1985). Measurement of PTH-stimulated cAMP synthesis in bone cells from rats subject to in vivo manipulation of vitamin D levels has provided further indication for such a mechanism; i.e., altered responsiveness to PTH in vitro reflected the results of PTH-induced homologous desensitization occurring in vivo; this was due to chronically elevated levels of PTH following loss of vitamin D (Crowell et al., 1981).

In many situations it may be difficult to experimentally distinguish between steroids affecting adenylate cyclase by acting on the receptive cell directly or by acting to cause increased or decreased secretion of a hormone. This is especially true in the CNS, where transmitter agonists are very closely opposed to post- and presynaptic adenylate cyclase systems. For example, the enhancement of VIP-stimulated cAMP synthesis in the hippocampus following adrenalectomy may reflect simply an adrenalectomy-induced drop in VIP release from hippocampal neurons and a subsequent up-regulation of postsynaptic VIP receptors (Harrelson et al., 1983, 1987; Harrelson and McEwen, 1985, 1987a,b).

Because of the almost universal presence of agonist-induced desensitization, as well as the possibility for heterologous desensitization, it will be important in the future that studies on steroid regulation of adenylate cyclase utilize preparations in which these phenomena can be sorted out from steroid effects mediated solely by adenylate cyclase itself. In this regard, studies on dispersed cells in culture may be especially helpful (Bilezikjian and Vale, 1983). However, even in dispersed cell preparations, the relationship between receptor desensitization and steroid control may be quite complex. For example, it has been reported that in cultured anterior pituitary cells, glucocorticoids are able to prevent agonist-induced desensitization by CRF (Ceda and Hoffman, 1986).

V. SUMMARY

We can draw a few conclusions from this brief survey of steroid effects on cyclic nucleotide second-messenger systems. First, changes in steroid status alter the ability of agents to affect adenylate cyclase in a wide variety of cell types in different tissues. There does not seem to be any restriction anatomically in the range of this steroid effect, in that it has been found in many different organ systems. Moreover, we can find evidence for steroid effects on cyclic nucleotide synthesis for each of the major steroid classes—glucocorticoids, mineralocorticoids, androgens, estrogens, progestins, and vitamin D. Thus, this effect is a quite general aspect of steroid hormone action.

Second, where steroids do alter adenylate cyclase, most often they do so in their target-cell populations in a very specific manner, i.e., the steroid hormone changes cyclase responsiveness to only one or a few agents, not to all of them. On the other hand, there are also cases in which the steroid alters basal adenylate cyclase activity and also the sensitivity to many or all agonists. Furthermore, in some tissues the steroid may increase sensitivity to the agonist, while diminishing sensitivity to the same agonist in other tissues.

Third, there is no single mechanism of the steroid effect, but rather a number of mechanisms. These include steroid-induced alterations in the number of specific receptors for certain agents, as well as alterations in the coupling of receptors to the adenylate cyclase catalytic subunit. Besides altering the amount of coupling proteins through gene expression, steroids have been inferred to affect the properties of the coupling proteins.

Fourth, steroids tend to modify the sensitivity of target-cell populations to hormones, which themselves are involved in the production of these steroids. For example, corticotroph cells in the anterior pituitary, which are responsible for synthesizing adrenocorticotrophic hormone and thus controlling the synthesis of glucocorticoids, are themselves targets for glucocorticoid regulation of their sensitivity to corticotropin-releasing factor. This pattern of steroid feedback to alter their own production can also be seen for PTH and vitamin D in bone-forming cells and for the gonadal steroids and pituitary sensitivity to gonadotropins.

Clearly, there is much still unknown as to how steroids regulate cAMP metabolism, and what role this regulation plays in the function of many different tissues throughout the body. As we move on from studying adenylate cyclase to looking at steroid regulation of cAMP-dependent enzymes and other second-messenger systems, we can anticipate new findings that will add to the complexity of what is already known, and that may help elucidate additional principles for the role of steroid regulation in cellular physiology.

ACKNOWLEDGMENTS

This work was supported by grants NS07080 from the National Institutes of Health and INT 8502384 and MH 41256 from the National Science Foundation to Bruce McEwen. Allan Harrelson is a Postdoctoral Fellow of the American Cancer Society.

REFERENCES

Abou-Samra, A.-B., Catt, K. H., and Aguilera, G. (1986). Involvement of protein kinase C in the regulation of adrenocorticotropin release from rat anterior pituitary cells. *Endocrinology* **118**, 212–217.

Aiba, H., Mori, K., Tanaka, M., Ooi, T., Roy, A., and Danchin, A. (1984). The complete nucleotide sequence of the adenylate cyclase gene of *Escherichia coli*. *Nucleic Acids Res.* **12**, 9427–9440.

Amechi, O. A., Butterworth, P. J., and Thomas, P. T. (1985). Effects of gonadal steroids on guanylate cyclase activity in the developing and adult brain. *Brain Res.* **342**, 158–161.

Barr, G. A., Ahn, H. S., and Makman, M. H. (1983). Dopamine-stimulated adenylate cyclase in hypothalamus: Influence of estrous cycle in female and castration in male rats. *Brain Res.* **277**, 299–303.

Bekiari, A. M., Saunders, R. B., Abulaban, F. S., and Yochim, J. M. (1984). Role of ovarian steroid hormones in the regulation of adenylate cyclase during early progestation. *Biol. Reprod.* **31**, 752–758.

Bell, J. D., Buxton, I. L. O., and Branton, L. L. (1985). Enhancement of adenylate cyclase activity in S49 lymphoma cells by phorbol ester. *J. Biol. Chem.* **260**, 2625–2628.

Bergamini, C. M., Pansini, F., Bettocchi, S., Jr., Segala, V., Dallocchio, F., Bagni, B., and Mollica, G. (1984). Hormonal sensitivity of adenylate cyclase from human endometrium: Modulation by estradiol. *J. Steroid Biochem.* **22**, 229–303.

Bilezikjian, L. M., and Vale, W. W. (1983). Glucocorticoids inhibit corticotropin-releasing factor-induced production of adenosine $3',5'$-monophosphate in cultured anterior pituitary cells. *Endocrinology* **113**, 657–662.

Birnbaumer, L., Codina, J., Mattera, R., Cerione, R. A., Hildebrandt, J. D., Sunyer, T., Rojas, F. J., Caron, M. G., Lefkowitz, R. J., and Iyengar, R. (1985). Regulation of hormone receptors and adenylyl cyclases by guanine nucleotide binding N proteins. *Recent Prog. Horm. Res.* **41**, 41–99.

Bitensky, M. W., Russell, V., and Blanco, M. (1970). Independent variation of glucagon and epinephrine responsive components of hepatic adenyl cyclase as a function of age, sex and steroid hormones. *Endocrinology* **86**, 154–159.

Blondeau, J.-P., and Baulieu, E.-E. (1984). Progesterone receptor characterized by photoaffinity labeling in the plasma membrane of *Xenopus laevis* oocytes. *Biochem. J.* **219**, 785–792.

Blondeau, J.-P., and Baulieu, E.-E. (1985). Progesterone-inhibited phosphorylation of an unique M_r 48,000 protein in the plasma membrane of *Xenopus laevis* oocytes. *J. Biol. Chem.* **260**, 3617–3625.

Bokoch, G. M., Katada, T., Northup, J. K., Hewlett, E. L., and Gilman, A. G. (1983). Identification of the predominant substrate for ADP-ribosylation by islet activating protein. *J. Biol. Chem.* **258**, 2072–2080.

Brostrom, M. A., Kon, C., Olson, D. R., and Breckenridge, B. McL. (1974). Adenosine $3',5'$-monophosphate in glial tumor cells treated with glucocorticoids. *Mol. Pharmacol.* **10**, 711–720.

Casperson, G. F., Walker, N., and Bourne, H. R. (1985). Isolation of the gene encoding adenylate cyclase in Saccharomyces cerevisiae. *Proc. Natl. Acad. Sci. U.S.A.* **82**, 5060–5063.

Catherwood, B. D. (1985). 1,25-Dihydrocholecalciferol and glucocorticosteroid regulation of adenylate cyclase in an osteoblast-like cell line. *J. Biol. Chem.* **260**, 736–743.

Ceda, G. P., and Hoffman, A. R. (1986). Glucocorticoid modulation of corticotropin-releasing factor desensitization in cultured rat anterior pituitary cells. *Endocrinology* **118**, 58–62.

Chan, T. M., Blackmore, P. F., Steiner, K. E., and Exton, J. H. (1979). Effects of adrenalectomy on hormone action on hepatic glucose metabolism. Reciprocal change in alpha and beta adrenergic activation of hepatic glycogen phosphorylase and calcium mobilization in adrenalectomized rats. *J. Biol. Chem.* **254**, 2428–2432.

Chen, T. L., and Feldman, D. (1978). Glucocorticoid potentiation of the adenosine 3',5'-monophosphate response to parathyroid hormone in cultured rat bone cells. *Endocrinology* **102**, 589–596.

Chen, T. L., and Feldman, D. (1979). Glucocorticoid receptors and actions in subpopulations of cultured rat bone cells. Mechanism of dexamethasone potentiation of parathyroid hormone-stimulated cyclic AMP production. *J. Clin. Invest.* **63**, 750–758.

Chen, T. L., Hauschka, P. V., and Feldman, D. (1986). Dexamethasone increases 1,25-dihydroxyvitamin D_3 receptor levels and augment bioresponses in rat osteoblast-like cells. *Endocrinology* **118**, 1119–1126.

Codina, J., Hildebrandt, J. D., Iyengar, R., Birnbaumer, L., Sekura, R. D., and Manclark, C. R. (1983). Pertussis toxin substrate, the putative Ni of adenylyl cyclases, is an alpha/beta heterodimer regulated by guanine nucleotide and magnesium. *Proc. Natl. Acad. Sci. U.S.A.* **80**, 4276–4280.

Coussen, F., Haiech, J., D'Alayer, J., and Monneron, A. (1985). Identification of the catalytic subunit of brain adenylate cyclase, a calmodulin-binding protein of 135 kDa. *Proc. Natl. Acad. Sci. U.S.A.* **82**, 6736–6740.

Crowell, J. A., Cooper, C. W., Toverud, S. U., and Boass, A. (1981). Influence of vitamin D on parathyroid hormone-induced adenosine 3',5'-monophosphate production by bone cells isolated from rat calvariae. *Endocrinology* **109**, 1715–1722.

Davies, A. O., de Lean, A., and Lefkowitz, R. J. (1981). Myocardial beta-adrenergic receptors from adrenalectomized rats: Impaired formation of high-affinity agonist–receptor complexes. *Endocrinology* **108**, 720–722.

de Souza, E. B., Insel, T. R., Perrin, M. H., Rivier, J., Vale, W. W., and Kuhar, M. H. (1985). Differential regulation of corticotropin-releasing factor receptors in anterior and intermediate lobes of pituitary and in brain following adrenalectomy in rats. *Neurosci. Lett.* **56**, 121–128.

Etgen, A. M., and Pettiti, N. (1986). Norepinephrine-stimulated cyclic AMP accumulation in rat hypothalamic slices: Effects of estrous cycle and ovarian steroids. *Brain Res.* **375**, 385–390.

Exton, J. H., Frienman, N., Wong, E. H.-A., Brineaux, J. P., Corbin, J. D., and Park, C. R. (1972). Interaction of glucocorticoids with glucagon and epinephrine in the control of gluconeogenesis and glycogenolysis in liver and lipolysis in adipose tissue. *J. Biol. Chem.* **247**, 3579–3588.

Finidori-Lepicard, J., Schorderet-Slatkine, S., Hanoune, J., and Baulieu, E.-E. (1981). Progesterone inhibits membrane-bound adenylate cyclase in *Xenopus laevis* oocytes. *Nature (London)* **292**, 255–257.

Forray, C., and Richelson, E. (1985). Glucocorticoids potentiate the prostaglandin E1-mediated cyclic AMP formation by a cultured murine neuroblastoma clone. *J. Neurochem.* **45**, 79–85.

Foster, S. J., and Harden, T. K. (1980). Dexamethasone increases beta-adrenoceptor density in human astrocytoma cells. *Biochem. Pharmacol.* **29**, 2151–2153.

Gilman, A. G., Jr. (1985). G proteins and dual control of adenylate cyclase. *Cell* **36**, 577–579.

Goodhardt, M., Ferry, N., Buscaglia, M., Baulieu, E.-E., and Hanoune, J. (1984). Does the guanine nucleotide regulatory protein Ni mediate progesterone inhibition of *Xenopus* oocyte adenylate cyclase? *EMBO J.* **3**, 2653–2657.

Granner, D., Chase, L. R., Aubach, G. D., and Tompkins, G. M. (1968). Tyrosine aminotransferase: Enzyme induction independent of adenosine 3',5'-monophosphate. *Science* **162**, 1018–1020.

Gunaga, K. P., and Menon, K. M. J. (1973). Effect of catecholamines and ovarian hormones on cyclic AMP accumulation in rat hypothalamus. *Biochem. Biophys. Res. Commun.* **54**, 440–448.

Gunaga, K. P., Kawano, A., and Menon, K. M. J. (1974). In vivo effect of estradiol benzoate on the accumulation of adenosine 3 ',5 '-cyclic monophosphate in the rat hypothalamus. *Neuroendocrinology* 16, 273–281.

Harrelson, A. L., and McEwen, B. S. (1985). Dependence of glucorticoid actions in hippocampus modulating VIP-stimulated cyclic AMP formation on in vivo presence of ACTH. *Abstr. Soc. Neurosci.* 11, 372–375.

Harrelson, A. L., and McEwen, B. S. (1987a). Hypophysectomy increases vasoactive intestinal peptide-stimulated cAMP generation in the hippocampus of the rat. *J. Neurosci.*, in press.

Harrelson, A. L., and McEwen, B. S. (1987b). Gonadal and adrenal steroid modulation of neurotransmitter-stimulated cAMP accumulation in the hippocampus of the rat. *Brain Res.* 404, 89–94.

Harrelson, A. L., McEwen, B. S., and Rostene, W. H. (1983). Adrenalectomy modifies neurotransmitter stimulated cyclic AMP accumulation in hippocampal slices. *Abstr. Soc. Neurosci.* 26, 87.

Harrelson, A. L., Rostene, W. H., and McEwen, B. S. (1987). Adrenocortical steroids modify neurotransmitter-stimulated cAMP accumulation in the hippocampus and limbic brain of the rat. *J. Neurochem.*, in press.

Hildebrandt, J. D., Codina, J., Risinger, R., and Birnbaumer, L. (1984). Identification of a gamma subunit associated with the adenylyl cyclase regulatory proteins Ns and Ni. *J. Biol. Chem.* 259, 2039–2042.

Johnson, G. S., and Jaworski, C. J. (1983). Glucocorticoids increase GTP-dependent cyclase activity in cultured fibroblasts. *Biochem. Pharmacol.* 23, 648–652.

Jordana, X., Allende, C. C., and Allende, J. E. (1981). Guanine nucleotides are required for progesterone inhibition of amphibian oocyte adenylate cyclase. *Biochem. Int.* 3, 527–532.

Kaku, K., Tsuchiya, M., Matsuda, M., Inoue, Y., Kaneko, T., and Yanaihara, N. (1985). Light and agonist alter vasoactive intestinal peptide binding and intracellular accumulation of adenosine 3 ',5 '-monophosphate in the rat pineal gland. *Endocrinology* 117, 2371–2375.

Katoaka, T., Broek, D., and Wigler, M. (1985). DNA sequence and characterization of the Saccharomyces cerevisiae gene encoding adenylate cyclase. *Cell* 43, 493–505.

Kent, G. N., Jilka, P. L., and Cohn, D. B. (1980). Homologous and heterologous control of bone cell adenosine 3 ',5 '-monophosphate response to hormones by parathormone, prostaglandin E2, calcitonin and 1,25-dihydroxycholecalciferol. *Endocrinology* 107, 1474–1481.

Knecht, M., Darbon, J.-M., Ranta, T., Baukal, A. J., and Catt, K. J. (1984). Estrogens enhance the adenosine 3 ',5 '-monophosphate-mediated induction of follicle-stimulating hormone and luteinizing hormone receptors in rat granulosa cells. *Endocrinology* 115, 41–49.

Kubata, M., Ng, K. W., and Martin, T. J. (1985). Effect of 1,25-dihydroxyvitamin D_3 on cyclic AMP responses to hormones in clonal oestogenic sarcoma cells. *Biochem. J.* 231, 11–17.

Kumakura, K., Hoffman, M., Cocchi, D., Trabucchi, M., Spano, P. F., and Mueller, E. E. (1979). Long-term effect of ovariectomy on dopamine-stimulated adenylate cyclase in rat striatum and nucleus accumbens. *Psychopharmacology* 61, 13–16.

Lai, E., Rosen, O. M., and Rubin, C. S. (1982). Dexamethasone regulates the beta-adrenergic receptor subtype expressed by 3T3-L1 preadipocytes and adipocytes. *J. Biol. Chem.* 257, 6691–6696.

Lee, T. P., and Reed, C. E. (1977). Effects of steroids on the regulation of the levels of cyclic AMP in human lymphocytes. *Biochem. Biophys. Res. Commun.* 78, 998–1004.

Londos, C., Honner, R. C., and Dhillon, G. S. (1985). cAMP-dependent protein kinase and kinase and lipolysis in rat adipocytes. III. Multiple modes of insulin regulation of lipolysis and regulation of insulin responsiveness by adenylate cyclase regulators. *J. Biol. Chem.* 260, 15139–15145.

McEwen, B. S., DeKloet, E. R., and Rostene, W. H. (1986). Adrenal steroid receptors and actions in the nervous system. *Physiol. Rev.* **66**, 1121–1188.

Manganiello, V., and Vaughan, M. (1972). An effect of dexamethasone on adenosine 3′,5′-monophosphate content and adenosine 3′,5′-monophosphate phosphodiesterase activity of cultured hepatoma cells. *J. Clin. Invest.* **51**, 2763–2767.

Manganiello, V., Breslow, J., and Vaughn, M. (1972). An effect of dexamethasone on the cyclic AMP content of human fibroblasts stimulated by catecholamines and prostaglandins. *J. Clin. Invest.* **51**, 60A.

Marone, G., Lichtenstein, L. M., and Plaut, M. (1980). Hydrocortisone and human lymphocytes increases in cyclic adenosine 3′:5′-monophosphate and potentiation of adenylate cyclase activating agents. *J. Pharmacol. Exp. Ther.* **215**, 469–478.

Mazancourt, P., Thotakura, N. R., and Giudicelli, Y. (1982). Does glucocorticoid deprivation promote the expression of adenosine receptor-sites stimulating adenylate cyclase in rat adipocyte membranes? *Biochem. Biophys. Res. Commun.* **108**, 987–994.

Miller, T. B., Exton, J. H., and Park, C. R. (1971). kA block in epinephrine-induced glucogenolysis in hearts from adrenalectomized rats. *J. Biol. Chem.* **246**, 3672–3678.

Mobley, P. L., and Sulser, F. (1980a). Adrenal corticoids regulate sensitivity of noradrenaline receptor-coupled adenylate cyclase in brain. *Nature (London)* **286**, 608–610.

Mobley, P. L., and Sulser, F. (1980b). Adrenal steroids affect the norepinephrine-sensitive adenylate cyclase system in the rat limbic forebrain. *Eur. J. Pharmacol.* **65**, 321–322.

Mobley, P. L., Mannier, H., and Sulser, F. (1983). Norepinephrine-sensitive adenylate cyclase system in rat brain: Role of adrenal corticosteroids. *J. Pharmacol. Exp. Ther.* **226**, 71–77.

Morrill, G. A., Ziegler, D., and Kostellow, A. B. (1981). The role of Ca^{++} and cyclic nucleotides in progesterone initiation of the meiotic divisions in amphibian oocytes. *Life Sci.* **29**, 1821–1835.

Morton, D. B., and Truman, J. W. (1985). Steroid regulation of the peptide-mediated increase in cyclic GMP in the nervous system of the hawkmoth *Manduca sexta. J. Comp. Physiol. A.* **157**, 423–432.

Nakamura, T., and Ui, M. (1985). Simultaneous inhibitions of inositol phospholipid breakdown, arachidonic acid release and histamine secretion in mast cells by islet-activating protein, *Pertussis* toxin. *J. Biol. Chem.* **260**, 3584–3593.

Ng, B., Hekkelman, J. W., and Heersche, J. N. M. (1979). The effect of cortisol on the adenosine 3′,5′-monophosphate response to parathyroid hormone of bone in vitro. *Endocrinology* **104**, 1130–1135.

Niehoff, D. L., and Mudge, A. W. (1985). Somatostatin alters beta-adrenergic receptor–effector coupling in cultured rat astrocytes. *EMBO J.* **4**, 317–321.

Northup, J. K., Sternweis, P. C., Smigle, M. D., Schleifer, L. S., Ross, E. M., and Gilman, A. G. (1980). Purification of the regulatory component of adenylate cyclase. *Proc. Natl. Acad. Sci. U.S.A.* **77**, 6516–6520.

Olate, J., Allende, C. C., Allende, J. E., Sekura, R. D., and Birnbaumer, L. (1984). Oocyte adenylyl cyclase contains Ni, yet the guanine nucleotide-dependent inhibition by progesterone is not sensitive to pertussis toxin. *FEBS Lett.* **175**, 25–30.

Parsons, B., Rainbow, T. C., MacLusky, N.J., and McEwen, B. S. (1982). Progestin receptor levels in rat hypothalamic and limbic nuclei. *J. Neurosci.* **2**, 1446–1452.

Paul, S. M., Axelrod, J., Saavedra, J. M., and Skolnick, P. (1979). Estrogen-induced efflux of endogenous catecholamines from the hypothalamus in vitro. *Brain Res.* **178**, 499–505.

Pfaff, D. W., and Keiner, M. (1973). Atlas of estradiol-concentrating cells in the central nervous system of the female rat. *J. Comp. Neurol.* **151**, 121–158.

Pfeuffer, E., Dreher, R.-M., Metzger, H., and Pfeuffer, T. (1985). Catalytic unit of adenylate cyclase: Purification and identification of affinity crosslinking. *Proc. Natl. Acad. Sci. U.S.A.* **82**, 3086–3090.

Rajerison, R., Marchetti, J., Roy, C., Bockaert, J., and Jard, S. (1974). The vasopressin-sensitive adenylate cyclase of the rat kidney. *J. Biol. Chem.* **249**, 6390–6400.

Richards, J. S., Ireland, J. J., Rao, M. C., Bernath, G. A., Midgely, A. R., and Reichert, L. E., Jr. (1976). Ovarian follicular development in the rat: Hormone receptor regulation by estradiol, follicle stimulating hormone and luteinizing hormone. *Endocrinology* **99**, 1562–1570.

Roberts, V. J., Singhal, R. L., and Roberts, D. C. S. (1984). Corticosterone prevents the increase in noradrenaline-stimulated adenyl cyclase activity in rat hippocampus following adrenalectomy or metopirone. *Eur. J. Pharmacol.* **103**, 235–240.

Rodan, S. B., Fischer, M. K., Egan, J. J., Epstein, P. M., and Rodan, G. A. (1984). The effect of dexamethasone on parathyroid hormone stimulation of adenylate cyclase in ROS 17/2.8 cells. *Endocrinology* **115**, 951–958.

Ross, P. S., Manganiello, V. C., and Vaughan, M. (1977). Regulation of cyclic nucleotide phosphodiesterases in cultured hepatoma cells by dexamethasone and N(6),O(2')-dibutyryl-adenosine 3',5'-monophosphate. *J. Biol. Chem.* **252**, 1448–1456.

Rostene, W. H. (1984). Neurobiological and neuroendocrine functions of the vasoactive intestinal peptide (VIP). *Prog. Neurobiol.* **22**, 103–129.

Rotsztejn, W. H., Dussaillant, M., Nobou, F., and Rosselin, G. (1981). Rapid glucocorticoid inhibition of vasoactive intestinal peptide-induced cyclic AMP accumulation and prolactin release in rat pituitary cells in culture. *Proc. Natl. Acad. Sci. U.S.A.* **78**, 7584–7588.

Rougon, G., Noble, M., and Mudge, A. W. (1983). Neuropeptides modulate adrenergic response of purified astrocytes in vitro. *Nature (London)* **305**, 715–717.

Rubin, J. E., and Catherwood, B. D. (1984). 1,25-Dihydroxyvitamin D causes attenuation of cyclic AMP responses in monocyte-like cells. *Biochem. Biophys. Res. Commun.* **123**, 210–215.

Sadler, S. E., Maller, J. L., and Cooper, D. M. F. (1984). Progesterone inhibition of *Xenopus* oocyte adenylate cyclase is not mediated via the *Bordetella pertussis* toxin substrate. *Mol. Pharmacol.* **26**, 526–531.

Schaeffer, L. D., Chenoweth, M., and Dunn, A. (1969). Adrenal cortocosteroid involvement in the control of liver glycogen phosphorylase activity. *Biochim. Biophys. Acta* **192**, 292–303.

Smigel, M., Katada, T., Northup, J. K., Bokoch, G. M., Ui, M., and Gilman, A. G. (1984). Mechanisms of guanine nucleotide-mediated regulation of adenylate cyclase activity. *Adv. Cyclic Nucleotide Protein Phosphorylation Res.* **17**, 1–18.

Stoff, J. S., Handler, J. S., and Orloff, J. (1972). The effect of aldosterone on the accumulation of adenosine 3':5'-cyclic monophosphate in toad bladder epithelial cells in response to vasopressin and theophylline. *Proc. Natl. Acad. Sci. U.S.A.* **69**, 805–808.

Stone, E. A., Herrera, A. S., Carr, K. D., and McEwen, B. S. (1986). Selective regulation of brain beta adrenergic receptors by the noradrenergic system and brain alpha adrenergic receptors by the pituitary adrenal system. *Fed. Proc., Fed. Am. Soc. Exp. Biol.* 839–3.

Thotakura, N. R., de Mazancourt, P., and Giudicelli, Y. (1982). Evidence for a defect in the number of beta adrenergic receptors and in the adenylate cyclase responsiveness to guanine nucleotides in fat cells after adrenalectomy. *Biochem. Biophys. Acta* **717**, 32–40.

Tilders, F., Tatmoto, K., and Berkenbosch, F. (1984). The intestinal peptide PHI-27 potentiates the action of corticotropin-releasing factor on ACTH release from rat pituitary fragments in vitro. *Endocrinology* **115**, 1633–1635.

Tornello, S., Orti, E., Weisenberg, L., and De Nicola, A. F. (1984). Regulation of adenosine-3',5'-monophosphate levels in rat anterior pituitary by adrenal corticoids. *Metabolism* **33**, 224–229.

Wagner, H. R., Crutcher, K. A., and Davis, J. N. (1979). Chronic estrogen treatment decreases beta-adrenergic responses in rat cerebral cortex. *Brain Res.* **171**, 147–151.

Watanabe, A. M., McConnaughey, M. M., Strawbridge, R. A., Fleming, J. W., Jones, L. R., and Besch, H. R., Jr. (1978). Muscarinic cholinergic receptor modulation of beta-adrenergic receptor affinity for catecholamines. *J. Biol. Chem.* **253**, 4833–4836.

Weissman, B. A., and Skolnick, P. (1975). Stimulation of adenosine 3′,5′-monophosphate formation in incubated rat hypothalamus by estrogenic compounds: Relationship to biologic potency and blockade by anti-estrogens. *Neuroendocrinology* **18**, 27–34.

Wolfe, B. B., Harden, T. K., and Molinoff, P. B. (1976). Beta-adrenergic receptors in rat liver: Effects of adrenalectomy. *Proc. Natl. Acad. Sci. U.S.A.* **73**, 1343–1347.

Wong, G. L., Luben, R. A., and Cohn, D. V. (1977). 1,25-Dihydroxycholecalciferol and parathormone: Effects on isolated osteoclast-like cells. *Science* **197**, 663–665.

Yamada, S., Yamamura, H. I., and Roeske, W. R. (1980). The regulation of cardiac alpha (1)-adrenergic receptors by guanine nucleotides and by muscarinic cholinergic agonists. *Eur. J. Pharmacol.* **63**, 239–241.

Yamamoto, K. R. (1985). Steroid receptor regulated transcription of specific genes and gene networks. *Annu. Rev. Genet.* **19**, 209–252.

Yeager, R. E., Heideman, W., Rosenberg, G. B., and Storm, D. R. (1985). Purification of the calmodulin-sensitive adenylate cyclase from bovine cerebral cortex. *Biochemistry* **24**, 3776–3783.

Yuwiler, A. (1983). Light and agonists alter pineal N-acetyltransferase induction by vasoactive intestinal polypeptide. *Science* **220**, 1082–1083.

Part III

Magnocellular Neurons

Expression of the Oxytocin and Vasopressin Genes

DIETMAR RICHTER AND HARTWIG SCHMALE

Institut für Zellbiochemie und Klinische Neurobiologie
Universität Hamburg
2000 Hamburg 20, Federal Republic of Germany

I. INTRODUCTION

Vasopressin and oxytocin are nonapeptide hormones synthesized in the magnocellular neurons of the hypothalamus. Except for two amino acids, both peptides are identical in their sequences, with the C-terminal residues being amidated. Because of their structural relationship, they can be considered to be members of a common hormone family of the hypothalamus though their functions are quite distinct: Vasopressin controls water retention in the kidney, while oxytocin regulates milk ejection and uterus contraction during birth. Besides their hormonal functions, the two nonapeptides appear to be involved in many other processes, such as learning and memory; responses to tolerance development and physical dependency on alcohol, opiate, or heroin; addiction; rewarded behavior; cardiovascular regulation; control of body temperature; and brain development (De Wied, 1983).

Both hormones are expressed predominantly in the same hypothalamic areas, e.g., the supraoptic nucleus (SON) and the paraventricular nucleus (PVN), yet there is no evidence that the two nonapeptides are present in one and the same hypothalamic neuron (Sofroniew, 1983).

The primary structure of the two hormones was determined more than 30 years ago by du Vigneaud and colleagues (Turner *et al.*, 1951); Du Vigneaud *et al.*, 1954). Subsequent studies of the biosynthetic pathway leading to the mature hormones revealed the presence of high-molecular-weight forms of immunoreactive oxytocin or vasopressin. These studies were initiated during the 1960s by Sachs and co-workers, who first proposed an elegant concept for the biosynthesis of vasopressin (Sachs and Takabatake, 1964; Sachs *et al.*, 1969). According to this hypothesis, the hormone is synthesized via a larger biologically inactive precursor, and is then processed into the mature hormone when transported from the hypothalamus to the posterior pituitary. This group also proposed the later confirmed precursor model that the hormone synthesis is linked to that of its carrier protein, known as neurophysin.

There are two neurophysins—one associated with vasopressin, the other with oxytocin (Acher, 1983)—which consist of roughly 90 amino acids, with the central part (∿60 amino acids) being strongly conserved in all known neurophysin sequences. Variable sequences are at the N- and C-termini of the proteins. The neurophysins are considered to function as carriers of the respective hormone within the axons of the magnocellular neurons (Acher, 1983).

Today the structure of vasopressin and oxytocin precursors and their corresponding genes from a number of vertebrates have been elucidated (Richter, 1987). This article concentrates on the molecular biology of the two peptide hormones and summarizes some of the structural/functional aspects of the precursors. Regulatory processes at the transcriptional and translational level in the hypothalamus and peripheral organs are discussed, including the genetic defect in the synthesis of vasopressin leading to central diabetes insipidus. For the physiological and clinical backgrounds of the two hormones the reader is referred to review articles (e.g., Rascher *et al.*, 1985).

II. PRECURSOR AND GENE STRUCTURE

The vasopressin precursor consists of the hormone, its carrier protein neurophysin, and a glycopeptide moiety of yet unknown function. The oxytocin precursor lacks the glycopeptide and consists only of the hormone and the respective carrier neurophysin (Fig. 1). As with other peptide hormone precursors, those for vasopressin and oxytocin include a signal peptide separated from the prohormone by a small neutral amino acid,, e.g., alanine. The putative signal sequences of human, cow and pig consist of 19 amino acids each. The gene sequence of the rat vasopressin precursor predicts a longer signal sequence, with 23 amino acid residues and three possible translation start sites (Fig. 2).

The peptide units within one polyprotein are separated either by paired (between hormone and carrier neurophysin) or by single basic amino acid residues (between carrier and glycopeptide), which may indicate the existence of different sets of processing enzymes discriminating between the two cleavage signals (Richter, 1983). Although the paired basic residues represent the most frequent form of processing signal, the one comprising a single residue is found only in a few cases (e.g., precursors to growth hormone release factor or somatostatin).

The nucleotide sequence encoding the calf, rat, and human oxytocin precursor (Rehbein *et al.*, 1986) predicts that there is a supernumerary basic amino acid at the carboxy terminus of the precursor which is not found in any of the isolated neurophysins and which is either an arginine (rat and human) or histidine (calf) residue. The extra basic amino acid that is apparently removed during or after translation is found at the same position as the single arginine separating the neurophysin and the glycopeptide in the vasopressin precursor. It is tempting to speculate that an ancestral oxytocin precursor also contained a C-terminal peptide, which, because there was no need for it, had been discarded during evolution leaving the terminal basic amino acid in today's precursor as a rudiment of the ancestral cleavage signal.

The vasopressin precursor from rat, cow, pig, and man includes a C-terminal glycopeptide with a glycosylation site Asn–Ala–Thr. The glycopeptide sequences so far known show remarkably high homologies, with the glycosylation site and the leucine-rich center part being well conserved. The consecutive leucine residues may represent alternative processing signals for converting the glycopeptide into subfractions (Smyth and Massey, 1979). The glycopeptide initially isolated from posterior pituitary extracts (Holwerda, 1972) has been localized in vasopressin-producing magnocellular neurons, yet its biological function remains to be resolved (Watson *et al.*, 1982).

Despite their obvious sequence homology, the two hormones are the products of independent genes that evolved divergently away from a common ancestral gene some 400 million years ago (Fig. 1). Vasopressin and oxytocin genes have been isolated and sequenced from a number of vertebrates. It appears that each of the vasopressin and oxytocin genes is present in a

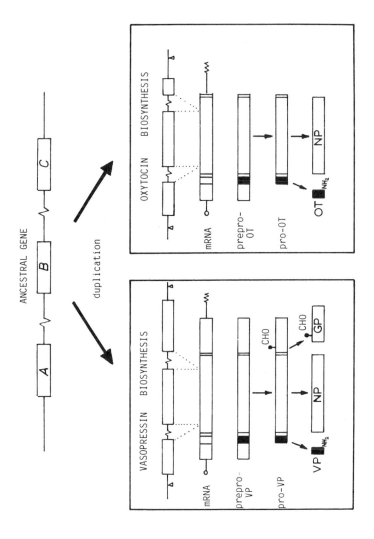

Fig. 1. Structural organization of the vasopressin and oxytocin precursors and their encoding genes. VP, Vasopressin; OT, oxytocin; CHO, carbohydrate chain; GP, glycopeptide; NP, neurophysin.

single copy in the mammalian genome (Schmale *et al.*, 1983; Schmale and Richter, 1984; Ivell and Richter, 1984a; Ruppert *et al.*, 1984; Sausville *et al.*, 1985). Each gene comprises three exons separated by two intervening sequences. Each exon encodes one of the principal functional domains of the polyprotein, i.e., the hormone, the carrier protein, and the glycopeptide. In the case of the oxytocin gene, the third exon comprises only the C-terminal variable part of the neurophysin plus an extra basic amino acid.

The human vasopressin- and oxytocin-encoding genes have both been mapped on chromosome 20 (Riddell *et al.*, 1985). There the two genes are located in close proximity, connected by ∿12 kb. The coding strands of the two genes are arranged inverted relative to one another (Fig. 3). In rats, the two genes also appear to be closely linked, as indicated by mapping of an ∿16-kb DNA fragment with specific probes for vasopressin and oxytocin(Mohr *et al., 1987*). It will be of considerable interest to study the regulatory sites of the two genes, e.g., promoter and enhancer sequences, and their effects on the tissue-specific expression of the two genes. Although the two human genes are in close proximity, only the vasopressin-encoding gene is specifically expressed in a human small cell lung cancer derived cell line (Sausville *et al.*, 1985). Also it is well documented that only one population of magnocellular neurons in the hypothalamus expresses either gene (Sofroniew, 1983).

The sequence of the vasopressin- and oxytocin-encoding precursors from hypothalamic cDNA libraries of various sources including human have been identified (Rehbein *et al.*, 1986). In Fig. 2, a lysine vasopressin precursor from pig is compared to those from other species. The Suinidae are unusual among mammals in being the only family to have lysine at position 8 of the nonapeptide. The change from arginine to lysine is due to an exchange of an A for a G nucleotide in the second position of the corresponding codon in the cDNA. The lower lines of Fig. 2 show the mRNA sequences encoding the oxytocin precursors, including that from human hypothalamus. The human oxytocin sequence agrees largely with a sequence predicted from the human gene (Sausville *et al.*, 1985), except that the cDNA also includes an additional valine codon. This extra amino acid is described in the neurophysin protein sequence obtained by Edman degradation (Acher, 1983), but was absent in the published gene sequence. It appears most likely that the presence or absence of this amino acid represents a polymorphism of the human oxytocin neurophysin precursor.

In general, the sequence of the vasopressin- and oxytocin-associated neurophysins predicted from the cDNA agree with those obtained by conventional amino acid sequence determination, with few exceptions. For instance, the DNA sequence predicts that the rat neurophysin of the oxytocin precursor contains a glutamic acid instead of a glutamine residue in position 92—a difference that may be due to a microheterogeneity between various rat strains (Ivell and Richter, 1984a).

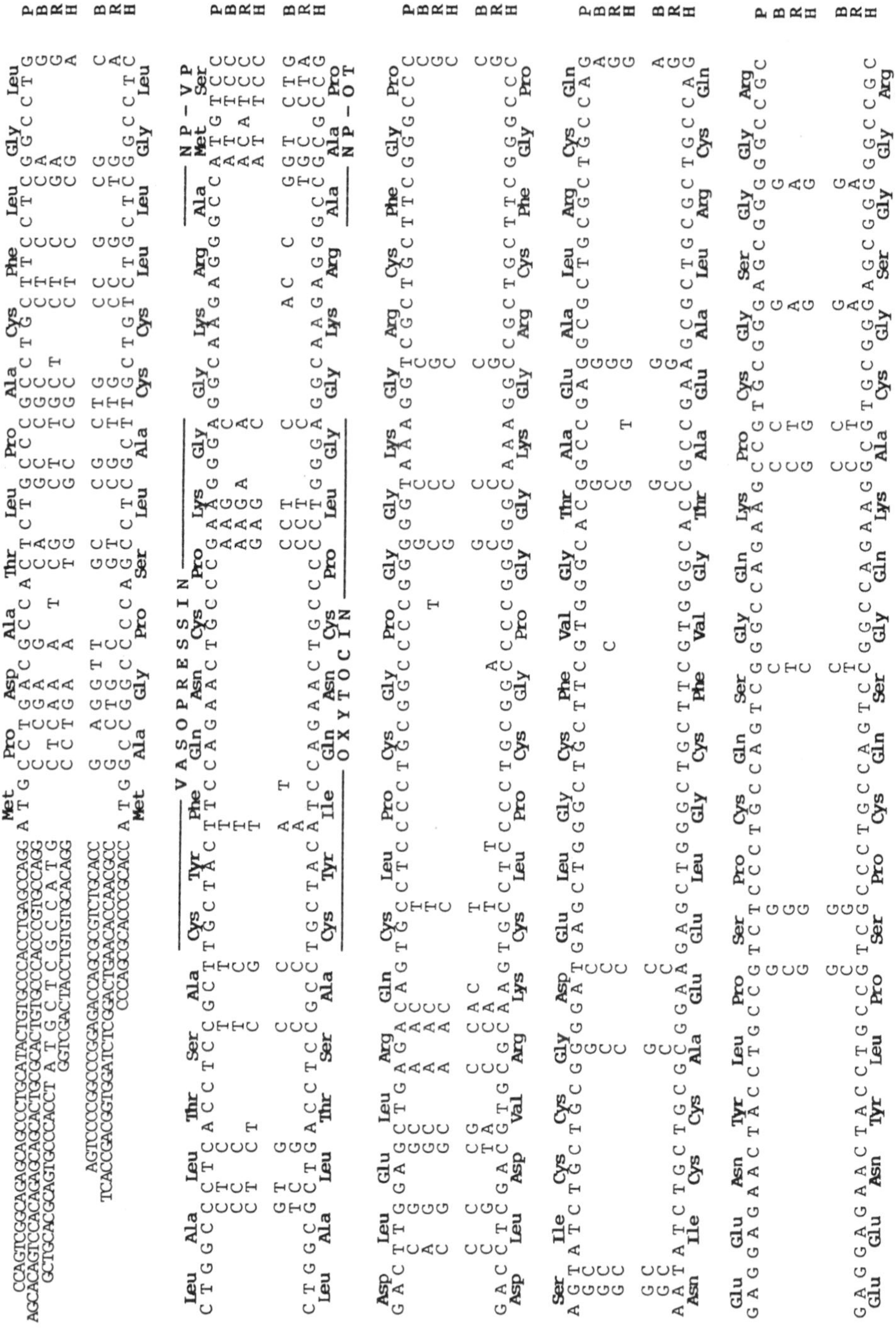

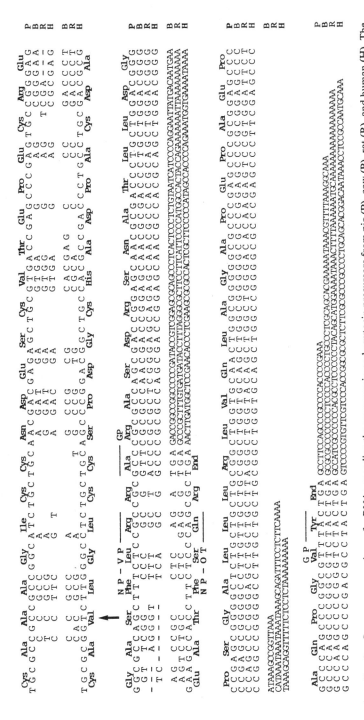

FIG. 2. Sequence comparison of mRNAs encoding the vasopressin and oxytocin precursors from pig (P), cow (B), rat (R), and human (H). The complete nucleotide sequences are given for the pig vasopressin mRNA and the human oxytocin mRNA. Complete homology between the seven sequences within the coding region is indicated by empty space, otherwise replacement by the respective nucleotide(s) are indicated. Dashed lines indicate positions where no corresponding sequence exists and which were introduced by the computer to get best fitting with the other protein sequences. The black dot in the vasopressin cDNA sequence encoding the mutated vasopressin precursor from Brattleboro rats indicates the position of the single-base deletion responsible for the mutant syndrome. NP-OT, Oxytocin-associated neurophysin; NP-VP, vasopressin-associated neurophysin; GP, glycopeptide; the arrow indicates the extra valine not included in the human gene sequence (Sausville et al., 1985).

257

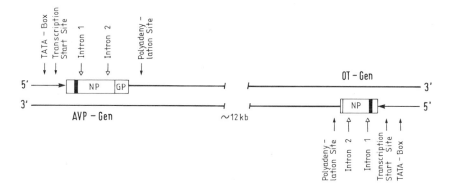

Fɪɢ. 3. Structural organization of the human vasopressin (AVP) and oxytocin (OT) genes (Sausville *et al.*, 1985). For abbreviations, see legend to Fig. 1.

III. GENE REGULATION

There is little information available concerning the regulation of the vasopressin- and oxytocin-encoding genes at the transcriptional or translational level. Tissue-specific expression of the human vasopressin gene has been reported (Sausville *et al.*, 1985; see also Section II). Because the two human genes differ particularly in their adjacent 5′-flanking region, differential expression may be due to specific recognition signals within these sequences (Supowit *et al.*, 1984). This would be in line with the high degree of sequence homology found in the 5′-flanking region of the vasopressin gene from rat, cow, and man (Ruppert *et al.*, 1984; Sausville *et al.*, 1985).

First attempts in studying the regulation of the two genes at the transcriptional level have been performed by liquid or blot hybridization assays. Another method, *in situ* hybridization, employs thin sections of hypothalamus, which are hybridized to specifically labeled DNA probes; the positive signals are quantified by counting the grains (McCabe *et al.*, 1986a,b). In the following paragraphs, examples are given for elevated mRNA levels under different physiological conditions estimated by these techniques. However, altered mRNA levels may not necessarily reflect the events at the transcriptional level, but just different turnover activities.

A. Osmotic Stress and Vasopressin-Encoding mRNA Levels

Rats placed under osmotic stress by drinking 2% saline respond with an increase of the mRNA encoding vasopressin (Table I; Majzoub *et al.*, 1983;

TABLE I

INFLUENCE OF OSMOTIC STRESS ON LEVELS OF mRNA ENCODING THE
VASOPRESSIN AND OXYTOCIN PRECURSOR[a]

Region	Control	Osmotic stress
Supraoptic nucleus	21.4 ± 5.6	97.9 ± 44.7
Paraventricular nucleus	16.4 ± 2.5	35.1 ± 2.1
Suprachiasmatic nucleus	19.8 ± 5.6	17.0 ± 4.5

[a]Data from Burbach et al. (1984). Rats were given 2% sodium chloride in tap water to drink for 14 days. mRNA levels are expressed as total pg vasopressin mRNA per microdissected region ± SEM. (For experimental details see Burbach et al., 1984.)

Burbach et al., 1984). This is in agreement with the in vivo finding that these animals respond with an elevated plasma osmolality and increased vasopressin levels (Rascher et al., 1985). Also, increased vasopressin levels are found in the supraoptic and paraventricular, but not in the suprachiasmatic, nucleus (Zerbe and Palkovits, 1984). Again these data coincide with the finding that the level for mRNA encoding vasopressin increases significantly in the SON, less in the PVN, but not in the suprachiasmatic nucleus (SCN) (Burbach et al., 1984). Both the immunological and hybridization experiments suggest that the vasopressin-producing neurons of the SON and PVN on the one hand, and the SCN on the other, are differentially regulated, and this would agree with the anatomical projections of the neurons. It is known that the neurons of the SON and PVN producing vasopressin project toward the posterior pituitary, whereas those of the SCN project toward other brain areas (Laczi et al., 1983).

It would be interesting to see whether the expression of the vasopressin gene in SON and PVN versus SCN is controlled by different—peripheral and/or central—signals. Whether the different levels of the vasopressin-coding mRNA in the three hypothalamic areas assayed reflect indeed differential regulation of the vasopressin gene has to be examined more rigorously, for instance, by run-on experiments in isolated hypothalamic nuclei. Also, other physiological parameters are needed that may affect specifically the vasopressin mRNA level in the SCN but not in the SON or PVN.

In a converse experiment the question has been asked whether an excess of vasopressin down-regulates the expression of the corresponding gene (Mohr et al., 1985; Rehbein et al., 1986). The data suggest that the levels of vasopressin mRNA in rats decreases compared to those in nontreated animals (Mohr, unpublished; Table II). Interestingly, there is also a

TABLE II
Influence of Excess Vasopressin on Regulation of Vasopressin- and Oxytocin-Encoding mRNA Levels[a]

	Water intake (ml)	Urine output (ml)	RNA extracted (μg)	Vasopressin mRNA	Oxytocin mRNA	β-Actin mRNA/ μg total RNA
Wistar						
Controls	41.5 ± 3.0	11.0 ± 0.8	47.0 ± 15.6	100.0 ± 12.8%	100.0 ± 28.3%	ND
Vasopressin-treated	34.4 ± 1.6	6.6 ± 3.3	64.0 ± 9.2	66.7 ± 8.9%	55.4 ± 3.4%	ND
Brattleboro						
Controls	348.8 ± 16.8	>150	54.5 ± 14.3	100.0 ± 24.2%	100.0 ± 34.1%	100.0 ± 26.5%
Vasopressin-treated	57.4 ± 12.5	8.5 ± 4.2	61.1 ± 9.4	72.1 ± 35.0%	79.5 ± 24.0%	95.3 ± 20.1%

[a]For experimental details, see Rehbein *et al.* (1986). mRNA levels are expressed as a percentage of the controls ± SEM. ND, Not determined.

reduction in the oxytocin mRNA, though the control, actin mRNA, appears not to be affected. Similar results are obtained for the mutant (Brattleboro) rat which lacks vasopressin. Whether these data are evidence for a feedback regulation via the hormone has to be studied.

Another method to quantify elevated or down-regulated mRNAs for the two peptide hormones is *in situ* hybridization, which combines a number of advantages. It allows not only the cells expressing the respective gene to be identified and also the mRNA levels to be quantified, but also permits one to study the effect of endogenously induced factors on the morphology of the respective cells (Uhl *et al.*, 1985; Fuller *et al.*, 1985; McCabe *et al.*, 1986a,b,). For instance, it has been shown that expression of the rat oxytocin gene in the SON is specifically raised in late pregnant as compared to nonpregnant rats (McCabe *et al.*, 1986a,b). With an alternative technique using the solution-hybridization method, other groups have come to similar conclusions (Burbach *et al.*, 1986). Also, when rats placed under osmotic stress are analyzed by the *in situ* hybridization technique, neurons specific for vasopressin synthesis contain increased grain numbers of the specific mRNA compared to cells without grains—presumably producing oxytocin (Uhl *et al.*, 1985; Fuller *et al.*, 1985; McCabe *et al.*, 1986a,b).

B. Expression and Regulation in Peripheral Tissues

Synthesis of both vasopressin and oxytocin has been shown to occur also in cells other than the hypothalamic magnocellular neurons. For example, vasopressin and oxytocin mRNAs have been identified in the rat cerebellum (Ivell *et al.*, 1986). Immunological studies suggest the presence of at least vasopressin in certain cells of the cerebellum (Caffe *et al.*, 1985). In the case of the oxytocin-encoding mRNA, two transcripts of 600 and 650 bp have been detected; the latter corresponding in size to the expected hypothalamic gene product (Ivell *et al.*, 1986). At present, the difference in size of the two oxytocin-encoding mRNAs is not understood, nor has any group shown the existence of more than one oxytocin gene.

Production of the two hormones has been reported for testes and ovarian corpus luteum as well as for lung carcinomas, thymus, and adrenal gland (reviewed by Ivell, 1986). Substantiation of these claims would involve a reconsideration of the relative role of the hypothalamus in peripheral functions depending upon circulating oxytocin or vasopressin.

Studies of mRNA from bovine corpus luteum have indicated that this organ indeed expresses the oxytocin gene (Ivell and Richter, 1984b). Depending upon the phase of the luteal cycle, up to 250 times more oxytocin mRNA per organ is produced than in a single hypothalamus. Dot blot analyses have shown that the onset of the expression of the oxytocin gene in the corpus luteum is correlated with the onset of ovulation (Ivell *et al.*,

1985). This result is clearly of considerable significance for the oxytocin-dependent control of reproductive function.

The sequence of the cloned luteal oxytocin mRNA is identical to that already described from the bovine hypothalamus. This is at present the only confirmation by sequence analysis of the peripheral oxytocin or vasopressin peptide. All other identifications have been done either by radioim-munoassay, high-performance liquid chromatography and/or bioassay. But in no case has direct sequence analysis of the peptides been carried out (Ivell, 1986).

Determination of oxytocin mRNA by *in situ* hybridization using cRNA probes shows the oxytocin gene to be transcribed exclusively in the large cells of the corpus luteum at the beginning of the estrous cycle (Fehr *et al.*, 1987). This result confirms the time of onset of oxytocin gene expression as determined by dot blot analysis.

IV. PROCESSING OF THE PRECURSORS

In the following some general remarks are made on the processing of the two hormone precursors. For further details the reader is referred to Chapter 9.

The first step in the processing of the oxytocin or vasopressin precursors is considered to take place during transport of the preprohormones across the membrane of the endoplasmic reticulum (Kreil, 1981). The signal peptide at the N-terminus of the precursors with its middle portion composed of hydrophobic amino acids plays an important role in this transport process across the membrane. After or during traversing the membrane, the signal peptide is removed by proteolytic cleavage (Kreil, 1981). The resulting pro-hormone is then, at least in the case for vasopressin, glycosylated (Gainer, 1982). Part of this glycosylation occurs cotranslationally (Schmale and Richter, 1980). Subsequently, the precursors are transferred to the Golgi ap-paratus, where they become concentrated and formation of neurosecretory granules begins with the precursors undergoing a number of maturation steps. In the case of the two hypothalamic precursors for vasopressin and oxytocin, there is ample time for these steps during their transport along the axons of the neuronal cells. Early kinetic studies indicate that in the dog this transport requires about 1.5 hours from the time of synthesis to release of the nonapeptide (Sachs and Takabatake, 1964).

Pulse–chase experiments *in vivo* in the rat show that cleavage of the nonapeptide precursors occurs continuously during axonal transport, with both cleaved and uncleaved products appearing in the neurohypophysis after a few hours (Gainer *et al.*, 1977). HPLC analysis of hypothalamic nuclear extracts also shows both hormones and the respective neurophysin in the proximal regions of the neurons (Swann *et al.*, 1982). Uncleaved pro-hormones have been found, however, in secretory granules of posterior pituitary (Russell *et al.*, 1980). Apparently, processing of the vasopressin and oxytocin precursors occurs throughout axonal transport between the cell body and the nerve terminals. The processing takes place in the neurosecretory granules, probably at neutral or slightly acidic pH. That the posttranslational processing is most likely dependent on time rather than place is indicated by experiments using colchicine, which blocks axonal transport of the precursor, and by doing so increases the proportion of end product to precursor in the hypothalamic cell bodies (Camier *et al.*, 1985).

Little information is available on intermediate forms of the vasopressin or oxytocin precursors. The glycopeptide has been isolated by concanavalin A Sepharose chromatography from neurosecretory granules, which contain a neurophysin but not a vasopressin antigenic moiety (North *et al.*, 1983). A similar intermediate has been detected in the guinea pig posterior pituitary, implying that in the neurohypophysis the first processing step occurs between vasopressin and the neurophysin (Robinson and Jones, 1983). The converse appears to be the case when frog oocytes are programmed with hypothalamic mRNA. Cleavage of the vasopressin precursor occurs here between neurophysin and the glycopeptide at the single arginine residue (Richter, 1983). The second cleavage step between hormone and neuro-physin has not been observed in the heterologous system. This suggests that the processing enzymes of the frog oocyte recognize single but not pairs of basic amino acids, supporting the notion that there are at least two different processing systems in the cell, one acting on pairs and the other on single basic amino acids (see Chapter 9).

A. Processing Signals and Modifications

A number of signals can be recognized within the vasopressin and oxy-tocin precursors that serve either for the proteolytic cleavage process (single or pairs of basic amino acids), for glycosylation (Asn–X–Thr), or for amidation (Gly). The significance of these signals for the processing event is stressed by the finding that the nucleotide sequence encoding glycine, lysine, and arginine—the three amino acids that separate the hormones from their respective carrier proteins—is highly conserved between the two genes.

Conversion of the two precursors into the active hormones has to include their amidation. Glycine has been found to serve as the signal in the amida-tion reaction of many oligopeptides (Richter, 1983). It is now well establish-ed that this amino acid serves as nitrogen donor leaving glyoxylate to be released as the other product. An enzyme catalyzing the amidation reaction has been isolated from soluble fractions of secretory granules of rat (Brad-bury *et al.*, 1982; Eipper *et al.*, 1983) and bovine posterior pituitary

(Kanmera and Chaiken, 1985). The enzyme requires a glycine residue with a free carboxyl group, suggesting that removal of the basic amino acids is a precondition before the amidation reaction can take place. Ascorbic acid stimulates, whereas divalent metal chelators such as EDTA inhibit, the amidation of model peptides (Mains *et al.*, 1984).

A significant difference between the vasopressin and oxytocin precursors is that the former becomes N-glycosylated, presumable at the Asn–X–Thr site within the C-terminal glycopeptide moiety. Glycosylation of the vasopressin precursor has been demonstrated *in vivo* and *in vitro* (Gainer, 1982). The purpose of the glycosylation is unclear, especially since the oxytocin precursor does not depend on this modifying step for synthesis or secretion.

Experiments with tunicamycin, an antibiotic known to block glycosylation, have suggested that glycosylation may be involved with transport mechanisms. Following intraventricular tunicamycin injection in the rat the vasopressin components do not appear to be properly axonally transported (Gonzalez *et al.*, 1981). The Golgi cisternae of the magnocellular neurons become grossly distended, pointing to an excessive accumulation in this organelle.

B. Processing Enzymes

It is generally assumed that the processing enzymes contain a broad substrate specificity, e.g., when a cDNA encoding human proinsulin is expressed in a tumor cell line normally secreting adrenal corticotropic hormone (ACTH), processing of the proinsulin precursor to insulin occurs correctly (Moore *et al.*, 1983).

Basically, two groups of processing enzymes have been proposed as candidates for catalyzing the conversion of polyproteins in general into mature peptides: Serine protases, including kallikreins and plasmin or plasminogen activator, and thiol proteases with cathepsin B-like properties.

Serine proteases appear to have a neutral or slightly basic pH optimum and one or more sensitive thiol groups. They may be involved in cleavage at single or at pairs of basic amino acids. Some of these enzyme preparations have been obtained from secretory granules of pituitaries (Smyth *et al.*, 1977) or adrenal chromaffin granules (Lindberg *et al.*, 1982). One serine protease isolated from yeast cleaves peptide precursors such as proenkephalin between two basic amino acids (Mizuno and Matsuo, 1984).

Thiol proteases have a slightly acidic pH optimum, with properties similar to cathepsin B, cleaving between pairs of basic amino acids. Most studies have been done with preparations obtained from secretory granules of pituitaries (Gainer, 1982). The model peptide used in many studies is proopiomelanocortin (Douglass *et al.*, 1984). It is claimed that after the first cleavage between the two basic residues a second, carboxypeptidase B-like enzyme, removes the C-terminal end, and an aminopeptidase-like enzyme removes the N-terminal arginine or lysine residue (Loh and Parish, 1984). Another thiol protease has been found in secretory granules from anglerfish pancreas; this enzyme cleaves proinsulin, prosomatostatin, and proglucagon, and has a preference for arginine residues (Noe, 1981).

Processing of a synthetic oxytocin precursor, consisting of the hormone, the extra glycine, the two basic amino acids, and the first six amino acids of the neurophysin, has been used as a model for studying the action of processing enzymes from secretory granules of bovine neurohypophysis (Clamagirand et al., 1986). There it is proposed that an endopeptidase opens the precursor chain, cleaving between the basic amino acid (Arg) and neurophysin (Ala); then a carboxypeptidase B-like enzyme removes Lys and Arg (Fig. 4). The C-terminal glycine serves finally as the signal for amidating the oxytocin (Kanmera and Chaiken, 1985). Unknown is the enzyme responsible for taking off the supernumary histidine found at the C-terminus of the oxytocin precursor, which is absent in authentic neurophysin (Acher, 1983).

Although information on processing enzymes is far from complete, existing data clearly indicate the complex nature of this posttranslational step. There may well be a protease gene family, expressing different but related gene products, which has become specialized as tissue-specific proteases. Analysis of the genes encoding the various proteases will doubtlessly shed more light in this field.

V. EXPRESSION OF THE MUTATED VASOPRESSIN GENE IN DIABETES INSIPIDUS (BRATTLEBORO) RATS

Diabetes insipidus is known as a genetic disorder affecting water metabolism. There are two types of diabetes insipidus, the neurohypophyseal and the neophrogenic form. The former lacks the hormone, the latter is due to vasopressin resistance of the kidney (Sokol and Valtin, 1982). In both forms, the symptoms–excretion of large volumes of dilute urine and intake of increased amounts of fluid–are similar. In man X-linked and autosomally dominant forms of vasopressin-sensitive diabetes insipidus have been reported (Green et al., 1967). The renal type appears to be X-linked inherited (Ten Bensel and Peters, 1970).

The vasopressin-sensitive diabetes insipidus has been studied extensively in mutant (Brattleboro) rats, in which vasopressin appears to be completely lacking from the nerve terminals of the posterior pituitary. In rats, the defect is autosomally recessive and seems only to affect the vasopressinergic neurons (Sokol and Valtin, 1982). The mutant rat lacks not only the

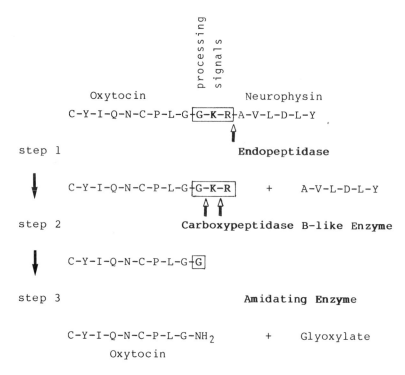

FIG. 4.   Processing model for preprooxytocin. (Modified according to Clamagirand *et al.*, 1986.)

hormone, but also its corresponding neurophysin as well as the glyco-peptide.

However, it has been shown that vasopressin may be expressed in peripheral tissues of the mutant rat (Lim *et al.*, 1983; Nussey *et al.*, 1984). Even more surprising, and in contrast to earlier reports (Watson *et al.*, 1982), are immunocytochemical studies that claim that the mutant rat not only contains vasopressin, but also the glycopeptide in a few hypothalamic neurons (Richards *et al.*, 1985; Mezey *et al.*, 1986). If correct, the data imply that the normal vasopressin precursor can be synthesized even in the Brattleboro rat. Because the specificity of the antiglycopeptide antibodies used in those studies is not yet proved, these findings must await further interpretation.

A. Structure of the Mutated Precursor as Deduced from Gene and mRNA Sequence

The mutated vasopressin gene from Brattleboro rats has been isolated and sequenced (Schmale and Richter, 1984; Schmale *et al.*, 1984). The mutated gene contains a deletion of a G residue in the second exon, which encodes most of the neurophysin protein (Figs. 3,5). Because of the mutation a frameshift occurs with an open reading frame which no longer contains a stop codon for protein termination, thus predicting a precursor with vasopressin and with most of the initial neurophysin intact, but with a modified C-terminus ending, theoretically, with a polylysine tail (Schmale and Richter, 1984).

In the following discussion, some of the C-terminal changes are listed. Starting from amino acid residue 76 of the precursor almost all of the remaining amino acid residues have been replaced. The new reading frame of the mutated gene would predict the replacement of a basic amino acid, which normally separates the neurophysin from the glycopeptide, as well as the loss of the typical sequence for glycosylation. Of 14 cysteine residues originally present in the neurophysin molecule, 5 are substituted by other amino acids—a change that should impact upon the folding of the mutated precursor. When a computer program is used to plot the secondary structure of the normal and mutant precursor, significant changes are predicted starting at the mutation, with more α-helical and β-turn structures in the mutated precursor (Richter, 1987; Rehbein *et al.*, 1986). This comparison of the secondary structure does not include the possible polylysine residues at the C-terminus of the precursor.

B. Transcription of the Mutated Gene

A number of experiments have shown that the mutated gene is transcribed.

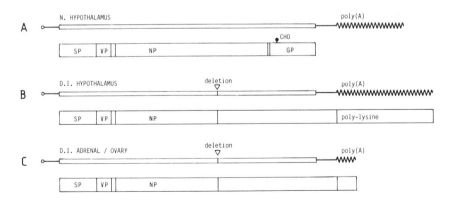

Fig. 5. Scheme comparing the structures of the predicted vasopressin precursors from (A) nonmutant rat hypothalamuses, (B) Brattleboro rat hypothalamuses, and (C) Brattleboro peripheral tissues. SP, Signal peptide; VP, vasopressin; NP, neurophysin; GP, glycopeptide; CHO, carbohydrate side chain.

1. Northern blot analysis of hypothalamic poly(A)$^+$ RNA from normal and Brattleboro rats yields a vasopressin-specific mRNA (Schmale and Richter, 1984; Majzoub et al., 1984; Majzoub, 1985; Ivell et al., 1986).

2. The primary transcripts from normal and mutant rats are correctly spliced, as indicated by microinjection experiments of the respective genes into the nucleus of frog oocytes or by transfection studies in 3T3 cells (Schmale and Richter, 1984; Schmale et al., 1984).

3. Sl mapping of mRNAs from normal and Brattleboro rats gives identical results, excluding the possibility of alternative transcription start sites (Schmale and Richter, 1984).

4. Finally, cloning and sequencing of hypothalamic mRNA from Brattleboro rats show no differences in the coding sequence compared to the corresponding gene. The mRNA lacks the single G nucleotide as expected (Schmale et al., 1984; Ivell et al., 1986).

The size of the vasopressin-encoding mRNA from hypothalamuses of the mutant and wild-type rats, initially thought to be more or less identical (Schmale and Richter, 1984), was subsequently determinated to be 720 bp for either Wistar or Long Evans rats and roughly 800 bp for the Brattleboro rat (Ivell et al., 1986). Because comparison of respective mRNAs by Sl mapping shows identical transcription start sites (Schmale and Richter, 1984), the variation in size has to be attributed to the 3'-end. When the poly(A) tails of the vasopressin mRNA from the wild-type and mutant rats were removed by digestion with RNase H, both hypothalamic mRNAs migrate identically at 580 bp (Ivell et al., 1986).

1. PERIPHERAL mRNA

As pointed out above, immunological data suggest that vasopressin is synthesized also in peripheral organs such as adrenal gland or testis from both wild-type and Brattleboro rats (Lim et al., 1984; Nussey et al., 1984). This agrees well with Northern blot analysis of these tissues (Ivell et al., 1986), which yield positive signals for the adrenal, ovary, and testes of wild-type (Long Evans or Wistar rats) and mutant (Brattleboro) rats, though the signals are approximately two orders of magnitude lower than those from hypothalamic mRNAs.

In all the peripheral tissues studied irrespective of rat strain, the vasopressin-encoding mRNA appears consistently shorter (620 bp) than its hypothalamic counterpart, though when treated with RNase H a 580-bp transcript is obtained (Ivell et al., 1986). Thus, the different sized RNAs from hypothalamic or peripheral organs are due to tissue-specific differences in the length of the poly(A) tails. Unclear at present is why the hypothalamic vasopressin mRNA from mutant rats has a longer poly(A) tail than its nonmutant counterpart. One possibility is that because the ribosomes read through into the noncoding 3 '-region, and presumably only leave the messenger slowly, this region including the poly(A) tail is more protected against cytoplasmic RNase activity. Hence, cytoplasmic clipping of the poly(A) tail would be inhibited in the case of the mutant mRNA.

When comparing the expression of the vasopressin gene in various tissues, the hypothalamus exhibits the highest levels of transcription, with a somewhat reduced level of hpothalamic vasopressin mRNA in the mutant rat (Rehbein et al., 1986; Table II). Surprisingly, both the adrenal and testicular mutant mRNAs are also considerably reduced in hybridization intensity in comparison to wild-type rats (Ivell et al., 1986).

2. In Situ HYBRIDIZATION IN TISSUES FROM NORMAL AND MUTANT RATS

Using the in situ hybridization technique with specific cDNA probes derived from the 3 '-end of the mRNA encoding the vasopressin precursor, it has been demonstrated by a number of groups that positive signals can be obtained in thin sections from the hypothalamus of the mutant rat (Fig. 6). The signals or grain numbers are more or less identical when compared to

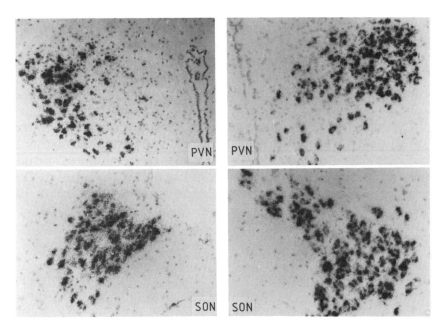

FIG. 6. In situ hybridization identifying specifically vasopressin-encoding mRNA in hypothalamic sections from Brattleboro (right panel) and Wistar rats (left panel). A 240-b vasopressin-specific RNA probe labeled with ^{35}S has been applied. (Courtesy of Dr. S. Fehr, Hamburg.) SON, Supraoptic nucleus; PVN, paraventricular nucleus.

the wild-type sections (Uhl *et al.*, 1985; Fuller *et al.*, 1985; S. Fehr, unpublished). In one case it has been observed that the number of grains in the Brattleboro rat is significantly reduced (McCabe *et al.*, 1986a,b). Whether this is due to technical problems or differences between animal colonies remains to be seen. In any event, it clearly demonstrates that with different techniques it can be shown that the vasopressin-mutated gene is expressed in the Brattleboro rat to more or less the same extent.

C. Identification of the Mutated Vasopressin

1. *In Vitro* SYNTHESIS OF THE MUTATED PRECURSOR

For identification of the predicted mutated precursor, antibodies raised against various parts of the precursor sequence have been used in cell-free translation systems programmed with hypothalamic mRNA from mutant and wild-type rats. With antibodies against a synthetic tetradecapeptide (CP-14) corresponding to part of the C-terminus of the mutated precursor, a product with a molecular weight of roughly 26,000 has been obtained in a cell-free translation system programmed with the mutated mRNA (Ivell *et al.*, 1986); no normal-sized vasopressin precursor (molecular weight 19,000) can be detected. The 26,000 molecular weight protein is absent among the translation products of nonmutant hypothalamic mRNA, and cross-reacts also with vasopressin and neurophysin antibodies. This is not surprising, because two-thirds of the mutated precursor is identical with the normal one.

The failure to synthesize authentic vasopressin precursor is due to the shifted reading frame, which no longer has the signal UGA for terminating protein synthesis, and thus allows ribosomes to proceed further on into the normally 3 '-untranslated region as well as into the poly(A) sequence. The expected translation product should consist of an additional 90 amino acids (including 70 lysine residues!) giving rise to the mutated product with a molecular weight between 26,000 and 28,000—a prediction that is confirmed by the studies cited above.

2. *In Vitro* PROCESSING OF THE MUTATED PRECURSOR

It has been shown earlier that *in vitro* the first processing step can be studied by the addition of microsomal membranes to a cell-free protein-synthesizing system with the result that the signal peptide of the normal vasopressin precursor is cleaved off and the provasopressin is glycosylated

(Schmale and Richter, 1980). In this case rat preprovasopressin has a molecular weight of 19,000, and that of the processed glycosylated proform is 23,000.

Similar experiments have been performed with the mutated mRNA, to determine whether the highly positively charged lysine tail at the C-terminus of the mutated vasopressin precursor interferes with ribosomal translation and/or the initial processing step by binding to structures such as microsomal membranes. For convenience, the mutated as well as the normal rat cDNA have been subcloned into SP6 vectors, which allow the *in vitro* generation of eukaryotic mRNAs.

The transcribed mRNAs have been used in the subsequent cell-free translation experiments (H. Schmale, unpublished data). Also, mRNAs have been constructed with poly(A) tails of various length by subcloning cDNAs encoding the mutated or the normal vasopressin precursor.

When mutated mRNAs are translated with no or very short poly(A) sequences (25 A residues), the mutated preprovasopressin is converted into an unglycosylated proform (Fig. 7). Translation of mutant mRNAs with longer poly(A) tails (80 A residues) obviously does not yield a longer protein precursor and processing by microsomal membranes seems to be at least less efficient. How far ribosomes proceed into the poly(A) sequence, and what precisely stops translation there, are unknown. Apparently, if a critical length of the poly(A) stretch is translated into polylysine, then further translation and perhaps membrane-directed processing become blocked.

That the length of the poly(A) sequence per se is not the reason for the processing defect is indicated by control experiments with normal rat mRNA containing different length poly(A) sequences. The resulting vasopressin precursor is processed normally.

The result that the mutated preprovasopressin is processed into the prohormone when derived by translation of mRNAs containing no or only few residues at the 3 '-end would explain the *in vivo* findings that, in adrenal and testis, low amounts of vasopressin can be detected in the mutant rat (Lim *et al.*, 1984; Nussey *et al.*, 1984). As shown recently, the mutated vasopressin-encoding mRNA from peripheral organs contains only a short poly(A) tail (Ivell *et al.*, 1986). Hence, this precursor with only a few lysine residues at its C-terminus could be sufficiently processed and packaged into the respective secretory vesicles. The *in vitro* data suggest that the vasopressin defect in the mutant rat must be at the posttranslational step and is due to the abnormal C-terminus.

3. IMMUNOCYTOCHEMICAL IDENTIFICATION

The mutated vasopressin precursor has been identified so far *in vivo* by immunocytochemical studies. With the CP-14 antibodies, positive staining

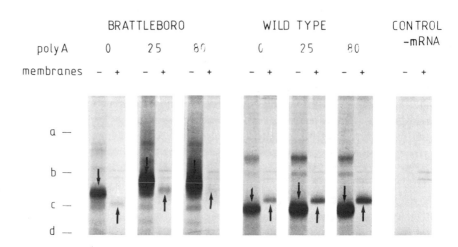

FIG. 7. *In vitro* processing of the mutated vasopressin precursor in the presence of the micro-somal membranes. Cell-free translation of mRNA generated *in vitro* in the SP 6 transcription system. Proteins were synthesized in the reticulocyte lysate with [^{35}S]cysteine as radioactive amino acid. mRNA was derived from clones carrying either the mutated (left panel) or wild-type cDNA (middle panel) with no or different lengths of poly(A) tails; the numbers of A residues are in-dicated. Microsomal membranes are derived from dog pancreas. Downward pointing arrows show the position of the unprocessed, the upward pointing arrows show that of the processed precursors. The mutated precursor varied in size depending on the poly(A) length, whereas the normal precursor showed constant molecular weights regardless of the length of the poly(A) se-quence. In the presence of membranes, the mutated precursor is processed into a form lower in size, whereas the normal one is core-glycosylated into a higher molecular weight form. The other bands in the upper part of the lanes are either endogenous proteins of the reticulocyte system or dimers of the respective precursor. The right panel shows control experiments without mRNAs. (a–d) Molecular weight markers of the size 43K, 25.7K, 18.4K, 14.3K.

has been found in the neurons of the SON of Brattleboro rats (Krisch *et al.*, 1986; Guldenaar *et al.*, 1986). Staining is predominantly in the perikarya of the SON, with somewhat less in those of the PVN and none evident in the SCN, the median eminence, or the neural lobe. In all animals, the SON always contains the greatest numbers of immunoreactive cells. The staining is preferentially restricted to the peripheral rim of the cytoplasm in a patchy manner. No positive staining reaction has been detected in sections from wild-type rats using CP-14 antibodies. In the Brattleboro rat a few cells can be stained also with antivasopressin antibodies; when it is observed, the subsequent section shows immunoreaction with CP-14 antiserum, sug-gesting a colocalization of both peptides within the same perikaryon.

At the electron-microscopic level, CP-14 immunoreactivity has been demonstrated on the secretory cisternae of the Golgi apparatus, on lysosome-like bodies, and on parts of the rough endoplasmic reticulum

(Krisch et al., 1986). With vasopressin antibodies positive signals have been confined to very limited areas of the rough endoplasmic reticulum. No positive immunoreactivity is observed in neurosecretory granules. Because the mutated vasopressin precursor contains a number of lysine residues at its C-terminus, this highly charged moiety is probably bound to the membrane surfaces and/or other structures, eventually triggering the scavenging mechanisms of the lysosomal system.

The observation (Richards et al., 1985; Mezey et al., 1986) that a few perikarya of the SON from Brattleboro rats can be stained not only with antivasopressin but also with antibodies raised against sheep or human glycopeptide antibodies is unexpected and difficult to explain. If one excludes the existence of a second vasopressin-encoding gene in rats—and there is no evidence for such an assumption—or an alternative splicing process, then one is left with a more technical problem (e.g., the specificity of the antibodies used in the various studies). Another possibility would be the observation of a viral system, in which a reversion of the initial reading frame has been postulated (Jacks and Varmus, 1985). Such a shift in the reading frame during translation has been described in bacteria and could represent another way to regulate gene expression in eukaryotes.

Taken together, the mutated vasopressin gene in the Brattleboro rat represents the rare case in which the genetic defect has been resolved to the molecular level. The single frameshift mutation is unusual because of the open reading frame, which excludes termination of protein synthesis. The resulting mutated precursor can be identified in vitro and in vivo by immunocytochemical studies. Whether the mutated precursor is rejected by the cell because of the abnormal polylysine structure at its C-terminus, or because of the altered C-terminal amino acid sequences that represent a "wrong tag" not allowing the correct packaging in neurosecretory granules, remains to be seen. Recombinant DNA methods should allow one to construct appropriate precursor models to study the posttranslational steps and hence to answer the question why there is no release of vasopressin from the mutated precursor.

VI. CONCLUSIONS AND PERSPECTIVES

Vasopressin and oxytocin precursors are typical representatives of cellular polyproteins—a definition initially introduced for viral polyproteins (Koch and Richter, 1980; Douglas et al., 1984). They are composed of several distinct entities, whereby the structural organization of those precursors may vary. Both hormones are encoded by single genes; the expression of the two precursors occurs not only in the hypothalamus, but also in peripheral organs such as testis or corpus luteum. This raises the question of the transcriptional control of these genes in the hypothalamus and in the peripheral organs.

Luteal cell systems should be helpful to study *in vitro* the regulation, for example, of the oxytocin gene expression, which *in vivo* appears to be linked to the hormonal status of the estrous cycle. Furthermore, it will be of interest to see how the numerous physiological functions proposed for the two peptides (de Wied, 1983) can be correlated to differential expression of the two genes, depending on the anatomical location of the neuronal cells and their projection to appropriate sites of the brain. Regulation will also occur at the level of posttranslational processing and release, and hence will require more detailed studies of the mechanisms involved. Finally, how are the signals relayed from the cell membrane to the nucleus? Are there feedback signals from the precursor synthesis toward the transcription of a gene? Genetic defects as found in the Brattleboro rats may be helpful in getting answers to some of these problems.

ACKNOWLEDGMENTS

We thank Dr. R. Ivell for editorial help and the Deutsche Forschungsgemeinschaft for financial support.

REFERENCES

Acher, R. (1983). *In* "Brain Peptides" (D. Krieger, M. Brownstein, and J. Martin, eds.), pp. 135–163. Wiley, New York.
Bradbury, A. F., Finnie, M. D. A., and Smyth, D. G. (1982). *Nature (London)* **298**, 686–688.
Burbach, J. P. H., De Hoop, M. J., Schmale, H., Richter, D., De Kloet, E. R., Ten Haaf, J. A., and De Wied, D. (1984). *Neuroendocrinology* **39**, 582–584.
Burbach, J. P. H., Van Tol, H. H. M., Bakkus, M. H. C., Schmale, H., and Ivell, R. (1986). *J. Neurochem.* **47**, 1814–1821.
Caffe, A. R., van Leeuwen, F. W., Buijs, R. M., De Vries, G. J., and Geffard, M. (1985). *Brain Res.* **338**, 160–164.
Camier, M., Barre, N., and Cohen, P. (1985). *Brain Res.* **334**, 1–8.
Clamagirand, C., Camier, M., Bousetta, H., Fahy, C., Morel, A., Nicolas, R., and Cohen, P. (1986). *Biochem. Biophys. Res. Commun.* **134**, 1190–1196.
De Wied, D. (1983). *Prog. Brain Res.* **60**, 115–167.
Douglass, J., Civelli,O., and Herbert, E. (1984). *Annu. Rev. Biochem.* **53**, 665–715.
Du Vigneaud, V., Ressler, C., Swan, J. M., Katsoyannis, P. G., and Roberts, C. W. (1954). *J. Am. Chem. Soc.* **76**, 3115–3121,
Eipper, B. A., Mains, R. E., and Glembotski, C. C. (1983). *Proc. Natl. Acad. Sci. U.S.A.* **80**, 5144–5148.
Fehr, S., Ivell, R., Koll, R., Schams, D., Fields, M., and Richter, D. (1987). *FEBS Lett.* **210**, 45–50.
Fuller, P. J., Clements, J. A., and Funder, J. W. (1985). *Endocrinology* **116**, 2366–2368.
Gainer, H. (1982). *Prog. Brain Res.* **60**, 205–215.
Gainer, H., Sarne, Y., and Brownstein, M. J. (1977). *J. Cell Biol.* **73**, 366–381.
Gonzalez, C. B., Swann, R. W., and Pickering, B T. (1981). *Cell Tissue Res.* **217**, 199–210.

Green, J. R., Buchan, G. C., Alvord, E. C., Jr., and Swanson, A. G. (1967). *Brain* **90**, 707–714.

Guldenaar, S. E. F., Nahke, P., and Pickering, B. T. (1986). *Cell Tissue Res.* **244**, 431–436.

Holwerda, D. A. (1972). *Eur. J. Biochem.* **28**, 340–346.

Ivell, R. (1986). *Curr. Top. Neuroendocrinol.* **6**, 1–18.

Ivell, R., and Richter, D. (1984a). *Proc. Natl. Acad. Sci. U.S.A.* **81**, 2006–2010.

Ivell, R., and Richter, D. (1984b). *EMBO J.* **3**, 2351–2354.

Ivell, R., Brackett, K. H., Fields, M. J., and Richter, D. (1985). *FEBS Lett.* **190**, 263–267.

Ivell, R., Schmale, H., Krisch, B., Nahke, P., and Richter, D. (1986). *EMBO J.* **5**, 971–977.

Jacks, T., and Varmus, H. E. (1985). *Science* **230**, 1237–1242.

Kanmera, T., and Chaiken, I. M. (1985). *J. Biol. Chem.* **260**, 10118–10124.

Koch, G., and Richter, D., eds. (1980). "Biosynthesis, Modification, and Processing of Cellular and Viral Polyproteins." Academic Press, New York.

Kreil, G. (1981). *Annu. Rev. Biochem.* **50**, 317–348.

Krisch, B., Nahke, P., and Richter, D. (1986). *Cell Tissue Res.* **244**, 351–358.

Laczi, F., Gaffori, D., de Kloet, E. R., and de Wied, D. (1983). *Brain. Res.* **260**, 342–346.

Lim, A. T., Lolait, S., Barlow, J. W., Wai Sum O, Zois, I., Toh, B. H., and Funder, J. W. (1983). *Nature (London)* **303**, 709–711.

Lindberg, I., Yang, H. Y. I., and Costa, E. (1982). *Biochem Biophys. Res. Commun.* **106**, 186–193.

Loh, Y. P., and Parish, D. (1984). *Int. Congr. Endocrinol. 7th; Quebec* No. S72.

McCabe, J. T., Morrell, J. I., Ivell, R., Schmale, H., Richter, D., and Pfaff, D. W. (1986a). *J. Histochem. Cytochem.* **34**, 45–50.

McCabe, J. T., Morell, J. I., Richter, D., and Pfaff, D. W. (1986b). *Front. Neuroendocrinol.* **9**, 149–167.

Mains, R. E., Glembotski, C. C., and Eipper, B. A. (1984). *Endocrinology* **114**, 1522–1530.

Majzoub, J. A. (1985). *In* "Vasopressin" (R. W. Schrier, ed.), pp. 465–474. Raven, New York.

Majzoub, J. A., Rich, A., v. Boom, J., and Habener, J. F. (1983). *J. Biol. Chem.* **258**, 14061–14064.

Majzoub, J. A., Pappey, A., Burg, R., and Habener, J. (1984). *Proc. Natl. Acad. Sci. U.S.A.* **81**, 5296–5299.

Mezey, E., Seidah, N. G., Chretien, M., and Brownstein, M. J. (1986). *Neuropeptides* **7**, 79–85.

Mizuno, K., and Matsuo, H. (1984). *Nature (London)* **309**, 558–560.

Mohr, E., Hillers, M., Ivell, R., Haulica, I. D., and Richter, D. (1985). *FEBS Lett.* **193**, 12–16.

Mohr, E., Schmitz, E., and Richter, D. (1987). *Biochimie*, in press.

Noe, B. D. (1981). *J. Biol. Chem.* **256**, 4940–4946.

North, W. G., Mitchell, T. I., and North, G. M. (1983). *FEBS Lett.* **152**, 29–34.

Nussey, S. S., Ang, V. T. Y., Jenkins, J. S., Chowdrey, H. S., and Bisset, G. W. (1984). *Nature (London)* **310**, 64–66.

Rascher, W., Lang, R. E., and Unger, T. (1985). *Curr. Top. Neuroendocrinol.* **4**, 101–136.

Rehbein, M., Hillers, M., Mohr, E., Ivell, R., Morley, S., Schmale, H., and Richter D. (1986). *Biol. Chem. Hoppe-Seyler* **367**, 695–704.

Rholam, M., Nicolas, P., and Cohen, P. (1982). *Biochemistry* **21**, 4968–4973.

Richards, S.-J., Morris, R. J., and Raisman, G. (1985). *Neuroscience* **16**, 617–623.

Richter, D. (1983). *Prog. Nucleic Acid Res. Mol. Biol.* **30**, 245–266.

Richter, D. (1987). *In* "The Peptides" (S. Udenfriend, J. Meienhofer, eds.), pp. 41–72, Academic Press, Orlando.

Riddell, D. C., Mallonee, R., Phillips, J. A., Parks, J. S., Sexton, L. A., and Hamerton, J. L. (1985). *Somat. Cell Mol. Genet.* **11**, 189–195.

Robinson, I. C. A. F., and Jones, P. M. (1983). *Neurosci. Lett.* **39**, 273–278.

Ruppert, S., Scherer, G., and Schuetz, G. (1984). *Nature (London)* **263**, 211–214.

Russell, J. T., Brownstein, M. J., and Gainer, H. (1980). *Endocrinology* **107**, 1880–1891.

Sachs, H., and Takabatake, Y. (1964). *Endocrinology* **75**, 943–948.

Sachs, H., Fawcett, P., Takabatake, Y., and Portanova, R. (1969). *Recent Prog. Horm. Res.* **25**, 447–491.

Sausville, E., Carney, D., and Battey, J. (1985). *J. Biol. Chem.* **260**, 10236–10241.

Schmale, H., and Richter, D. (1980). *FEBS Lett.* **121**, 358–362.

Schmale, H., and Richter, D. (1984). *Nature (London)* **308**, 705–709.

Schmale, H., Ivell, R., Breindl, M., Darmer, D., and Richter, D. (1984). *EMBO J.* **3**, 3289–3293.

Smyth, D. G., and Massey, D. E. (1979). *Biochem. Biophys. Res. Commun.* **87**, 1006–1010.

Smyth, D. G., Austen, B. M., Bradbury, A. F., Geisow, M. J., and Snell, C. R. (1977). *In* "Centrally Active Peptides" (J. Hughes, ed.), pp. 231–239. Macmillan, London.

Sofroniew, M. V. (1983). *Prog. Brain Res.* **60**, 101–114.

Sokol, H. W., and Valtin, H., eds. (1982). *Ann. N.Y. Acad. Sci.* **394**, 1–828.

Supowit, S., Potter, E., Evans, R. M., and Rosenfeld, M. G. (1984). *Proc. Natl. Acad. Sci. U.S.A.* **81**, 2975–2979.

Swann, R. W., Gonzales, C. B., Birkett, S. D., and Pickering, B. T. (1982). *Biochem. J.* **208**, 339–349.

Ten Bensel, R. W., and Peters, E. R. (1970). *J. Pediat.* **77**, 439–443.

Turner, R. A., Pierce, J. G., and Du Vigneaud, V. J. (1951). *J. Biol. Chem.* **191**, 21–28.

Uhl, G. R., Zingg, H. H., and Habener, J. F. (1985). *Proc. Natl. Acad. Sci. U.S.A.* **22**, 5555–5559.

Watson, S. J., Seidah, N. G., and Chretien, M. (1982). *Science* **217**, 853–855.

Zerbe, R. L., and Palkovits, M. (1984). *Neuroendocrinology* **38**, 285–289.

CURRENT TOPICS IN MEMBRANES AND TRANSPORT, VOLUME 31

The Secretory Vesicle in Processing and Secretion of Neuropeptides

JAMES T. RUSSELL

Unit on Neuronal Secretory Systems
National Institutes of Child Health and Human Development
National Institutes of Health
Bethesda, Maryland 20892

I. INTRODUCTION

Since the discovery of proinsulin by Steiner and Oyer (1967), experimental evidence accumulated over the past two decades has clearly shown that most secreted proteins and peptides are synthesized as larger precursor proteins (proproteins or prohormones) and subsequently modified into biologically active peptides by a variety of posttranslational processing steps (see Steiner *et al.*, 1974; Koch and Richer, 1980; Zimmermann *et al.*, 1980; Freedman and Hawkins, 1980; Habener, 1981; Docherty and Steiner, 1982; Loh and Gainer, 1983; Loh *et al.*, 1984; Gainer *et al.*, 1985a, for reviews). These posttranslational events consist of specific proteolytic cleavages of the precursor proteins as well as other enzymatic events (e.g., acetylation, amidation) and are highly organized within specific intracellular membrane-bound compartments. It has become clear that the biosynthesis of neuropeptides and other polypeptides such as albumin, immunoglobulin, mellitin, and various membrane and viral proteins also follows this sequence of cellular biochemical events.

It is reasonable to assume that the general mechanism of biosynthesis of secretory proteins elucidated in other more accessible experimental systems (e.g., pancreatic acinar cells; see Palade, 1975) is operative in neuronal systems in the genesis of secretory polypeptides. This process begins with the synthesis of the prepropeptide on the rough endoplasmic reticulum membranes, its transport across the membrane into the cisternae, and subsequent translocation to the Golgi apparatus, where it is ultimately packaged into secretory vesicles. Various posttranslational processing steps occur during this translocation period through the various membrane systems. These include removal of the signal peptide from the prepropeptide, asparagine-linked glycosylation of the precursor in the endoplasmic reticulum, disulfide bond formation, and more complex glycosylation events in the Golgi apparatus (for reviews see Koch and Richer, 1980; Zimmermann et al., 1980; Farquhar and Palade, 1981; Castel et al., 1984). However, the precursor polypeptide arriving at the Golgi is still in the propeptide form and needs to be further processed into the final peptide products. These processing events seem to occur in the fully formed secretory vesicles within which the propeptides are packaged together with other components. The conclusion that the final posttranslational processing events occur in secretory vesicles would imply that the secretory vesicles should contain not only the enzymes necessary, but also the cofactors, and that the vesicles must maintain their internal milieu conducive for the enzyme activities. The scope of this article is to focus on the biochemical processes by which the secretory vesicle maintains its contents chemically and energetically ideal for propeptide processing and exocytotic secretion.

II. STRUCTURE OF PRECURSORS AND CLEAVAGE SITES

Over the last decade the use of recombinant DNA technology has yielded valuable data on the amino acid sequences of a large number of prohormones and propeptides (Nakanishi et al., 1979; Noda et al., 1982; Kakidani et al., 1982; Gubler et al., 1982; Land et al., 1982, 1983; Ruppert et al., 1984; Schmale and Richter, 1984). This information was valuable in deducing the specific structural domains within the propeptides, that are the sites at which processing enzymes must act. It has also been learned that a series of enzymatic events must take part in the conversion of propeptides to final products. These include endopeptidases to initially cleave the propeptides, followed by exopeptidases to trim the terminal amino acids, and finally specific modification of NH_2- and/or COOH-termini (e.g., acetylation amidation). Almost all the propeptides contain pairs of basic amino acid residues (Arg or Lys) flanking the peptide sequences that are to be excised (see Table I). The most common sequence pair is Lys-Arg, although Arg-Lys, Arg-Arg, and Lys-Lys are also found at the putative cleavage sites. Several reviews on processing enzymes have dealt extensively with the amino acid sequence signal requirements in detail (Habener, 1981; Loh and Gainer, 1983; Loh et al., 1984; Gainer et al., 1985a).

I will point out certain salient points here for the sake of completeness and to emphasize certain chemical requirements the enzymatic reactions impose on the secretory vesicles. Although this pattern of cleavage sites seems to be characteristic of propeptides, it is by no means universal. In a significant number of cases, propeptides are processed at sites that do not have dibasic amino acid residues (Docherty and Steiner, 1982; Gainer et al., 1985a). Furthermore, some cleavage sites contain single basic amino acids, and not all dibasic residues within the propeptide structure serve as cleavage sites (see Table I). Another point to consider is the diversity in the actual cleavage site within the dibasic amino acid signal, i.e., whether the cleavage occurs at the N-terminal side, in between, or at the C-terminal side of the dibasic pair. Analysis of peptides occurring in situ shows that all three possibilities occur in different cell types (Gainer et al., 1985a, for review). It is also clear that, in addition to the dibasic residue-specific endopeptidase, exopeptidases (aminopeptidase and/or carboxypeptidase) are required to trim the remaining N- and/or C-terminal basic amino acid residues following endopeptidase cleavage to produce the biologically active peptide.

Another structural requirement for posttranslational processing is the presence of a glycine residue immediately distal to the amino acid to be amidated at the C-termius in the biologically active peptide (Table I). This was recognized by Smyth and co-workers (Bradbury et al., 1982; Bradbury and Smyth, 1983a) while studying the amidating reaction and the enzyme responsible for it, and is borne out by the sequence data now available through recombinant DNA technology. The glycine residue is utilized as the amide donor in this hydroxylating monooxygenase reaction (Bradbury et al., 1982; Eipper et al., 1983a–c; Glembotski et al., 1984; see later).

It is clear even in the case of one of the simplest of propeptides (e.g., pro-oxyphysin, precursor for oxytocin and its neurophysin) that a large number (four) of different processing enzymes is required before the final products can be fashioned (see Fig. 1 in Gainer et al., 1985a).

III. THE PROCESSING ENZYMES

The three classes of processing enzymes involved in posttranslational modifications of propeptides to be considered are (1) endopeptidases that

TABLE I
SCHEMATIC REPRESENTATION OF PROPEPTIDE-PROCESSING SITES[a]

Peptide	Prohormone	Processing site
Vasopressin	Propressophysin	AVP ⊢Gly-Lys-Arg⊣ NP-VP ⊢Arg⊣ Glyco Ⓢ
Oxytocin	Prooxyphysin	OT ⊢Gly-Lys-Arg⊣ NP-OT ⊢Arg
α-MSH	POMC	-Lys-Arg⊣ α-MSH ⊢Gly-Lys-Lys-Arg-Arg
γ₁-MSH	POMC	-Arg⊣ γ₁-MSH ⊢Gly-Arg-Arg
Met-Enkephalin	Pro-Enkephalin	-Lys-Arg⊣ Tyr-Gly-Gly-Phe-Met ⊢Arg-Arg
		-Lys-Arg⊣ Tyr-Gly-Gly-Phe-Met ⊢Lys-Lys
Dynorphin	Prodynorphin	-Lys-Arg⊣ Dynorphin ⊢Lys-Arg
Dynorphin₁₋₁₃	Prodynorphin	-Lys-Arg⊣ Tyr-Gly------Leu-Lys ⊢Trp-Asp
Substance P	Protachykynin	-Ala-Arg⊣ Arg-Pro------Met ⊢Gly-Lys-Arg

[a]The biologically active peptides are shown with rectangles, and the processing sites flanking the amino acids at the N- and C- termini are obtained from the sequences of propeptides. Underlined glycine residues are utilized for the amidation of the peptide (see text). AVP, Arginine[8] vasopressin; NP-VP, vasopressin-associated neurophysin; Glyco, glycopeptide in precursor to vasopressin; OT, oxytocin; NP-OT, oxytocin-associated neurophysin; MSH, melanocyte-stimulating hormone.

cleave the large propeptide into fragments, (2) exopeptidases that remove amino acids from the N- and/or C-terminus of the resulting fragments, and (3) nonproteolytic enzymes that modify the N- or C-terminus of the peptide (e.g., acetylation and amidation). As pointed out in Section I, because the processing appears to occur within secretory vesicles, these enzymatic activities also should be localized within them. Steiner and colleagues (Docherty and Steiner, 1982) working with proinsulin set forth the criteria to identify an authentic processing enzyme. These criteria have been revised and augmented since then, particularly as more and more experimental data have been collected in the field of opioid peptides, neurohypophysical peptides, and adrenal chromaffin cell systems (see Loh and Gainer, 1983; Loh et al., 1984a; Gainer et al., 1985a, for review). I will attempt to give a concise summary of the state of our knowledge to date.

A. Endopeptidases Specific for Cleavage at Dibasic Residues

Since 1982, several laboratories have succeeded in demonstrating enzymatic activities in secretory tissue that cleave prohormones into peptide fragments specifically at dibasic residues (Fletcher et al., 1980, 1981; Noe, 1981; Chang et al., 1982; Loh and Gainer, 1982). These include enzymatic activities that convert proinsulin, proglucagon, and prosomatostatin from secretory vesicles from pancreatic islets (Fletcher et al., 1980, 1981; Noe, 1981), and enzymatic activities obtained from secretory vesicles isolated from bovine anterior, intermediate, and neural lobes of the pituitary that convert intact POMC into appropriate peptide products (Chang et al., 1982; Loh and Chang, 1982; Loh and Gainer, 1982). The distinctive features of these enzymes were that they were active at a pH range of 4 to 6, and were inactive against smaller peptide substrates even though they contained dibasic amino acid residues. The latter property may suggest that the processing enzymes have conformational requirements in addition to the cleavage signals in the propeptide sequences. The inhibitor profile was distinct from trypsin or cathepsin B-like endopeptidases, suggesting that they are a separate class of endopeptidases localized within secretory vesicles.

Over 10 years passed after the first suggestion by Steiner and co-workers (Steiner et al., 1974) describing the necessity of an endopeptidase in prohormone processing before the first one was purified to homogeneity from bovine pituitary intermediate lobe secretory vesicles (Loh et al., 1985; Loh, 1986a). These studies have led to the tentative classification of the dibasic residue-specific endopeptidase from the intermediate lobe and initial kinetic analyses. In the literature there still exists a controversy on the actual classification of the endopeptidase. Whereas the enzyme obtained from the pituitary intermediate and neural lobes appears to be aspartyl proteases (Loh et al., 1985; Loh, 1986a), the chromaffin vesicle enzyme, which cleaves fragments of proenkephalin, is classified as a serine protease (Lindberg et al., 1984; Mizuno and Matsuo, 1985; Mizuno et al., 1985) and the pancreatic islet enzyme as a thiol protease (Kemmler et al., 1973; Fletcher et al., 1980, 1981; Docherty and Steiner, 1982). It is not clear whether this discrepancy is due to different cell types utilizing different enzymes or heterogeneity in enzyme preparations. However, the pituitary enzyme isolated from secretory vesicles in the intermediate or neural lobe of the pituitary by Loh and co-workers can successfully convert proinsulin and proenkephalin (Loh et al., 1985; Loh, 1986a,b). However, in gene transfection experiments Drucker et al. (1986) introduced a metallothionein–glucagon fusion gene into a pituitary cell line (GH4) and an islet cell line (RN38) and found that the pituitary cell line was able to process proglucagon only partially, whereas the cell line from the pancreas, where glucagon is normally made, was able to secrete almost completely processed glucagon. Most of these enzyme activities have pH optima in the acidic range (pH 4.0 to 5.5) (Fletcher *et al.*, 1980, 1981; Docherty and Steiner, 1982; Chang and Loh, 1984; Loh *et al.*, 1985; Loh, 1986a), with a few exceptions (Habener *et al.*, 1977; Lindberg *et al.*, 1984; Mizuno and Matsuo, 1985; Cromlish *et al.*, 1985; Gomez *et al.*, 1985).

B. Endopeptidases Specific for Cleavage at Single Basic Residues

The propeptide structures reveal that some of the endopeptidases must cleave at single basic amino acid residues (e.g., prohormones of vasopressin, oxytocin, and prodynorphin). Two such enzyme activities have been reported (Wallace *et al.*, 1984; Devi and Goldstein, 1984, 1985), one isolated from the atrial gland of *Aplysia californica* (Wallace *et al.*, 1984), and the other from rat brain membranes (Devi and Goldstein, 1984, 1985). The former is an acidic protease (pH optimum 6.2) and is activated under sulfhydryl reagents, and the latter is a neutral protease. These activities do not appear to be due to similar enzymes. The endopeptidase required for making vasopressin-associated neurophysin from propressophysin, which involves cleavage at a single arginine residue, has not been identified. Such an enzyme must be present in vasopressinergic secretory vesicles.

C. Exopeptidases

The peptide fragments resulting from the action of endopeptidases on propeptides need to be further processed to remove the basic amino acid residues at the N- and C-termini. These include a carboxypeptidase B-like activity and an aminopeptidase. Carboxypeptidase B-like enzymes have been identified in secretory vesicles from various tissues (Fricker and Snyder 1982, 1983; Docherty and Hutton, 1983; Hook *et al.*, 1982; Wallace *et al.*, 1982; Hook and Loh, 1984). The enzyme from chromaffin vesicles was isolated to homogeneity (Fricker and Snyder, 1982, 1983; Fricker *et al.*, 1983a,b; Suppattapone *et al.*, 1984), and shown to be a metalloprotease stimulated by Co^{2+} and inhibited by Cu^{2+} and Cd^{2+}. The pH optimum of this enzyme was 5.4–5.8.

As pointed out earlier (Table I), because the endopeptidase specific for dibasic amino acid sequences can cleave the propeptide in between the two dibasic residues (e.g., Lys-Arg, Arg-Arg, or Arg-Lys) (Loh *et al.*, 1985; Loh, 1986a) the resultant peptide fragments could contain basic amino acid residue extensions at both the N- and the C-termini. In this case, an aminopeptidase activity would be required to trim the amino acid extension at the N-terminus. Such an enzymatic activity was detected in isolated secretory vesicles from the intermediate and neural lobes of the pituitary (Gainer *et al.*, 1985b). This enzymatic activity was activated by Co^{2+} and Zn^{2+} and was maximally active at pH 6.0. It has also been reported that the precursor for mellitin in the honeybee and for α-mating factor in yeast requires a dipeptidyl aminopeptidase for complete processing (Julius *et al.*, 1983; Kriel *et al.*, 1977).

D. Nonproteolytic Enzymes

A number of biologically active peptides are amidated at the carboxyl terminus and acetylated at the amino terminus. These terminal modifications afford resistance against degradative enzymes upon secretion, thereby increasing the biological half-life, and in some instances are required for activation of the receptor and biological activity (Eberle and Schwyzer, 1975; Lowry *et al.*, 1977; Smyth *et al.*, 1979; Eipper *et al.*, 1983a,b; Glembotski, 1982a,b; Manning *et al.*, 1984). Corticotropin-releasing factor, vasopressin, gastrin, and cholecystokinin (CCk) require the C-terminal amide for biological activity, and acetylation of the N-terminus of α-melanocyte-stimulating hormone (α-MSH) is necessary for its activity (Lowry *et al.*, 1977). N-Terminal acetylation of β-endorphin, however, inhibits its opioid potency (Smyth *et al.*, 1979).

1. N-TERMINAL ACETYLATION

An acetyltransferase activity that can catalyze acetyl-group transfer from acetyl-CoA to the N-terminus of desacetyl α-MSH has been found and extensively characterized in bovine intermediate lobe secretory vesicles (Glembotski, 1982a,b; Chappell *et al.*, 1982), and in homogenates of rat hypothalamus (Barnea and Cho, 1983; O'Donohue, 1983). The acetylation occurred at the serine residue (α-MSH) or tyrosine residue (β-endorphin). This acetyltransferase enzyme was soluble and was active at a pH range of 5.7–8.0.

2. C-TERMINAL AMIDATION

Secretory vesicles isolated from pituitary homogenates have been shown to contain enzymes that will catalyze the amidation of carboxyl termini of model peptides (Bradbury *et al.*, 1982; Bradbury and Smyth, 1983a; Eipper *et al.*, 1983a,b; Mains *et al.*, 1984). A series of reports from Smyth's laboratory in London gave the impetus for studies of this enzyme (Bradbury *et al.*, 1982; Bradbury and Smyth, 1983a). Their studies showed that secretory vesicle preparations from porcine pituitaries contained a specific enzyme, which amidated model peptides with a C-terminal glycine. It was shown that the activity required the presence of Cu^{2+}, and was inhibited by EDTA. Furthermore, the presence at the C-terminus of a glycine residue was mandatory for this enzymatic activity in order to amidate the penultimate amino acid residue. This requirement for a glycine residue at the C-terminus of the peptide intermediate to be amidated was consistent with the fact that the amino acid sequences of precursor molecules contained a glycine residue prior to the dibasic residues at the C-terminus of the peptide to be amidated (see Table I). Subsequent studies have elegantly shown that this enzyme required molecular oxygen, ascorbic acid, and Cu^{2+} (Eipper et al., 1983a,b; Bradbury and Smyth, 1983a,b; Glembotski et al., 1984). The reaction was recognized to be a monooxygenation and the enzyme was named peptidyl-glycine α-amidating monooxygenase (PAM) (Eipper et al., 1983a,b). In its characteristics this enzyme resembles the dopamine β-hydroxylating monooxygenase (dopamine β-hydroxylase, DBH), found in noradrenergic and adrenergic secretory vesicles; but unlike DBH, PAM was active in the pH range 6.0–8.0. However, the pH optimum of PAM activity in in vitro assays was somewhat dependent on the synthetic peptide substrate utilized (Bradbury and Smyth, 1983b). PAM has been purified to homogeneity and was shown to exist in two distinct forms (Murthy et al., 1986).

IV. THE SECRETORY VESICLE AS THE PROCESSING ENVIRONMENT

A. Evidence for Propeptide Processing in Secretory Vesicles

Various lines of experimental evidence indicate that the secretory vesicle is the site of posttranslational processing of propeptides. Early studies on the biosynthesis of insulin (Kemmler et al., 1973; Steiner et al., 1974) and vasopressin (Sachs et al., 1969) suggested that processing of these prohormones occurred within secretory vesicles. Compelling evidence that neuropeptides were processed in neurosecretory vesicles came from experiments in neurosecretory systems. In the hypothalamo-neurohypophysial system both provasopressin and prooxytocin were shown to be axonally transported intact before conversion to the peptide products (Gainer et al., 1977a,b; 1982a). Because the axonal transport of the newly synthesized proteins occurs within neurosecretory vesicles (Kent and Williams, 1974; Haddad et al., 1980), these data strongly suggested that neurosecretory vesicles contained the machinery for posttranslational processing of these propeptides. Similar results were obtained in experiments using the neurosecretory neuron R_{15} from A. californica (Gainer et al., 1977c, 1982b).

In studies using cDNA for human proinsulin, Moore *et al.* (1983a,b) showed that injection of this cDNA into fibroblasts or adrenal corticotropic hormone (ACTH)-secreting AtT-20 cells resulted in two opposite results; whereas the AtT-20 cells, which normally process and secrete ACTH, were able to process proinsulin into insulinlike peptides, the fibroblasts secreted unprocessed proinsulin. The fibroblasts, which lack the cellular apparatus to package and store secretory products for regulated secretion, secreted proinsulin via a nonregulated, constitutive pathway. These elegant experiments suggested that packaging mechanisms and processing enzymes in the tumor cell line could sort and process proinsulin as well as POMC, and that the presences of the cellular machinery for secretory vesicle formation was necessary for prohormone processing. In another set of experiments using preproglucagon gene, Drucker *et al.* (1986) found that when a metallothionein–glucagon fusion gene was introduced into a fibroblast cell line, the cells secreted unprocessed proglucagon. However, when the same construct was introduced into two endocrine cell lines (GH4, pituitary; RN38, islet cells), the cells were able to process the propeptide to varying degrees. Almost complete processing occurred in the islet cell line.

From a cell biology perspective, the secretory vesicle would be an advantageous site to localize the apparatus for processing of precursors to secretory peptides for the following reasons. (1) The secretory cell is committed to secreting large amounts of the peptides for any given stimulus and it would be more efficient. (2) Many precursors [e.g., proenkephalin, prodynorphin, and thyrotropin-releasing hormone (TRH)] are multivalent, in that they contain several copies of the similar biologically active peptides on the same propeptide, hence the sorting apparatus has only to package the precursor molecule before cleavage as opposed to having to sort many peptide products required by the secretory cell. (3) A similar case could be made for precursors that contain multiple peptide products for different physiological or cell biological reasons. The two examples are POMC, which contains ACTH, α-MSH, β-lipotropin hormone (β-LPH), and β-endorphin, and the precursors of the neurohypophysial peptides, oxytocin and vasopressin, which contain their associated neurophysins in their sequence. In the case of the former, the secretory cell is committed to secrete all these products concomitantly for physiological reasons; and in the latter case stoichiometric quantities of neurohypophysial peptides are stored within neurosecretory vesicles as a complex (Thorn *et al.*. 1982; Chaiken *et al.*, 1983; Castel *et al.*, 1984), perhaps to maintain the integrity of the vesicles (see later). (4) Another argument might be that because prohormone processing requires highly specific cleavage events at precise locations on the macromolecule (precursor), it has to be protected from other general proteolytic enzymes to which they may be susceptible (cathepsins, and trypsin-like activities), which are found in other cellular compartments. The processing event, unlike degradative processes, needs to be highly precise. (5) In many secretory cells, stimulation of secretory activity is associated with an increase in the rate of posttranslational processing of propeptides (Russell et al., 1981). It may be more efficacious for the secretory cell to compartmentalize the processing events for such physiological regulation. (6) The secretory vesicle membrane has been shown to be highly specialized, and low in permeability to most solutes, including H^+, unlike other intracellular membrane compartments such as the endoplasmic reticulum. This property allows for the vesicle to maintain a very stable intravesicular environment (e.g., pH 5.0–6.0) and provides the necessary cofactors for enzymatic activities.

B.   The Vesicular Processing Environment

From the foregoing discussion it is clear that the intravesicular space has to be organized in a manner most suited for the cell biological events taking place inside, in terms of specific contents (i.e., enzymes, ions, and redox state) (see Table II). Furthermore, the secretory vesicle must possess machinery to store all the components in very high concentrations in a stable state. Table II summarizes the requirements for optimal enzymatic activities involved in the posttranslational processing of propeptides. Two types of secretory vesicles that contain precursor-derived neuropeptides, the chromaffin vesicles from the adrenal medulla (which contain proenkephalin-derived peptides) and the neurosecretory vesicles from the nerve endings of the posterior pituitary (which contain vasopressin and oxytocin), have been extensively studied with respect to their properties. Many excellent reviews have been written on the subject of the intravesicular contents and membrane physicochemical properties of these secretory vesicles (Pollard et al., 1979a; Winkler and Westhead, 1980; Njus et al., 1981; Thorn et al., 1982; Winkler and Carmichael, 1982; Nordmann and Morris, 1982). In this article I will discuss only those areas that directly relate to processing and secretion of peptides from secretory cells and neuronal systems.

1.   THE SECRETORY VESICLE CORE

Both chromaffin and neurosecretory vesicles contain very high concentrations of secreted molecules. The chromaffin vesicles contain about 0.55 M catecholamines and about 0.0002 M Leu- and Met-enkephalins. The neurosecretory vesicles on the other hand contain about 0.1–0.2 M vasopressin or oxytocin and 0.001 M dynorphin peptides (vasopressin vesicles) and CCK peptides (oxytocin vesicles). Furthermore, both vesicles contain macromolecules that are secreted, chromogranins in the case of the chromaffin vesicles, and neurophysins (in equimolar amounts to the peptides)

TABLE II

REQUIREMENTS AND SOME PROPERTIES OF PROCESSING ENZYMES

Enzyme	Requirements	pH optimum	Classification	Tissue
Endopeptidases				
Dibasic residue-specific enzyme		4.0–5.0	Aspartyl protease	Neural and intermediate pituitary
Single basic residue-specific enzyme	Co^{2+}	5.0–6.5	Thiol protease	Brain
Exopeptidases				
Carboxypeptidase B	Co^{2+}, Zn^{2+}	5.0–6.0	Metalloprotease	Adrenal medulla pituitary
Aminopeptidase	Co^{2+}, Zn^{2+}	5.0–6.0	Metalloprotease	Neural and intermediate pituitary
Nonproteolytic enzymes				
Peptidyl-α-amidating monooxygenase	Cu^{2+}, ascorbate, O_2	6.0–8.0	Monooxygenase	Pituitary, adrenal medulla
N-Acetyltransferase	Acetyl-CoA	6.5–8.0		Pituitary

in the case of the neurosecretory vesicles. In addition to the secretory products, both types of vesicles have been shown to contain various other small molecules and ions (i.e., ATP, GTP, Ca^{2+}, Mg^{2+}, ascorbic acid). The composition of chromaffin vesicle contents has been extensively reviewed (Winkler and Westhead, 1980; Njus et al., 1981; Winkler and Carmichael, 1982). Table III lists the components shown to be present in neurosecretory vesicles isolated from bovine neurohypophyses. Both in the case of the chromaffin vesicles and neurosecretory vesicles very little is known about the organization of the vesicle core, which results in such high concentrations of solutes with relatively low overall osmotic pressure, or about the function of many of these molecules. In order to achieve this functional organization, the vesicle must satisfy a number of conflicting requirements. The charged molecules within the vesicles must be neutralized with opposing charges (e.g., cations for anions), and the molecules must interact with each other strongly enough to reduce the osmotic concentration. Yet the interaction cannot be so strong that it will interfere with the dispersal of the secretory products upon exocytosis. In the case of the chromaffin vesicles, the vesicle membrane must maintain a large concentration gradient for catecholamines with respect to the cytoplasm, yet the catecholamines must pass back and forth freely across the membrane for amine biosynthesis to occur. Finally, both types of secretory vesicles must protect their matrix from oxidation and degradation. We now understand, at least in general, how some of these different objectives are achieved.

The total concentration of the intravesicular solutes is approximately three times higher than the cytosolic osmolarity. Since the isolated secretory vesicles behave as perfect osmometers (Johnson and Scarpa, 1976; Morris and Schovanka, 1977; Nordmann and Morris, 1982; Sudhof, 1982), it is apparent that the intravesicular components interact so that the osmolarity of the matrix is no higher than that of the cytosol. Intravesicular interactions between different components have been identified in chromaffin vesicles (catecholamines and ATP) (Berneis et al., 1970, 1971; Colburn and Maas, 1965), insulin-containing vesicles (insulin–zinc complex) (Howell, 1974), and neurosecretory vesicles (vasopressin or oxytocin and neurophysins) (see Pickering and Jones, 1978). It is interesting that the interaction between neurohypophysial peptides and their neurophysins is strictly pH sensitive and is optimal at a pH range of 5.2–5.8 (Ginsburg and Ireland, 1964; Pickering and Jones, 1978; Breslow, 1975,1979). There are two ways to account for the low osmotic activity of the secretory vesicle core. The entire matrix may be nonideal but rather a homogeneous solution in which all components have a similar osmotic activity which is less than one. Alternatively, a fraction of the solutes may exist with an activity coefficient close to one and the remainder in an osmotically inactive complex. Studies using NMR, X-ray diffraction, and fluorescence techniques in chromaffin

TABLE III
COMPOSITION OF NEUROSECRETORY VESICLES[a]

Component	nmol/mg protein	Reference
Vasopressin	66	Gratzl et al. (1980); Russell (1981)
Oxytocin	42	Gratzl et al. (1980); Russell (1981)
Neurophysins	Equimolar amounts to vasopressin and oxytocin	Nordmann and Morris (1982); see Castel et al. (1984)
Glycopeptide	Equimolar amounts to vasopressin alone in vasopressin vesicles	See Castel et al. (1984)
Dynorphin	(vasopressin vesicles)	Watson et al. (1983); Whitnall et al. (1983)
Leu-enkephalin	(oxytocin vesicles)	Martin et al. (1983)
Cholecystokinin	(oxytocin vesicles)	Martin et al. (1983)
Ascorbic acid	70	Russell et al. (1985)
Calcium	35	Thorn et al. (1975)
Hexosamines	29 (μg/mg protein)	Hvas and Thorn (1973)
Soluble protein	0.82 (mg/mg protein)	Russell (1981); Gratzl et al. (1980)
Prohormone-converting enzyme		Chang et al. (1982)
Carboxypeptidase B		Hook and Loh (1984)
Aminopeptidase		Gainer et al. (1985a)
Peptidyl-α-amidating monooxygenase		Eipper et al. (1983b)

[a] The concentrations of the minor secretory products (dynorphin, corticotropin-releasing factor, enkephalins, and cholecystokinin) have been found to be approximately 1000 times less than the major products, vasopressin and oxytocin. See Winkler and Westhead (1980) and Winkler and Carmichael (1982) for the composition of chromaffin vesicles for comparison.

vesicles suggest no signs of crystalline structure or solute immobilization (Sharp and Richards, 1977a,b; Daniels *et al.*, 1978; Sharp and Sen, 1978; see also Njus *et al.*, 1981 for review). In addition, because the secretory vesicle behaves as a perfect osmometer, and because the osmotically active and inactive phases can coexist only at a critical concentration, on theoretical grounds the argument that the vesicle core would exist in two phases is untenable. Thus, the intravesicular matrix is probably isotropic and homogeneous, but is a nonideal solution in which the intravesicular components interact weakly to give an osmotic coefficient considerably less than in free solution.

2. THE SECRETORY VESICLE MEMBRANE

Attention has been focused on the secretory vesicle membrane properties with respect to their transport properties, ionic permeabilities, and protein and lipid compositions. The membrane of chromaffin vesicles is composed of 33% protein, 51% phospholipid, and 16% cholesterol (Winkler 1976). The neurosecretory vesicle membranes have a very similar composition (Thorn *et al.*, 1982; Nordmann and Morris, 1982).

a. Lipid Composition. The lipid composition of the chromaffin vesicles and the neurosecretory vesicle membranes has been analyzed (Blaschko *et al.*, 1967; Vilhardt and Hømer, 1972; de Oliveria Filgueiras *et al.*, 1979). The vesicle membranes contain relatively low concentrations of cholesterol compared with the plasma membranes. Otherwise, the phospholipid compositions are not remarkably different from that of the respective plasma membranes. One interesting aspect of chromaffin vesicles is the high concentration of lysophosphatidylcholine (16.7% of the total phospholipids) Blaschko *et al.*, 1967). However, it is possible that this high concentration of lysophosphatidylcholine was generated by the action of phospholipases during the isolation procedures. Neurosecretory vesicle membranes, on the other hand, contain very little lysophospholipids (Vilhardt and Hølmer, 1972).

b. Protein Composition. The protein compositions of the chromaffin and neurosecretory vesicles have been analyzed using sodium dodecyl sulfate–polyacrylamide gel electrophoresis (Abbs and Phillips, 1980; Zinder *et al.*, 1978; Cahill and Morris, 1979; Vilhardt *et al.*, 1975).In a comprehensive study of the chromaffin vesicle membranes, Abbs and Phillips (1980) resolved over 60 polypeptides, 13 of which are glycoproteins.

Analysis of neurosecretory vesicle membrane proteins on polyacrylamide gels was carried out by Vilhardt and co-workers (1975). They identified 19 major protein bands, 3 of which were stained by periodate and were identified as glycoproteins. Studies in our laboratory using two-dimensional gels revealed over 150 individual spots (Fig. 1.). However, very few of the

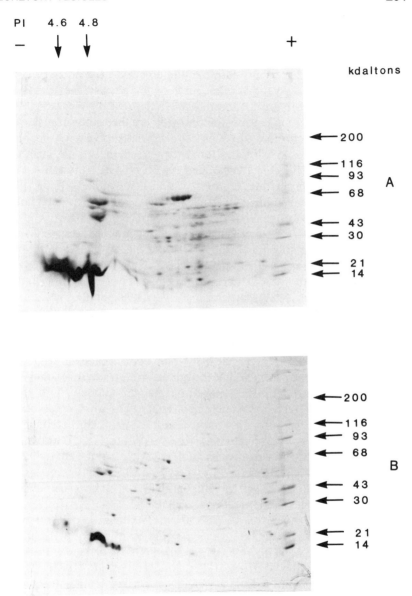

FIG. 1.   Proteins of neurosecretory vesicles. Two-dimensional polyacrylamide gel electrophoresis profiles of soluble proteins (A) and membrane proteins (B) from isolated neurosecretory vesicles. Neurosecretory vesicles were isolated according to Russell (1981) and the soluble components and membranes were separated by sonication in hypotonic solution followed by centrifugation. The proteins were solubilized in sodium dodecyl sulfate sample buffer and loaded on isoelectrofocusing gels and subsequently on polyacrylamide slab gels. p*I* values 4.6 and 4.8 indicated represent isoelectric points (p*I*) of neurophysins.

proteins are identified, except for the acidic intravesicular proteins, the neurophysins. Even under chaotropic extraction conditions (0.6 M KI), 7–10% of the neurophysins appears to be membrane bound and cannot be removed except by detergent treatment (unpublished experiments).

One interesting observation is that some of the intrinsic membrane proteins are found in both types of secretory vesicles (chromaffin vesicles and neurosecretory vesicles). The most noteworthy are the proton-translocating ATPase, and cytochrome b_{561}. In the chromaffin vesicles dopamine β-hydroxylase forms 25% of the total protein on the vesicle membrane (Winkler, 1976; Saxena et al., 1985), and cytochrome b_{561} forms 10% (Flatmark et al., 1971; Silsand and Flatmark, 1974; Duong and Fleming, 1985). A fraction of the propeptide-converting enzymes is also membrane bound. Approximately 40% of the dibasic residue-specific converting enzyme (Loh et al., 1985), and 30% of the peptidyl α-amidating monooxygenase (Murthy et al., 1986) appear to be associated with the membrane. In addition, a number of proteins have been found to be associated with the secretory vesicle membranes on the cytoplasmic face. These include calmodulin (Bader et al., 1984; Olsen et al., 1983), calelectrin (Walker, 1982; Sudhof et al., 1984), tubulin (Bernier-Valentin et al., 1983), actin (Burridge and Phillips, 1975; Meyer and Burger, 1979; Bader and Aunis, 1983), synexin (Creutz et al., 1978), etc. It is not clear, however, whether some of these are artifactually associated with the secretory vesicle membrane upon homogenization of the tissue. A few of the secretory vesicle membrane proteins are phosphorylated when incubated with [^{32}P]ATP, suggesting that the kinase activity is also associated with the secretory vesicle membrane. The most notable is the cyclic AMP-dependent protein kinase (Holz et al., 1980; Treiman and Gratzl, 1981; Amy and Kirshner, 1981; Treiman et al., 1980; Burgoyne and Geisow, 1982). In the case of the chromaffin vesicles, phosphatidylinositol kinase (Muller and Kirshner, 1975), and an NADH:(acceptor) oxidoreductase (Flatmark et al., 1971; Terland and Flatmark, 1972) have also shown to be present in purified vesicle fractions.

  c. *ATP-Dependent Proton Translocation.* The interior of most of the secretory vesicles has been found to be acidic by a variety of techniques (Johnson and Scarpa, 1976; Holz, 1978, 1979; Njus et al., 1978; Pollard et al., 1979a,b; Russell and Holz, 1981; Anderson et al., 1982). The intravesicular pH has been calculated using weak base accumulation techniques in most of these cases (Rottenberg et al., 1972; Schuldiner et al., 1972), and found to be between pH 5.5 and 5.8 when the vesicles are suspended in sucrose medium at pH 7.0. This large pH gradient over several hours in isolated vesicles suspended in isotonic media suggests that the vesicle membrane is very poorly permeable to protons. The one exception to this rule is the amylase-containing vesicles in parotid acinar cells, whose intravesicular pH was found to be close to 6.8 (Arvan et al., 1984). The pH gradient across the vesicle membrane is established by a proton pump present on the vesicle membranes. The proton pump has been recognized as an adenosine triphosphatase, which hydrolyzes ATP on the external face of the membrane and translocates protons into the vesicles (Njus et al., 1981; Beers et al., 1982; Rudnick, 1986). This H^+-pumping ATPase is electrogenic in nature, in that proton translocation is associated with a large increase in the internal positive membrane potential ($+70$ to $+100$ mV). Thus, when ATP is added to chromaffin vesicles or neurosecretory vesicles in the absence of permeant anion, a large change in $\Delta\Psi$ is measured without any changes in the ΔpH; however, if a permeant anion species (e.g., Cl^-) is present, a large change in ΔpH occurs without any change in the $\Delta\Psi$ (Holz, 1979; Johnson et al., 1979; Njus et al. 1978; Knoth et al., 1980). The total transmembrane electrochemical potential for H^+ ($\Delta\mu H^+$) varies little during this manipulation. These observations also suggest that the intravesicular space is buffered very well by the contents. Indeed, the buffering capacity of the intravesicular contents in chromaffin vesicles has been measured by indirect techniques to be approximately 300 μmol H^+/pH unit per gram dry weight (Johnson and Scarpa, 1976; Njus et al., 1978).

In all these respects the neurosecretory vesicles isolated from bovine posterior pituitaries are identical to chromaffin vesicles (Russell and Holz, 1981; Russell, 1984; Scherman et al., 1981; Scherman and Nordmann, 1982). The ΔpH measured using the accumulation of the weak base methylamine appears not to be due to a Donnan equilibrium across the vesicle membrane, as has been suggested in some studies (Scherman et al., 1981; Scherman and Nordmann, 1982), because the magnitude of the calculated ΔpH remains constant in low-ionic-strength medium (e.g., sucrose) and in isotonic potassium methylsulfate or potassium glutamate (Russell and Holz, 1980, and unpublished experiments in the author's laboratory). Thus all available evidence indicates that an electrogenic ATPase acidifies the interior of secretory vesicles in every tissue type studied (see also Njus et al., 1981; Rudnick, 1986, for reviews).

The electrogenic proton-pumping ATPase on secretory vesicles responsible for acidifying the vesicles appears to be a unique enzyme distinct from the $F_1 F_0$ ATPase found in mitochondria. Antibodies raised against the mitochondrial enzyme do not cross-react with the secretory-vesicle ATPase (Cidon and Nelson, 1983; Cidon et al., 1983). The inhibitor susceptibility of the secretory-vesicle enzyme is very different from that of the mitochondrial enzyme. Whereas oligomycin, azide, efrapeptin, and vanadate are potent inhibitors of the mitochondrial enzyme, they have very little effect on the secretory-vesicle proton pump. Furthermore, the secretory-vesicle H^+-ATPase is inactivated by sulfhydryl reagents, such as N-ethylmaleimide, NBD-CI (4-chloro-7-dinitrobenzo-2oxa-1,3-diazole), and 5,5'-dithiobis-(2-nitrobenzoic acid) (DTNB) in concentrations that have no effect on the F_1F_0 ATPase (Dean *et al.*, 1984; see also Rudnick, 1986, for review). The stoichiometry of H^+ transport by the secretory-vesicle ATPase is two H^+ for every ATP molecule hydrolyzed (Njus *et al.*, 1978; Beers *et al.*, 1982); it is three in the case of the mitochondrial ATPase. The secretory-vesicle ATPase has not been isolated in a functional form. Preliminary evidence suggests that it is a multisubunit enzyme (see Rudnick, 1986).

What physiological functions does the intravesicular acidity caused by H^+-pumping play in secretory vesicles? A variety of cell biological processes have been hypothesized and identified to depend upon the vesicular acid environment.

1. The most well-characterized function is intravesicular solute transport and storage in aminergic vesicles. In catecholaminergic chromaffin vesicles, amine accumulation proceeds using the ΔpH and the $\Delta \Psi$ generated by the H^+-pumping ATPase. These vesicles contain, in addition to the ATPase, a reserpine-sensitive amine translocator that exchanges H^+ for cytoplasmic amines utilizing both the ΔpH and the $\Delta \Psi$. Artificial imposition of ΔpH, $\Delta \Psi$, or both drives catecholamine accumulation (see Njus *et al.*, 1981; Rudnick, 1986, for reviews). Acetylcholine transport in cholinergic synaptic vesicles also depends upon the $\Delta \mu H^+$ across the membrane (Rudnick, 1986).

2. The second possibility is that the acid environment is necessary for the storage of high concentrations of solute within secretory vesicles. In the neurosecretory vesicles the peptides (oxytocin and vasopressin) are known to be stoichiometrically bound to the neurophysins. This neurophysin-nonapeptide binding is optimum at pH 5.5 (Ginsburg and Ireland, 1964; Breslow, 1975, 1979), and hence the pH of 5.7 found in neurosecretory vesicles could contribute to the stability of this complex. It has also been shown that the self-association between neurophysin molecules themselves is augmented by 100-fold when the neurophysin is bound to the peptides (Chaiken *et al.*, 1983), and because hormone binding is pH sensitive, the core complex forms that, in order to reduce the osmotic activity of the contents, may depend upon the acidic intravesicular environment. Reducing the osmotic activity of the vesicle core not only serves to maintain secretory vesicle stability, but also provides the vesicle with a very high osmotic potential if the interaction between the core components is ever disrupted. It is conceivable that this potential osmotic gradient could be utilized in the exocytotic process (see below). In many secretory systems exocytotic secretion is known to be sensitive to osmotic pressure increases in the external medium (Knight and Baker, 1982; Hampton and Holz, 1983; Pollard *et al.*, 1984).

3. The third possibility is that precursor processing depends on the pH of the vesicular environment. Indeed, in *in vitro* assays most of the enzymes

involved in propeptide processing are maximally active at acidic pH (pH 4.0–6.0)(Loh *et al.*,1984a, 1985; Gainer *et al.*, 1985a). It is conceivable that the processing steps might be prevented prior to packaging in the secretory vesicles simply because the pH in the endoplasmic reticulum and Golgi cisternae is not conducive for enzymatic activities. When packaging of the propeptides together with the enzymes in secretory vesicles is completed, the H^+ -pump could then acidify the vesicle interior and hence processing could commence.

4. Another possible function of the low pH in the secretory vesicles is to protect the stored products from oxidation. In chromaffin vesicles, catecholamines and ascorbic acid, and in neurosecretory vesicles, the disulfide-rich proteins, peptides, and ascorbic acid, may be protected by the acid environment. Other uses including intracellular signaling, energy transduction, and regulation of processing will undoubtedly be discovered for the proton pump and resulting ΔpH.

d. Electron Transfer across Secretory Vesicle Membranes. As pointed out earlier, chromaffin vesicles and neurosecretory vesicles utilize ascorbic acid as a cofactor for two distinct enzyme systems; dopamine β-hydroxylating monooxygenase in the former and peptidyl α-amidating monooxygenase in the latter. Both vesicles contain high concentrations of ascorbate (20–25 mM). Ascorbate is the preferred electron donor for DBH (Rosenberg and Lovenberg, 1980), and PAM activity appears to be identical to DBH in all respects, although the exact nature of the redox requirements of this enzyme is not fully worked out (Eipper *et al.*, 1983b; Murphy *et al.*, 1986). Because these enzymes utilize ascorbate in a one-electron reaction (i.e., two molecules of ascorbate-free radical are generated for each molecule of dopamine hydroxylated or peptide amidated) (Ljones and Skotland 1979; Skotland and Ljones 1980; Diliberto and Allen, 1980, 1981; Wakefield *et al.*, 1986a,b), and because the vesicles contain 0.6 M catecholamines or 0.1–0.2 M amidated peptides, stoichiometric considerations show that each molecule of ascorbic acid must be used 30–50 times in the generation of the vesicles' full complement of mature products. This is particularly relevant, because neither vesicle membrane is capable of transporting ascorbic acid. In addition, the chromaffin vesicle faces another major redox task, in that it must prevent oxidation and inactivation of the readily autooxidizable catecholamines.

The major redox-active protein that is common to both the secretory vesicles is cytochrome b_{561} (Silsand and Flatmark, 1974; Apps *et al.*, 1980; Duong and Fleming, 1982; Duong *et al.*, 1983; Wakefield *et al.*, 1984), which forms 10% of the chromaffin vesicle-membrane proteins and approximately 2%of the neurosecretory-vesicle protein (Duong *et al.*,1984). Cytochrome b_{561} was initially identified as the only heme-containing transmembrane protein in chromaffin vesicles, and was shown to have a high midpoint oxidation–reduction potential ($+140$mV) (Flatmark and Terland, 1971). It has been proposed that this cytochrome is involved in transport of electrons across the vesicle membrane to intragranular ascorbate-free radical (Wakefield *et al.*, 1982; Njus *et al.*, 1981, 1983; Srivastava *et al.*, 1984; Kelley and Njus, 1986). The experimental evidence for this conclusion is the following. (1) Cytochrome b_{561} is efficiently reduced by ascorbate, which is found in high concentration within the vesicles (Ingebretson *et al.*, 1980; Diliberto and Allen, 1980, 1981; Levine and Morita, 1985). (2) In chromaffin vesicle ghosts loaded with ascorbic acid, Njus *et al.* (1981, 1983; Hardenak *et al.*, 1985) demonstrated the passage of electrons from the intraghost ascorbate to an external electron acceptor, such as ferricyanide or cytochrome c. (3) Studies on spectral changes associated with electron transfer in cytochrome b_{561} showed that the cytochrome can reduce semidehydroascorbate directly (Kelley and Njus, 1986). (4) In a series of elegant experiments using electron paramagnetic resonance (EPR) to monitor and quantitate semidehydroascorbate, Wakefield and colleagues (1986a,b) showed that, in intact chromaffin vesicle, DBH activity was associated with the conversion of intravesicular ascorbate to semidehydroascorbate, which is reduced back to ascorbate via electron transfer from extravesicular ascorbic acid by cytochrome b_{561}. These experiments also showed that the semidehydroascorbate formed in the cytoplamic side of the vesicle could be reduced to ascorbate by the mitochondrial NADH:semidehydroascorbate oxidoreductase. DBH activity in isolated chromaffin vesicle ghosts as measured by the conversion of tyramine to octopamine was associated with intravesicular utilization of ascorbic acid, and extravesicular ascorbate was found to keep the intravesicular ascorbate in the reduced state (Beers *et al.*, 1986), presumably by providing electrons via cytochrome b_{561}. That cytochrome b_{561} is responsible for electron transfer was demonstrated in reconstitution experiments in which the isolated cytochrome was found to transfer electrons from intravesicular ascorbic acid to an external electron acceptor in phospholipid vesicles (Srivatsava *et al.*, 1984).

It has become evident that the neurosecretory vesicles also contain cytochrome b_{561} and high concentrations of ascorbic acid (Duong *et al.*, 1984; Russell *et al.*, 1985). On the other hand, no DBH activity was detected in neurosecretory vesicles by immunochemical techniques or by more sensitive enzymatic assays (Duong *et al.*, 1984). The cytochrome b_{561} on neurosecretory vesicles was shown to function as an electron carrier from intravesicular ascorbic acid to extravesicular ferricyanide or cytochrome c (Russell *et al.*, 1985). It is conceivable that the PAM activity in peptidergic vesicles, like DBH activity in chromaffin vesicles, is supported by an identical biochemical mechanism. Cytochrome b_{561} is found in all the tissues tested so far in which peptide amidation may be occurring (Duong *et al.*, 1984). Although the requirement of PAM activity for ascorbic acid has been amply documented (Bradbury *et al.*, 1982; Eipper *et al.*, 1983a,b; Murthy *et al.*, 1986), direct experimental evidence that PAM activity generates ascorbate free radical within the vesicle is still lacking. However, available experimental evidence suggests that identical biochemical components in two different secretory vesicles from different cell types may function to support different biochemical reactions. Cytochrome b_{561} has been identified in secretory vesicles from platelets (Wilkins and Salgainicoff, 1981; Johnson and Scarpa, 1981), although it appears to be immunologically distinct from the chromaffin-vesicle or the neurosecretory-vesicle cytochrome (Duong *et al.*, 1984).

Thus the evidence accumulated is consistent with the model that cytochrome b_{561}, by transferring electrons from ascorbate on the cytoplasmic side to ascorbate free radical on the inside of the vesicle, can couple the ascorbate-consuming reaction of DBH or PAM inside the vesicle to the ascorbate-regenerating reaction of the NADH:semidehydroascorbate oxidoreductase on the outer mitochondrial membrane (Fig 2; Diliberto and Allen, 1980, 1981; Njus *et al.*, 1981; Diliberto *et al.*, 1982; Wakefield *et al.*, 1982, 1986a,b). The proton electrochemical potential generated by the

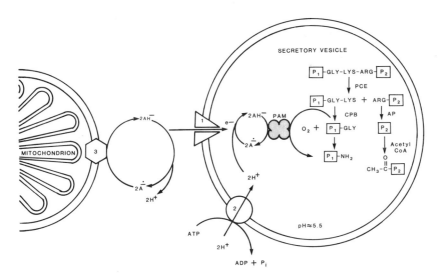

FIG. 2. Diagrammatic representation of some of the secretory-vesicle processes involved in posttranslational processing of propeptides. PCE, Dibasic residue-specific prohormone-converting enzyme; CPB, carboxpeptidase B; AP, aminopeptidase; PAM, peptidyl-glycine α-amidating monooxygenase; 1, cytochrome b_{561}; 2, H + -translocating Mg^{2+}ATPase; 3, mitochondrial NADH:semidehydroascorbate oxidoreductase; P1 and P2, model peptides; AH, ascorbate; A⁻, semidehydroascorbate. See text and Table II for other cofactor requirements for the different processing enzymes.

H^+-ATPase on the vesicle membrane drives the flow of electrons in the direction from the cytosol to the vesicle interior, and replenishes protons consumed by the turnover of the enzymes inside the vesicle. Furthermore, the low pH within the vesicle could conceivably favor regeneration of the ascorbate free radical, because the redox potential of ascorbate is pH dependent (Clark, 1960). One missing link in this model is that we do not understand the mechanism by which ascorbate is packaged in the secretory vesicles. It is possible that ascorbate is packaged into nascent vesicles at the Golgi apparatus, but the concentrations found inside the secretory vesicles are approximately four times higher than in the cytosol (Ingebretsen *et al.*, 1980; Levine and Morita, 1985).

e. Permeability Characteristics and Ionic Conductances on Secretory-Vesicle Membranes. Because most secretory-vesicle membranes have a proton-translocating ATPase, and some of them a transmembrane electron-transport system, ion permeabilities are a functionally important characteristic. The secretory-vesicle membranes are virtually impermeant to H^+, K^+, Na^+, and Mg^{2+} (Johnson and Scarpa, 1976a). Evidence has been obtained for a Ca^{2+}–Na^+ antiport system in both chromaffin and neurosecretory vesicles, but its functional significance has not been identified (Krieger-Brauer and Gratzl, 1983). Anion permeability of chromaffin vesicles determined by the measurement of vesicle swelling (optical density changes), or lysis in the presence of $Mg^{2+}ATP$, follows the order $SCN- > I- > CCl_3CO_2^- > BR^- > Cl^- > SO_4^{3-} > CH_3CO_2^-$, F^-, PC_4^{3-} (Casey *et al.*, 1976; Phillips, 1977). The Cl^- conductance calculated from these experiments is approximately 3×10^{-7} S per cm^2, which is similar to the Cl^- conductances found in artificial phospholipid bilayer membranes. Neurosecretory vesicles exhibit identical anion permeability characteristics (Russell, 1984). Both isolated chromaffin and neurosecretory vesicles are relatively stable in isosmotic KCl, or NaCl solutions in the absence of MgATP (Johnson and Scarpa, 1976a). However, when incubated in solutions containing potassium salts in the presence of the potassium ionophore valinomycin, both vesicles lyse and release their contents (Dolais-Kitabgi and Perlman, 1975; Russell, 1984). The presence of both cation and anion channels on secretory vesicles has been identified (see below).

Isolated secretory vesicles are relatively stable in isotonic (300 mOsm) sucrose solutions, suggesting that the membrane is impermeant to sucrose, but undergo lysis when the osmolarity is lowered to 200 mOsm (Holz, 1966; Normann and Morris, 1982). However, chromaffin vesicles lyse rapidly in isosmolar solutions of other nonelectrolytes, such as erythritol, arabitol, and glucose. Thus, nonelectrolytes smaller than sucrose seem to be more permeant across the vesicle membrane (Holz, 1986).

It has been proposed that the secretory-vesicle membrane may contain ionic channels, some of which may be gated by Ca^{2+} (Stanley and Ehrenstein, 1985). According to this model, the function of Ca^{2+} in initiation of secretion is to open Ca-activated cation channels present on the vesicle membrane. Opening of these channels results in an influx of cations, followed by anions via anion channels in order to preserve electroneutrality. Consequently, the vesicle-core osmotic activity is altered due to the increased ion concentrations or by interfering with the organization of the vesicle core. According to this model, it is this osmolarity increase that is the rate-limiting step for Ca-dependent exocytosis (see Stanley and Ehrenstein, 1985). This model is largely consistent with the experimental evidence available and can explain very rapid secretory events, and facilitation of neurotransmitter release in nerve endings. The presence of a Ca-activated cation channel with large conductances (>400 pS) and calcium activation at submicromolar concentration ranges has been identified in isolated neurosecretory vesicles from the posterior pituitary in lipid-bilayer-incorporation experiments (Stanley *et al.*, 1986a). Furthermore, an anion channel also has been identified in similar experiments in neurosecretory-vesicle membranes (Stanley *et al.*, 1986b). In a series of experiments using chromaffin vesicles fused with large lipsomes, patch-clamp data showed the presence of ionic channels with large conductances of ∿ 100–500 pS (Picaud *et al.*, 1984). Biochemical evidence for the presence of anion channels on both chromaffin vesicles and neurosecretory vesicles has existed for some time (Casey *et al.*, 1976, 1977; Pazzoles *et al.*, 1980; Pazzoles and Pollard, 1978).

As mentioned earlier the secretory mechanism may utilize the large potential osmotic energy in secretory vesicles in the exocytotic secretory process, and it is conceivable that the ionic channels may function to trigger such an event. Secretion from intact chromaffin cells initiated by acetylcholine (Hampton and Holz, 1983; Pollard *et al.*, 1981; Holz, 1986), and in permeabilized cells initiated by increasing pCa (Knight and Baker, 1982; Holz and Senter, 1986; Knight, 1986) is inhibited by increasing the osmolarity of the bathing medium, suggesting a possible role for osmotic changes during the secretory process. In these experiments, one caveat is that the increased osmolarity also causes shrinkage of the cells themselves, and not only the chromaffin vesicles. However, experiments using large-molecular-weight polysaccharides, which do not cause cell shrinkage, also prevented secretion (see Holz, 1986, for review). In some experiments secretion seems to proceed in media deficient in permeant ionic species (e.g., sucrose solutions; Knight and Baker, 1982; Knight, 1986). Furthermore, additional experiments are needed to unequivocally demonstrate that the vesicle membrane, in fact, is the source for the ionic channels identified in lipid bilayer experiments of Stanley and co-workers (1986a,b). It is also conceivable that the ionic channels on vesicle membranes provide a mechanism for regulating the intravesicular environment (e.g., pH, or ionic concentrations) and thereby regulate the activities of processing enzymes involved in the posttranslational processing of propeptides. Thus, the ionic-channel hypothesis of Stanley and Ehrenstein (1985) for secretion needs to be examined in much more detail.

V. CONCLUDING REMARKS

In this article I have given a general overall picture of the secretory vesicle as a crucial cellular compartment, which participates not only in the biosynthesis, processing, and storage of secretory products, but also in the process of secretion. The fact that the secretory vesicle is the site of processing of propeptides would imply that on exocytosis all of the contents of the vesicle would be secreted, including the enzymes, all the products, and cofactors. Indeed experimental evidence to date supports this contention. Processing enzymes, such as PAM (Eipper *et al.*, 1985), dibasic residue-specific enzymes (Y.P. Loh, personal communication), DBH (see Pollard *et al.*, 1984), and carboxypeptidase B (Fricker *et al.*, 1983a,b), have been shown to be secreted in response to stimulation. Ascorbic acid is secreted by chromaffin cells in culture in response to stimulation with acetylcholine (Levine *et al.*, 1983; Daniels *et al.*, 1982, 1983; Levine and Morita, 1985). Chromogranins have also been shown to be secreted by adrenal chromaffin cells together with catecholamines (O'Connor and Deftos, 1986). The delivery of multiple components to a target cell in high concentrations, particularly in a synapse, raises the possibility of modulation of the function of the major transmitter substance by another component present in the secreted cocktail. On the other hand, some of the secreted products may feed back on the secretory cell itself to modify its activity. For example, dynorphin cosecreted with vasopressin is thought to have modulatory effects on the posterior pituitary neurosecretory nerve endings, which are endowed with opiate receptors (Bicknell *et al.*, 1985). Furthermore, since the secretory-vesicle compartment is the processing site, different cell types could use the same precursor protein in order to arrive at different biologically active peptides and thus different utilities for the same gene product in different tissues and/or at different stages of embryonic development (POMC in hypothalamus, anterior pituitary, and intermediate lobe of the pituitary) (Gainer *et al.*, 1985a; Loh, 1986b).

Another question to be considered is are the processing enzymes involved in posttranslational processing of different propeptides in different cell types identical, or are they separate enzymes belonging to a large gene family? The evidence accumulated seems to suggest that the carboxypeptidase B-like enzyme and PAM are similar in different tissues. The evidence for universality of the endopeptidases (dibasic or single basic residue-specific enzymes) is still preliminary and conflicting.

The major questions outstanding in secretory-vesicle organization concern the physical nature of the intravesicular contents, and the vesicle-membrane properties *in situ*. Approaches to study the interactions of vesicular contents have to be developed in different model systems for an integrated picture to emerge. Furthermore, the vesicle membrane specializations in terms of specific ionic channels and membrane-associated proteins have yet to be described in detail. Understanding of these properties of the vesicle membrane will be crucial for understanding the molecular mechanisms of vesicle transport and exocytosis. The mechanisms of regulation of intravesicular processes is also not clearly understood. Thus, concerted effort is needed to fully understand how a secretory cell puts together this fascinating array of molecules in the secretory-vesicle compartment to serve in its multiplicity of functions.

It is clear from the foregoing that the secretory vesicle is a highly specialized organelle with multiple cell biological roles, and it is energetically, and in its biochemical composition, ideal to support the various processes taking place in them. It is also clear that our knowledge of secretory-vesicle physiology is still in embryonic stages. Much more experimental work is needed in areas of propeptide processing, ionic conductances, and in understanding the nature of the vesicle core in order to complete the picture. Techniques of classical molecular biology (development of secretory and processing mutants) and modern molecular biology (gene transfer) have yet to be significantly applied to studies in the above areas. It is likely that when such approaches are applied, our understanding of the organization and regulation of the peptide-secretory processes will be greatly enhanced.

ACKNOWLEDGMENTS

I would like to thank Mr. Robert Moy for expert help in preparing the art work, and valuable help in typing the manuscript. Thanks are due to Dr. James Garbern, and Mrs. Maxine Schafer for help in proofreading. I would also like to thank Drs. Harold Gainer and Y. Peng Loh for discussions.

REFERENCES

Abbs, M. T., and Phillips, J. H. (1980). Organization of proteins of the chromaffin granule membrane. *Biochim. Biophys. Acta* **595**, 200–221.

Amy, C. M., and Kirshner, N. (1981). Phosphorylation of adrenal medulla cells proteins in conjunction with stimulation of catecholamine secretion. *J. Neurochem.* **36**, 847–854.

Anderson, D. C., King, S. C., and Parsons, S. M. (1982). Proton gradient linkage to active uptake of [³H]acetylcholine by Torpedo electric organ synaptic vesicles. *Biochemistry* **21**, 3037–3043.

Apps, D. K., Pryde, J. G., and Phillips, J. H. (1980). Cytochrome *b*-561 is identical with chromomembrin B, a major polypeptide of chromaffin granule membranes. *Neuroscience* **5**, 2279–2287.

Arvan, P., Rudnick, G., and Castle, J. D. (1984). Osmotic properties and internal pH of isolated rat parotid secretory granules. *J. Biol. Chem.* **259**, 13567–13577.

Bader, M. F., and Aunis, D. (1983). The 97-kD α-actinin-like protein in chromaffin granule membranes from adrenal medulla: Evidence for localization on the cytoplasmic surface and for binding to actin filaments. *Neuroscience* **8**, 165–181.

Bader, M. F., Hikita, T., and Trifaro, J. M. (1984). Calcium-dependent calmodulin binding to chromaffin granule membranes: Presence of a 65-kilodalton calmodulin-binding protein. *J. Neurochem.* **44**, 526–539.

Barnea, A., and Cho, G. (1983). Acetylation of adrenocorticotropin and β-endorphin by hypothalamic and pituitary acetyl transferases. *Neuroendocrinology* **37**, 434–439.

Beers, M. F., Carty, S. E., Johnson, R. G., and Scarpa, A. (1982). H⁺-ATPase and catecholamine transport in chromaffin granules. *Ann. N.E. Acad. Sci.* **402**, 116–133.

Beers, M. F., Johnson, R. G., and Scarpa, A. (1986). Evidence for an ascorbate shuttle for the transfer of reducing equivalents across chromaffin granule membranes. *J. Biol. Chem.* **261**, 2529–2535.

Berneis, K.H., Pletscher, H., and DaPrada, M. (1970). Phase separation in solutions of noradrenaline and adenosine triphosphate: Influence of bivalent cations and drugs. *Br. J. Pharmacol.* **39**, 382–389.

Berneis, K. H., DaPrada, M., and Pletcher, A. (1971). A possible mechanism for uptake of biogenic amines by storage organelles: Incorporation into nucleotide–metal aggregates. *Experienta* **27**, 917–918.

Bernier-Valentin, F., Aunis, D., and Rousset, B. (1983). Evidence for tubulin-binding sites on cellular membranes: Plasma membranes, mitochondrial membranes, and secretory granule membranes. *J. Cell Biol.* **97**, 209–216.

Bicknell, R. J., Chapman, C., and Leng, G. (1985). Effects of opioid agonists and antagonists on oxytocin and vasopressin release in vitro. *Neuroendocrinology* **41**, 142–148.

Blaschko, H., Firemark, H., Smith, A. D., and Winkler, H. (1967). Lipids of the adrenal medulla: Lysolecithin, a characteristic constituent of chromaffin granules. *Biochem. J.* **104**, 545–549.

Bradbury, A. F., and Smyth, D. G. (1983a). Amidation of synthetic peptides by a pituitary enzyme: Specificity and mechanisms of the reaction. *In* "Peptides 1982" (K. Blaha and P. Mahlon, eds.), pp. 383–386. De Gruyter, Berlin.

Bradbury, A. F., and Smyth, D. G. (1983b). Substrate specificity of an amidating enzyme in porcine pituitary. *Biochem. Biophys. Res. Commun.* **112**, 372–377.

Bradbury, A. F., Finnie, M. D. A., and Smyth, D. F. (1982). Mechanism of C-terminal amide formation by pituitary enzymes. *Nature (London)* **298**, 686–688.

Breslow, E. (1975). On the mechanism of binding of neurohypophysical hormones and analogs of neurophysin. *Ann. N. Y. Acad. Sci.* **248**, 423–441.

Breslow, E. (1979). Chemistry and biology of the neurophysins. *Annu. Rev. Biochem.* **48**, 251–274.

Burgoyne, R. D., and Geisow, M. J. (1982). Phosphoproteins of the adrenal chromaffin granule membrane. *J. Neurochem.* **39**, 1387–1396.

Burridge, K., and Phillips, J. H. (1975). Association of actin and myosin in the secretory granule membranes. *Nature (London)* **254**, 526–529.

Cahill, A. L., and Morris, S. J. (1979). Soluble and membrane lectin-binding glycoproteins of the chromaffin granule. *J. Neurochem.* **32**, 855–867.

Casey, R. P., Njus, D., Radda, G. K., and Sehr, P. A. (1976). Adenosine triphosphate-evoked catecholamine release in chromaffin granules, osmotic lysis as a consequence of proton translocation. *Biochem. J.* **158**, 583–588.

Casey, R.P., Njus, D., Radda, G. K., and Sehr, P. A. (1977). Active proton uptake by chromaffin granules: Observation by amine distribution and phosphorus-31 nuclear magnetic resonance techniques. *Biochemistry* **16**, 972–977.

Castel, M., Gainer, H., and Dellmann, H. D. (1984). Neuronal secretory systems. *Int. Rev. Cytol.* **88**, 303–459.

Chaiken, I. M., Tamakoki, H., Brownstein, M. H., and Gainer, H. (1983). Onset of neurophysin self association upon neurophysin/neuropeptide hormone precursor synthesis. *FEBS Lett.* **164**, 361–365.

Chang, T.-L., and Loh, Y. P. (1984). *In vitro* processing of proopicortin by membrane-associated and soluble converting enzymes activities from rat intermediate lobe secretory granules. *Endocrinology* **114**, 2092–2099.

Chang, T.-L., Gainer, H., Russell, J. T., and Loh, Y. P. (1982). Pro-opiocortin converting enzyme activity in bovine neurosecretory granules. *Endocrinology* **11**, 1607–1614.

Chappell, M., Loh, Y. P., and O'Donohue, T. L. (1982). Evidence for an optiomelanotropin acetyltransferase in the rat pituitary neurointermediate lobe. *Peptides* **3**, 405–410.

Cidon, S., and Nelson, N. (1983). A novel ATPase in the chromaffin-granule membrane. *J. Biol. Chem.* **258**, 2892–2998.

Cidon, S., Ben-David, H., and Nelson, N. (1983). ATP-driven proton fluxes across membranes of secretory organelles. *J. Biol. Chem.* **258**, 11684–11688.

Clark, W. M. (1960). "Oxidation–Reduction Potentials of Organic Systems," pp. 467–471. Williams & Wilkins, Baltimore.

Colburn, R. W., and Maas, J. W. (1965). Adenosinetriphosphate–metal–norepinephrine ternary complexes and catecholamine binding. *Nature (London)* **208**, 37–41.

Creutz, C. E., Pazoles, C. J., and Pollard, H. B. (1978). Identification and purification of an adrenal medullary protein (synexin) that causes calcium-dependent aggregation of isolated chromaffin granules. *J. Biol. Chem.* **253**, 2858–2866.

Cromlish, J. A., Seidah, N. G., and Chretien, M. N. (1985). Isolation and characterization of four proteases from porcine pituitary neuro-intermediate lobes: Relationship to the maturation enzyme of prohormones. *Neuropeptides* **5**, 493–496.

Daniels, A. J., Williams, R. J. P., and Wright, P. E.(1978). The character of stored molecules in chromaffin granules of the adrenal medulla: A nuclear magnetic resonance study. *Neuroscience* **3**, 573–585.

Daniels, A. J., Dean, G., Viveros, H. O., and Diliberto, E. J. (1982). Secretion of newly taken-up ascorbic acid by adrenomedullary chromaffin cells. *Science* **216**, 737–739.

Daniels, A. J., Dean, G., Viveros, O. H., and Diliberto, E. J. (1983). Secretion of newly taken up ascorbic acid by adrenomedullary chromaffin cells originates from a compartment different from the catecholamine storage vesicle. *Mol. Pharmacol.* **23**, 437–444.

Dean, G. E., Fishkes, H., Nelson, P. J., and Rudnick, G. (1984). The hydrogen ion-pumping adenosine triphosphate of platelet dense granule membrane. Differences from F_1F_0- and phosphoenzyme-type ATPases. *J. Biol. Chem.* **259**, 9569–9574.

deOliveria Filgueiras, O. M., Van den Beaselaar, A. N. P. H., and Van den Bosch, H. (1979). Localization of lysophosphatidylcholine in bovine chromaffin granules. *Biochim. Biophys. Acta* **558**, 73–84.

Devi, L., and Goldstein, A. (1984). Dynorphin converting enzyme with unusual specificity from rat brain. *Proc. Natl. Acad. Sci. U.S.A.* **81**, 1892–1896.

Devi, L., and Goldstein, A. (1985). Neuropeptide processing by single-step cleavage conversion of leumorphin (dynorphin B-29) to dynorphin-B. *Biochem. Biophys. Res. Commun.* **130**, 1168–1176.

Diliberto, E. J., Jr., and Allen, P. L. (1980). Semidehydroascorbate as a product of the enzymic conversion of dopamine to norepinephrine. *Mol. Pharmacol.* **17**, 421–426.

Dilberto, E. J., Jr., and Allen, P. L. (1981). Mechanism of dopamine β-hydroxylation. *J. Biol. Chem.* **256**, 3385–3393.

Diliberto, E. J., Jr., Dean, G., Carter, C., and Allen, P. L. (1982). Tissue, subcellular and submitochondrial distributions of semidehydroascorbate reductase. *J. Neurochem.* **39**, 563–568.

Docherty, K., and Hutton, J. C. (1983). Carboxypeptidase activity in the insulin secretory granule. *FEBS Lett.* **162**, 137–141.

Docherty, K., and Steiner, D. (1982). Post-translational proteolysis in polypeptide hormone biosynthesis. *Rev. Physiol.* **44**, 625–638.

Dolais-Kitabgi, J., and Perlman, R. L. (1975). The stimulation of catecholamine release from chromaffin granules by valinomycin. *Mol. Pharmacol.* **11**, 745–750.

Drucker, D. J., Mojsov, S., and Habener, J. F. (1986). Cell-specific post-translational processing of preproglucagon expressed from a metallothionein–glucagon fusion gene. *J. Biol. Chem.* **261**, 9637–9643.

Duong, L.-T., and Fleming, P. J. (1982). Isolation and properties of cytochrome b_{561} from bovine adrenal chromaffin granules. *J. Biol. Chem.* **257**, 8561–8564.

Duong, L.-T., and Fleming, P. J. (1983). The asymmetric orientation of cytochrome b_{561} in bovine chromaffin granule membranes. *Arch. Biochem. Biophys.* **228**, 332–341.

Duong, L.-T., Fleming, P. J., and Russell, J. T. (1984). An identical cytochrome b_{561} is present in bovine adrenal chromaffin vesicles and posterior pituitary neurosecretory vesicles. *J. Biol. Chem.* **259**, 4885–4889.

Eberle, A., and Schwyzer, R. (1975). Hormone receptor interactions. Demonstration of two message sequences (active sites) in melanotropin. *Helv. Chim. Acta* **58**, 1528–2535.

Eipper, B. A., Glembotski, C. C., and Mains, R. E. (1983a). Bovine intermediate pituitary α-amidation enzyme: Preliminary characterization. *Peptides* **4**, 921–928.

Eipper, B. A., Mains, R. E., and Glembotski, C. C. (1983b). Identification in pituitary tissues of a peptide α-amidation activity that acts on glycine-extended peptides requires molecular oxygen, copper, and ascorbic acid. *Proc. Natl. Acad. Sci. U.S.A.* **80**, 5144–5148.

Eipper, B. A., Myers, A. C., and Mains, R. E. (1983c). Selective loss of α-melanotropin-amidating enzyme activity in primary cultures of rat intermediate pituitary cells. *J. Biol. Chem.* **258**, 7292–7298.

Eipper, B. A., Myers, A. C., and Mains, R. E. (1985). Peptidyl-glycine-α-amidation activity in tissues and serum of the adult rat. *Endocrinology* **116**, 2496–2504.

Farquhar, M. G., and Palade, G. E. (1981). The Golgi apparatus (complex)—(1954–1981)—From artifact to center stage *J. Cell Biol.* **91**, 77s–103s.

Flatmark, T., and Terland, O. (1971). Cytochrome b_{561} of the bovine adrenal chromaffin granules; a high potential b-type cytochrome. *Biochim. Biophys. Acta* **253**, 487–491.

Flatmark, T., Terland, O., and Helle, K. B. (1971). Electron carriers of the bovine adrenal chromaffin granules. *Biochim. Biophys. Acta* **226**, 9–19.

Fletcher, D. J., Noe, B. D., Bauer, G. E., and Quigley, J. P. (1980). Characterization of the conversion of a somatostatin precursor to somatostatin by islet secretory granules. *Diabetes* **29**, 593–599.

Fletcher, D. J., Quigley, J. P., Bauer, G. E., and Noe, B. D. (1981). Characterization of pro-insulin and proglucagon-converting activities in isolated islet secretory granules. *J. Cell Biol.* **90**, 312–322.

Freedman, R. B., and Hawkins, H. C. (1980). "The Enzymology of Post-translational Modification of Proteins," Vol. 1, p. 456. Academic Press, London.

Fricker, L. D., and Snyder, S. H. (1982). Enkephalin convertase: Purification and characterization of a specific enkephali-synthesizing carboxypeptidase localized to adrenal chromaffin granules. *Proc. Natl. Acad. Sci. U.S.A.* **79**, 3886–3890.

Fricker, L. D., and Snyder, S. H. (1983). Purification and characterization of enkephalin convertase, an enkephalin-synthesizing carboxypeptidase. *J. Biol. Chem.* **258**, 10950–10955.

Fricker, L. D., Plummer, P. H., and Snyder, S. H. (1983a). Enkephalin convertase: Potent, selective, and irreversible inhibitors. *Biochem. Biophys. Res. Commun.* **111**, 994–1000.

Fricker, L. D., Suppattapone, S., and Snyder, S. H. (1983b). Enkephalin convertase: A specific enkephalin synthesizing carboxypeptidase in adrenal chromaffin granules, brain, and pituitary gland. *Life Sci.* **31**, 1841–1844.

Gainer, H., Sarne, Y., and Brownstein, M. J. (1977a). Neurophysin biosynthesis: Conversion of a putative precursor during axonal transport. *Science* **195** 1354–1356.

Gainer, H., Sarne, Y., and Brownstein, J. M. (1977b). Biosynthesis and axonal transport of rat neurohypophyseal proteins and peptides. *J. Cell Biol.* **73**, 366–381.

Gainer, H., Loh, Y. P., and Sarne, Y. (1977c). Biosynthesis of neuronal peptides. In "Peptides in Neurobiology" (H. Gainer, ed.), pp. 183–219. Plenum, New York.

Gainer, H., Russell, J. T., and Brownstein, M. J. (1982a). Axonal transport, neurosecretory vesicles, and the endocrine neuron. In "Axoplasmic Transport in Physiology and Pathology" (D. G. Weiss and A. Gorio, eds.), pp. 44–50. Springer-Verlag, Berlin.

Gainer, H., Loh, Y. P., and Neale, E. A. (1982b). The organization of post-translational precursor processing inpeptidergic neurosecretory cells. In "Proteins in the Nervous System: Structure and Function", pp. 131–145. Liss, A. R., New York.

Gainer, H., Russell, J. T., and Loh, Y. P. (1985a). The enzymology and intracellular organization of peptide precursor processing: The secretory vesicle hypothesis. *Neuroendocrinology* **40**, 171–184.

Gainer, H., Russell, J. T., and Loh, Y. P. (1985b). An aminopeptidase activity in bovine pituitary secretory vesicles that cleaves the N-terminal arginine from β-lipotropin$_{60-65}$. *FEBS Lett.* **175**, 135–139.

Ginsburg, M., and Ireland, M. (1964). Binding of vasopressin and oxytocin to protein in extracts of bovine and rabbit neurohypophyses. *J. Endocrinol.* **30**, 131–145.

Glembotski, C. C. (1982a). Acetylation of α-melanotropin and β-endorphin in the rat intermediate pituitary. *J. Biol. Chem.* **257**, 10493–10500.

Glembotski, C. C. (1982b). Characterization of the peptide acetyltransferase activity in bovine and rat intermediate pituitaries responsible for the acetylation of β-endorphin and α-melanotropin. *J. Biol. Chem.* **257**, 10501–10509.

Glembotski, C. C., Eipper, B. A., and Mains, R. E. (1984). Characterization of a peptide α-amidation activity from rat anterior pituitary. *J. Biol. Chem.* **259**, 6385–6392.

Gomez, S., Gluschankof, P., Morel, A., and Cohen, P. (1985). The somatostatin-28 convertase of rat brain cortex is associated with secretory granule membranes. *J. Biol. Chem.* **260**, 10541–10545.

Gratzl, M., Torp-Petersen, C., Dartt, D. A., Treiman, M., and Thorn, N. A. (1980). Isolation and characterization of secretory granules from bovine neurohypophyses. *Hoppe-Seyler's Z. Physiol.* **361**, 1615–1628.

Gubler, U., Seeburg, P., Hoffman, B. J., Gage, L. P., and Udenfriend, S. (1982). Molecular cloning establishes pro-enkephalin as precursor of enkephalin-containing peptides. *Nature (London)* **295**, 206–208.

Habener, J. F. (1981). Regulation of parathyroid hormone secretion and biosynthesis. *Rev. Physiol.* **43**, 211–223.

Habener, J. F., Chang, H. T., and Potts, J. T., Jr. (1977). Enzymic processing of parathyroid hormone by cell free extracts of parathyroid glands. *Biochemistry* **16**, 6385–6392.

Haddad, J. F. (1981). Regulation of parathyroid hormone secretion and biosynthesis. *Rev. Physiol.* **43**, 211–223.

Haddad, A., Guaraldo, S. P. M., Pelletier, G., Brasiliero, I. L. G., and Marchi, F. (1980). Glycoprotein secretion in the hypothalamoneurophypophysical system of the rat. *Cell Tissue Res.* **209**, 399–422,

Hampton, R. Y., and Holz, R. W. (1983). The effects of osmolality on the stability and function of cultured chromaffin cells and the role of osmotic forces in exocytosis. *J. Cell Biol.* **96**, 1082–1088.

Hardenak, G. J., Callahan, R. E., Barone, A. R., and Njus, D. (1985). An electron transfer dependent membrane potential in chromaffin-vesicle ghosts. *Biochemistry* **24**, 384–389.

Holz, R. W. (1978). Evidence that catecholamine transport into chromaffin vesicles is coupled to vesicle membrane potential. *Proc. Natl. Acad. Sci. U.S.A.* **75**, 5190–5194.

Holz, R. W. (1979). Measurement of membrane potential of chromaffin granules by the accumulation of triphenylmethylphosphonium cation. *J. Biol. Chem.* **254**, 6703–6709.

Holz, R. W. (1986). The role of osmotic forces in exocytosis from adrenal chromaffin cells. *Annu. Rev. Physiol.* **48**, 175–189.

Holz, R. W., and Senter, R. A. (1986). The effects of osmolality and ionic strength on secretion from adrenal chromaffin cells permeabilized with digitonin. *J. Neurochem.* **46**, 1835–1842.

Holz, R. W., Rothwell, G. E., and Ueda, T. (1980). Cholinergic agonist-stimulated phosphorylation of two specific proteins in bovine chromaffin cells: Correlation with catecholamine secretion. *Neurosci. Abstr.* **6**, 177.

Hook, V. Y. H., and Loh, Y. P. (1984). Carboxypeptidase B-like converting enzyme activity in secretory granules of rat pituitary. *Proc. Natl. Acad. Sci. U.S.A.* **81**, 2777–2780.

Hook, V. Y. H., Eiden, L. E., and Brownstein, M. J. (1982). A carboxypeptidase processing enzyme for enkephalin precursor. *Nature (London)* **295**, 341–342.

Howell, S. L. (1974). The molecular organization of the β granule of the islets of Langerhans. *In* "Advances in Cytopharmacology" (B. Ceccarelli, F. Clementi, and J. Meldolesi, eds.), Vol. 2, pp. 319–327. Raven, New York.

Hvas, S., and Thorn, N. A. (1973). Hexosamine and heparin in homogenate and subcellular fractions from bovine neurohypophyses. *Acta Endocrinol.* **74**, 209–214.

Ingebretsen, O. C., Terland, O., and Flatmark, T. (1980). Subcellular distributrion of ascorbate in bovine adrenal medulla. Evidence for accumulation in chromaffin granules against a concentration gradient. *Biochim. Biophys. Acta* **628**, 182–189.

Johnson, R. G., and Scarpa, A. (1976a). Ion permeability of isolated chromaffin vesicles. *J. Gen. Physiol.* **68**, 601–631.

Johnson, R. G., and Scarpa, A. (1976b). Internal pH of isolated chromaffin vesicles. *J. Biol. Chem.* **251**, 2189–2191.

Johnson, R. G., and Scarpa, A. (1981). The electron transport chain of serotonin-dense granules of platelets. *J. Biol. Chem.* **256**, 11966–11969.

Johnson, R. G., Pfister, D., Carty, S. E., and Scarpa, A. (1979). Biological amine transport in chromaffin ghosts. Coupling to the transmembrane proton and potential gradients. *J. Biol. Chem.* **254**, 10963–10972.

Julius, D., Blair, L., Brake, A., Sprague, G., and Thorner, J. (1983). Yeast α-factor is processed from a larger precursor polypeptide: The essential role of a membrane-bound dipeptidyl aminopeptidase. *Cell* **32**, 839–852.

Kakidani, H., Furutani, Y., Takahashi, H., Noda, M., Morimoto, Y., Hirose, T., Asai, M., Inayama, S., Nakanashi, S., and Numa, S. (1982). Cloning and sequence analysis of cDNA for porcine β-endorphin/dynorphin precursor. *Nature (London)* **298**, 245–249.

Kelley, P. M., and Njus, D. (1986). Cytochrome b_{561} spectral changes associated with electron transfer in chromaffin-vesicle ghosts. *J. Biol. Chem.* **261**, 6429–6432.

Kemmler, W., Steiner, D. F., and Borg, J. (1973). Studies on the conversion of proinsulin to insulin. *J. Biol. Chem.* **248**, 4544–4551.

Kent, C., and Williams, M. A. (1974). The nature of the hypothalamo-neurohypophysial neurosecretion in rat. A study by light-, and electron microscope autoradiography. *J. Cell Biol.* **60**, 554–570.

Knight, D. E. (1986). Calcium and exocytosis. (1986 Calcium and the Cell). *Ciba Found. Symp. Ser.* **122**, 250–269.

Knight, D. E., and Baker, P. F. (1982). Calcium-dependence of catecholamine release from bovine adrenal medullary cells after exposure to intense electric fields. *Membr. Biol.* **68**, 107–140.

Knoth, J., Handloser, K., and Njus, D. (1980). Electrogenic epinephrine transport in chromaffin granule ghosts. *Biochemistry* **19**, 2938–2942.

Koch, G., and Richter, D. (eds.) (1980). "Biosynthesis, Modification, and Processing of Cellular and Viral Polyproteins". Academic Press, New York.

Krieger-brauer, H., and Gratzl, M. (1983). Effects of monovalent and divalent cations on Ca^{++} fluxes across chromaffin secretory membrane vesicles. *J. Neurochem.* **41**, 1269–1276.

Kriel, G., Suchanek, G., and Kindas-Mugge, I. (1977). Biosynthesis of a secretory peptide in honeybee venom gland: Intermediates detected in vivo and in vitro. *Fed. Proc.* **36**, 2081–2086.

Land, H., Grez, M., Ruppert, S., Schmale, H., Rehbein, M., Richter, D., and Schutz, G. (1983). Deduced amino acid sequence from the bovine oxytocin–neurophysin I precursor cDNA. *Nature (London)* **302**, 342–344.

Land, H., Schutz, G., Schmale, H., and Richter, D. (1982). Nucleotide sequence of cloned cDNA encoding bovine arginine vasopressin–neurophysin II precursor. *Nature (London)* **295**, 299–303.

Levine, M., and Morita, K. (1985). Ascorbic acid in endocrine systems. *Vitam. Horm.* **42**, 1–64.

Levine, M., Asher, A., Pollard, H. B., and Zinder, O. (1983). Ascorbic acid and catecholamine secretion in cultured chromaffin cells. *J. Biol. Chem.* **258**, 13111–13115.

Lindberg, I., Yang, H.-Y. T., and Costa, E. (1984). An enkephalin generating enzyme in bovine adrenal medulla. *Biochem. Biophys. Res. Commun.* **106**, 186–193.

Ljones, T., and Skotland, T. (1979). Evidence from the acceleration of cytochrome c reduction for the formation of ascorbate free radical by dopamine β-monooxygenase. *FEBS Lett.* **108**, 25–27,

Loh, Y. P. (1986a). Kinetic studies on the processing of β-lipotropin by bovine pituitary intermediate lobe pro-opiomelanocortin converting enzyme. *J. Biol. Chem.* **261**, 11949–11955.

Loh, Y. P. (1986b). Peptide precursor processing enzymes within secretory vesicles. *Ann. N.Y. Acad. Sci.*, **493**, 292–307.

Loh, Y. P., and Chang, T.-L. (1982). Proopiocortin converting activity in rat intermediate and neural lobe secretory granules *FEBS Lett.* **137**, 57–62.

Loh, Y. P., and Gainer, H. (1982). Processing of normal and non-glycosylated forms of toad pro-opiocortin by rat intermediate (pituitary) lobe pro-opiocortin converting enzyme activity. *Life Sci.* **31**, 3043–3050.

Loh, Y. P., and Gainer, H. (1983). Biosynthesis and processing of neuropeptides. *In* "Brain Peptides" (D. Kreiger, M. J. Brownstein, and J. Martin, eds.), pp. 79–116. Wiley, New York.

Loh, Y. P., and Parish, D. C. (1986). Processing of neuropeptide precursors. *In* "Neuropeptides and their Peptidases" (A.J. Turner, ed), pp. 65–84, Ellis-Horwood.

Loh, Y. P., Tam, W. H. H., and Russell, J. T. (1984). Measurement of ΔpH and membrane potential in secretory vesicles isolated from bovine pituitary intermediate lobe. *J. Biol. Chem.* **259**, 8238–8245.

Loh, Y. P., Parish, D. C., and Tuteja, R. (1985). Purification and characterization of a paired basic residue-specific pro-opiomelanocortin converting enzyme from bovine pituitary intermediate lobe secretory vesicles. *J. Biol. Chem.* **260**, 7194–7205.

Lowry, P. T., Silman, R. E., Hope, J., and Scott, A. P. (1977). Structure and biosynthesis of peptides related to corticotropins and β-endorphins. *Ann. N.Y. Acad. Sci.* **297**, 49–53.

Mains, R. E., Glembotski, C. C., and Eipper, B. A. (1984). Peptide-α-amidation activity in mouse anterior pituitary AtT-20 cell granules: Properties and secretion. *Endocrinology* **114**, 1522–1530.

Manning, M., Olma, A., Klis, W., Kolodziejczyk, A., Nawrocka, E., Misicka, A., Seto, J., and Sawyer, W. (1984). Carboxy terminus of vasopressin required for activity but not binding. *Nature (London)* **308**, 652–653.

Martin, R., Geis, R., Holl, R., Schafer, M., and Voigt, K. H. (1983). Coexistence of unrelated peptides in oxytocin and vasopressin terminals of rat neurohypophyses: Immunoreactive methioning^5-enkephalin, leucine^5-enkephalin, and cholecystokinin-like substances *Neuroscience* **8**, 213–227.

Meyer, D., and Burger, M. M. (1979). Chromaffin granule surface: The presence of actin and the nature of its interaction with the membrane. *FEBS Lett.* **101**, 129–133.

Mizuno, K., and Matsuo, H. (1985). Proenkephalin processing enzyme with specificity toward paired basic residues purified from bovine adrenal chromaffin granules. *Neuropeptides* **5**, 489–492.

Mizuno, K., Kojima, M., and Matsuo, H. (1985). A putative prohormone processing protease in bovine adrenal medulla specifically cleaving in between Lys–Arg sequences. *Biochem. Biophys. Res. Commun.* **128**, 884–891.

Moore, H. P., Gumbiner, B., and Kelley, R. B. (1983a) Chloroquine diverts ACTH from a regulated to a constitutive pathway in AtT-20 cells. *Nature (London)* **302**, 434–436.

Moore, H. P., Walker, M., Lee, F., and Kelley, R. B. (1983b). Expressing a human proinsulin cDNA in a mouse ACTH secreting cell. Intracellular storage, processing, and secretion on stimulation. *Cell* **35**, 531–538.

Morris, S. J., and Schovanka, T. (1977). Some physical properties of adrenal medulla chromaffin granules isolated by a new continuous iso-osmotic density gradient method. *Biochim. Biophys. Acta* **464**, 53–64.

Muller, T. W., and Kirshner, N. (1975). ATPase and the phosphatidylinositolkinase activities of adrenal chromaffin vesicles. *J. Neurochem.* **24**, 1155–1161.

Murthy, A. S., Mains, R. E., and Eipper, B. A.(1986). Purification and characterization of peptidylglycine α-amidating monooxygenase from bovine neurointermediate pituitary. *J. Biol. Chem.* **261**, 1815–1822.

Nakanishi, S., Ionue, A., Kita, T., Nakamura, M., Chang, A. C. Y., Cohen, S. N., and Numa, S. (1979). Nucleotide sequence of cloned cDNA for bovine corticotropin-β-lipotropin precursor. *Nature (London)* **278**, 423–427.

Njus, D., Sehr, P. A., Radda, G. K., Ritchie, G. A., and Seeley, P. J. (1978). Phosphorus-31 nuclear magnetic resonance studies of active proton translocation in chromaffin granules. *Biochemistry* **17**, 4337–4343.

Njus, D., Knoth, J., and Zallakian, M. (1981). Proton-linked transport in chromaffin granules. *Curr. Top. Bioenerg.* **11**, 107–147.

Njus, D., Knoth, J., Cook, C., and Kelley, P. M. (1983). Electron transfer across the chromaffin granule membrane. *J. Biol. Chem.* **258**, 27–30.

Noda, M., Teranishi, Y., Takahashi, H., Toyosato, M., Notake, M., Nakanishin, S., and Numa, S. (1982). Isolation and structural organization of the human preproenkephalin gene. *Nature (London)* **297**, 432–434.

Noe, B. D. (1981). Inhibition of islet prohormone to hormone conversion by incorporation of arginine and lysine analogs. *J. Biol. Chem.* **256**, 4940–4946.

Nordmann, J. J., and Morris, J. F. (1982). Neurosecretory granules. *In* "Neurotransmitter Vesicles" (R. L. Klein, H. Langercrantz, and H. Zimmerman, eds.), pp. 41–64. Academic Press, London.

O'Connor, D. T., and Deftos, L. J. (1986). Secretion of chromogranin A by peptide-producing endocrine neoplasms. *N. Engl. J. Med.* **314**, 1145–1151.

O'Donohue, T. L. (1983). Identification of endorphin acetyltransferase in rat brain and pituitary gland. *J. Biol. Chem.* **258**, 2163–2167.

Olsen, S. F., Slaninova, J., Treiman, M., Saermark, T., and Thorn, N. A. (1983). Calmodulin binding to secretory granules isolated from bovine neurohypophyses. *Acta Physiol. Scand.* **118**, 355–358.

Palade, G. (1975). Intracellular aspects of the process of protein synthesis. *Science* **189**, 347–358.

Pazzoles, C. J., and Pollard, H. B. (1978). Evidence for stimulation of anion transport in ATP-evoked transmitted release from isolated secretory vesicles. *J. Biol. Chem.* **253**, 3962–3969.

Pazzoles, C. J., Creutz, C E., Ramu, A., and Pollard, H. B. (1980). Permeant anion activation of MgATPase activity in chromaffin granules. *J. Biol. Chem.* **255**, 7863–7869.

Phillips, J. H. (1977). Passive ion permeability of the chromaffin granule membrane. *Biochem. J.* **186**, 289–297.

Picaud, S., Marty, A., Trautmann, O. G.-W., and Henry, J.-P. (1984). Incorporation of chromaffin granule membranes into large-size vesicles suitable for patch-clamp recording. *FEBS Lett.* **178**, 20–24.

Pickering, B. T., and Jones, C. W. (1978). The neurophysins. *In* "Hormonal Proteins and Peptides," Vol. 5, pp. 103–158. Academic Press, New York.

Pollard, H. B., Pazoles, C. J., Creutz, C. E., and Zinder, O. (1979a). The chromaffin granule and possible mechanisms of exocytosis. *Int. Rev. Cytol.* **58**, 159–197.

Pollard, H. B., Shindo, H., Creutz, C. E., Pazoles, C. J., and Cohen, J. S. (1979b). Internal pH and state of ATP in adrenergic chromaffin granules determined by ^{31}P nuclear magnetic resonance spectroscopy. *J. Biol. Chem.* **254**, 1170–1177.

Pollard, H. B., Pazoles, C. J., and Creutz, C. E. (1981). Mechanism of calcium action and release of vesicle-bound hormones during exocytosis. *Recent Prog. Horm. Res.* **37**, 299–332.

Pollard, H. B., Pazoles, C. J., Creutz, C. E., Scott, J. H., Zinder, O., and Hotchkiss, A. (1984). An osmotic mechanism for exocytosis from dissociated chromaffin cells. *J. Biol. Chem.* **259**, 1114–1121.

Rosenberg, R. C., and Lovenberg, W. (1980). *In* "Essays in Neurochemistry and Neuropharmacology" (M. B. H. Youdim, W. Lovenberg, D. F. Sharman, and J. R. Lagnabo, eds.), Vol. 4, pp. 163–209. Wiley, Chichester.

Rottenberg, H., Grunwald, T., and Avron, M. (1972). Determination of ΔpH in chloroplasts 1. Distribution of [^{14}C]methylamine. *Eur. J. Biochem.* **25**, 54–63.

Rudnick, G. (1986). ATP-driven H^+ pumping into intracellular organelles. *Annu. Rev. Physiol.* **48**, 403–413.

Ruppert, S., Scherer, G., and Shutz, G. (1984). Recent gene conversion involving bovine vasopressin and oxytocin precursor genes suggested by nucleotide sequence. *Nature (London)* **308**, 554–557.

Russell, J. T. (1981). Isolation of purified neurosecretory vesicles from bovine neurohypophyses using isoomolar density gradients. *Anal. Biochem.* **113**, 229–238.

Russell, J. T. (1984). ΔpH, H^+ diffusion potentials, and Mg^{++} ATPase in neurosecretory vesicles isolated from bovine neurohypophyses. *J. Biol. Chem.* **259**, 9496–9507.

Russell, J. T., and Holz, R. (1981). Measurement of ΔpH and membrane potential in isolated neurosecretory vesicles from bovine neurohypophyses. *J.Biol Chem.* **256**, 5950–5953.

Russell, J. T., Brownstein, M. J., and Gainer, H. (1981). Time course of appearance and release of [^{35}S]cysteine labelled neurophysins and peptides in the neurohypophysis. *Brain Res.* **205**, 299–311.

Russell, J. T., Levine, M., and Njus, D. (1985). Electron transfer across posterior pituitary neurosecretory vesicle membranes. *J. Biol. Chem.* **260**, 226–231.

Sachs, H., Fawcett, P., Takabatake, Y., and Portanova, R. (1969). Biosynthesis and release of vasopressin and neurophysin. *Recent Prog. Horm. Res.* **25**, 447–491.

Saxena, A., Hensley, P., Osborne, J. C., Jr., and Fleming, P. J. (1985). The pH-dependent subunit dissociation and catalytic activity of bovine dopamine β-hydroxylase. *J. Biol. Chem.* **260**, 3386–3392.

Scherman, D., and Nordmann, J. J. (1981). Internal pH of isolated newly formed and aged neurohypophysial granules. *Proc. Natl. Acad. Sci. U.S.A.* **79**, 476–479.

Scherman, D., Nordmann, J., and Henry, J. (1982). Existence of an adenosine-5'-triphosphate dependent proton translocase in bovine neurosecretory granule membrane. *Biochemistry* **21**, 687–694.

Schmale, H., and Richter, D. (1984). Single base deletion in the vasopressin gene is the cause of diabetes insipidus in Brattleboro rats. *Nature (London)* **308**, 705–709.

Schuldiner, R., Rottenberg, H., and Avron, M. (1972). Determination of ΔpH in chloroplasts. *Eur. J. Biochem.* **25**, 64–70.

Sharp, R. R., and Richards, E. P. (1977a). Analysis of the carbon-13 and proton NMR spectra of bovine chromaffin granules. *Biochim. Biophys. Acta* **497**, 14–28.

Sharp, R. R., and Richards, E. P. (1977b). Molecular mobilities of soluble components in the aqueous phase of chromaffin granules. *Biochim. Biophys. Acta* **497**, 260–271.

Sharp, R. R., and Sen, R. (1978). Molecular mobilities in chromaffin granules. Magnetic field dependence of proton T_1 relaxation times. *Biochim. Biophys. Acta* **538**, 155–163.

Silsand, T., and Flatmark, T. (1974). Purification of cytochrome b-561. An integral heme protein of the adrenal chromaffin granule membrane. *Biochim. Biophys. Acta* **359**, 257–266.

Skotland, T., and Ljones, T. (1980). Direct spectrophotometric detection of ascorbate free radical formed by dopamine β-monoxygenase and by ascorbate oxidase. *Biochim. Biophys. Acta.* **630**, 30–35.

Smyth, D. G., Massey, D., Zakarian, S., and Finnie, M. (1979). Endorphins are stored in biologically active and inactive forms: Isolation of α-N-actyl peptides. *Nature (London)* **279**, 252–254.

Srivastava, M., Duong, L.-T., and Fleming, P. J. (1984). Cytochrome b_{561} catalyses transmembrane electron transfer. *J. Biol. Chem.* **259**, 8072–8075.

Stanley, E. F., Ehrenstein, G., and Russell, J. T. (1986a). Evidence for calcium-activated potassium channels in vesicles of pituitary cells. *Biophys. J.* **49**, 19a.

Stanley, E. F., Ehrenstein, G., and Russell, J. T. (1986b). Evidence for anion channels in secretory vesicles *Neurosci. Abstr.*

Stanley, E. S., and Ehrenstein, G. (1985). A model for exocytosis based on the opening of calcium-activated potassium channels in vesicles. *Life Sci.* **37**, 1985–1995.

Steiner, D. F., and Oyer, P. E. (1967). The biosynthesis of insulin by a human islet adenoma. *Proc. Natl Acad. Sci. U.S.A.* **57**, 473–480.

Steiner, D. F., Kemmler, W., Tager, H. S., and Peterson, J. D. (1974). Proteolytic processing in the biosynthesis of insulin and other proteins. *Fed. Proc., Fed. Am. Soc. Exp. Biol.* **33**, 2105–2115.

Sudhof, T. C. (1982). Core structure internal osmotic pressure and irreversible structural changes of chromaffin granules during osmometer behavior. *Biochim. Biophys. Acta* **684**, 27–39.

Sudhof, T. C., Ebbecke, M., Walker, J. H., Fritsche, U., and Boustead, C. (1984). Isolation of mammalian calelectrins: A new class of ubiquitous Ca^{2+}-regulated proteins. *Biochemistry* **23**, 1103-1109.

Suppattapone, S., Fricker, L. D., and Snyder, S. H. (1984). Purification and characterization of a membrane-bound enkephalin-forming carboxypeptidase, "enkephalin convertase." *J. Neurochem.* **42**, 1017-1023.

Terland, O., and Flatmark, T. (1973). NADH (NADPH): (Acceptor) oxidoreductase activities of the bovine adrenal chromaffin granules. *Biochim. Biophy. Acta* **305**, 206-218.

Thorn, N. A., Russell, J. T., and Vilhardt, H. (1975). Hexosamines, calcium, and neurophysin in secretory granules and the role of calcium in hormone release. *Annu. N.Y. Acad. Sci.* **248**, 202-217.

Thorn, N. A., Russell, J. T., and Treiman, M. (1982). The neurosecretory granule. *In* "The Secretory Granule" (A. M. Poisner and J. M. Trifaro, eds.), pp. 119-151. Elsevier, Amsterdam.

Treiman, M., and Gratzl, M. (1981). Endogenous protein phosphorylation activity in bovine chromaffin granules purified on a Percoll gradient (tentative identification of a major granule membrane phosphoprotein as a subunit of tyrosine hydroxylase). *Abstr. Meet. Int. Soc. Neurochem.* **8**, 327.

Treiman, M., Worm-Petersen, S., and Thorn, N. A. (1980). Complex phosphorylation activity in neurosecretosomal membranes isolated from ox neurohypophyses. *Biochem. J.* **188**, 657-666.

Vilhardt, H., and Hølmer, G. (1972). Lipid composition of membranes of secretory granules and plasma membranes from bovine neurohypophyses. *Acta Endocrinol.* **71**, 638-648.

Vilhardt, H., Baker, R. V., and Hope, D. B. (1975). Isolation and protein composition of membranes of neurosecretory vesicles and plasma membranes from the neural lobe of the bovine pituitary gland. *Biochem. J.* **148**, 57-65.

Wakefield, L. M., Cass, A. E. G., and Radda, G. K. (1982). Functional coupling between chromaffin granules and mitochondria via the granule cytochrome B. *Fed. Proc., Fed. Am. Soc. Exp. Biol.* **41**, 893.

Wakefield, L. M., Cass, A. E. G., and Radda, G. K. (1984). Isolation of a membrane protein by chromatofocussing: Cytochrome *b*-561 of the adrenal chromaffin granule. *J. Biochem. Biophys. Methods* **9**, 331-341.

Wakefield, L. M., Cass, A. E. G., and Radda, G. K. (1986a). Functional coupling between enzymes of the chromaffin granule membrane. *J. Biol. Chem.* **261**, 9739-9745.

Wakefield, L. M., Cass, A. E. G., and Radda, G. K. (1986b). Electron transfer across the chromaffin granule membrane: Use of a spin-label-spin-probe EPR method to distinguish intra- and extra-vesicular compartments. *J. Biol. Chem.* **261**, 9746-9752.

Walker, J. H. (1982). Isolation from cholinergic synapses of a protein that binds to membranes in a calcium-dependent manner. *J. Neurochem.* **39**, 815-823.

Wallace, E. F., Evans, C. F., Jurik, S. M., Mefford, I. N., and Barchas, J. D. (1982). Carboxypeptidase B activity from adrenal medulla—Is it involved in the processing of proenkephalin? *Life Sci.* **31**, 1793-1796.

Wallace, E. F., Weber, E., Barchas, I. D., and Evans, C. J. (1984). A putative processing enzyme from *Aplysia* that cleaves dynorphin A at the single arginine residue. *Biochem. Biophys. Res. Commun.* **119**, 415-422.

Watson, S. J., Akil, H., Fischli, W., Goldstein, A., Zimmerman, E., Nilaver, G., and Van Wimersma Greidanus, T. B. (1983). Dynorphin and vasopressin: Common localization in magnocellular neurons. *Science* **216**, 85-87.

Whitnall, M. H., and Gainer, H. (1986). Ultrastructural studies of peptide coexistence in corticotropin-releasing factor- and arginine-vasopressin-containing neurons. *In* "Neural and Endocrine Peptides and Receptors" (T. W. Moody, ed.), pp. 159-175. Plenum, New York.

Whitnall, M. H., Gainer, H., Cox, B. M., and Molineaux, C. J. (1983). Dynorphin-A-(1-8) is contained within vasopressin neurosecretory vesicles in rat pituitary. *Science* **222**, 1137-1139.

Whitnall, M. H., Mezey, E., and Gainer, H. (1985). Co-localization of corticotropin releasing factor and vasopressin in median eminence neurosecretory vesicles. *Nature (London)* **317**, 248-250.

Wilkins, J. A., and Salgainicoff, L. (1981). Participation of a transmembrane proton gradient in 5-hydroxytryptamine transport by platelet dense granules and dense-granule ghosts. *Biochem J.* **198**, 113-123.

Winkler, H. (1976). The composition of adrenal chromaffin granules: An assessment of controversial results. *Neuroscience* **1**, 65-80.

Winkler, H., and Carmichael, S. W. (1982). The chromaffin granule. *In* "The Secretory Granule" (A. M. Poisner and J. M. Trifaro, eds.), pp. 3-79. Elsevier, Amsterdam.

Winkler, H., and Westhead, E. W. (1980). The molecular organization of adrenal chromaffin granules. *Neuroscience* **5**, 1803-1823.

Zimmermann, M., Mumford, R. A., and Steiner, D. F. (1980). Precursor processing in the biosynthesis of proteins. *Ann. N.Y. Acad. Sci.* **343**, 1-449.

Zinder, O., Hoffman, P. G., Bonner, W. M., and Pollard, H. B. (1978). Comparison of chemical properties of purified plasma membranes and secretory vesicle membranes from the bovine adrenal medulla. *Cell Tissue Res.* **188**, 153-170.

CURRENT TOPICS IN MEMBRANES AND TRANSPORT, VOLUME 31

Mammalian Neurosecretory Cells: Electrical Properties *in Vivo* and *in Vitro*

D. A. POULAIN AND J. D. VINCENT

Unité de Neurobiologie des Comportements
Institut National de la Santé et de la Recherche Médicale U.176
Université de Bordeaux II
33077 Bordeaux Cedex, France

I. INTRODUCTION

In the mammalian hypothalamo-neurohypophysial system, neurons secreting vasopressin and oxytocin display contrasting electrophysiological behaviors, which are related to the release of either hormone. Thus, these neurons represent excellent models for the study of the relationship between patterns of electrical activity and hormone release, the endogenous mechanisms underlying their specific electrical activity, and finally the interplay between endogenous activity and afferent input.

Vasopressin neurons constitute the final common pathway of several neuroendocrine reflexes involved essentially in hydromineral and cardiovascular homeostasis. Electrophysiologically, the main feature of these neurons is their ability to evolve a phasic pattern, consisting of alternating periods of action potentials (bursts) and electrical silence (Fig. 1). Notwithstanding the essential role of afferent input, evidence has highlighted

313

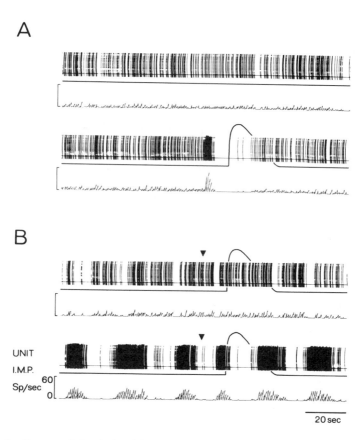

FIG. 1. Patterns of electrical activity of oxytocin and vasopressin neurons. Antidromically identified neurosecretory cells were recorded in the supraoptic nucleus of anesthetized lactating rats during suckling. On these polygraph records, action potentials represented by the polygraph pen deflections (UNIT) are drawn alongside the rate meter output (Spike/second) and the intramammary pressure recording (I.M.P.) from a cannulated mammary gland. (A) Oxytocin neuron. Upper trace shows background activity during suckling (note the flat intramammary pressure recording). Lower trace shows the same oxytocin neuron at the time of a milk ejection. The rise in intramammary pressure corresponds to 20 mmHg, equivalent to the response elicited by intravenous injection of 1 mU oxytocin. Note the brief high-frequency discharge of action potentials occurring 14 seconds before milk ejection (peak firing rate over a 0.5 second period, 65 spikes/second). (B) Vasopressin neurons: Example of one neuron displaying a slow irregular pattern of electrical activity (top) and of another neuron displaying a phasic mode of firing (bottom). In the few seconds preceding the milk ejection induced by suckling, no high-frequency discharge occurred (*arrowheads*). Note the similarity between the slow irregular activity of the oxytocin neuron outside the period of milk ejection and of the first vasopressin neuron.

the importance of endogenous membrane properties of the neurons for the generation of such bursting activity. Vasopressin neurons have thus become a fruitful model for the investigation of endogenous membrane properties of mammalian neurosecretory cells, an aspect of the physiology of these cells that will be discussed in detail in this article.

Oxytocin neurons, on the other hand, have been extensively studied in relation to lactation. During the milk-ejection reflex evoked by suckling, the cells display a characteristic electrical activity, which consists of a high-frequency discharge of action potentials, highly synchronized throughout the whole neuronal population (Fig. 1). This activity leads to the release of a pulse of oxytocin, which causes milk ejection. The possible endogenous mechanisms underlying this activity are unknown. In this article, we will therefore deal with the organization of the afferent input and, in particular, with two of its features, namely, (1) that the electrical activity of the cells is accompanied by a structural reorganization of the magnocellular centers, which constitutes a unique model of plasticity within the adult central nervous system (CNS); and (2) that the neurons appear to control their own activity by the release of their hormone.

II. VASOPRESSIN NEURONS

A. General Electrophysiological Properties

The study of the electrophysiological properties of vasopressin cells began with the demonstration that they can be identified by antidromic stimulation from the neurohypophysis (Yagi et al., 1966). Then, it was shown that, like other neurons, they can generate spontaneous action potentials, and can be synaptically driven by various pathways and stimuli (osmotic, cardiovascular, noxious, etc.). A major step in the understanding of their behavior was the discovery that the phasic pattern occasionally observed in these cells (Wakerley and Lincoln, 1971) is related to enhanced vasopressin release (Arnauld et al., 1975) and that it is specific to vasopressin neurons (Poulain et al., 1977).

The phasic pattern has now been described in the rat (Poulain et al., 1977; Brimble and Dyball, 1977), sheep (Jennings et al., 1978), rabbit (Paisley and Summerlee, 1984), and monkey (Hayward and Jennings, 1973; Arnauld et al., 1975), and appears to be a general property of vasopressin cells in mammals (Fig. 1). It consists of bursts of action potentials (duration 5–100 seconds, intraburst firing rate 5–15 Hz) alternating more or less regularly with periods of electrical silence (lasting also 5–100 seconds). The phasic pattern is rarely seen (in less than 10% of the cells) under basal conditions of hydromineral or cardiovascular equilibrium when plasma vasopressin concentration is minimal. Under stimulation for vasopressin release, there is a progressive recruitment of the number of cells in phasic activity (Arnauld et al., 1975; Wakerley et al., 1978). For each neuron, the bursting pattern appears whenever its general level of activity is enhanced. Thus, under conditions of basal hormone release, the mean firing rate is usually below 2 Hz; when the firing rate exceeds 2.5–3 Hz, most cells display a phasic activity (Poulain and Wakerley, 1982).

We will only briefly mention certain aspects of the afferent mechanisms that modulate the electrical activity of vasopressin neurons. Osmotic stimulation activates vasopressin neurons, but there is still a lot of controversy as to the nature and location of osmoreceptors: Are the cells themselves osmoreceptors? Or, are osmoreceptors part of other areas of the CNS? (For review see Bie, 1980; Leng et al., 1982; Sladek and Armstrong, 1985.) Vasopressin neurons are also influenced by cardiovascular input originating from baro-, chemo-, and voloreceptors in the peripheral vascular bed (see Morris, 1983; Poulain, 1983). Powerful inhibitory mechanisms can also affect the neurons, as, for example, oropharyngeal input during drinking (Arnauld and Du Pont, 1982).

The coincidence of phasic activity with enhanced hormone release has raised the question of the functional significance of this pattern. The relationship between mean firing rate of the cells and the amount of hormone released is not linear. A phasic pattern of stimulation in vitro releases more hormone that a regular pattern at the same mean frequency (Dutton and Dyball, 1979). This highlights the facilitatory effect on hormone release of spikes clustered at short intervals, a process that facilitates Ca^{2+} entry in the terminals. On the other hand, the refractory period between bursts appears to allow for the restoration of inactivated Ca^{2+} conductance at the end of a burst, an inactivation that leads to a decrease in hormone release if the cells fire for a very long period of time (Shaw et al., 1984; Cazalis et al., 1985; for review see Nordmann, 1983; Shaw et al., 1983).

B. Endogenous Electrophysiological Characteristics

By analogy with other neuronal systems (as in invertebrates), the existence of a bursting pattern of electrical activity in vasopressin neurons suggested that it could be at least partly due to endogenous properties of the neuronal membrane. This contention prompted the development of various in vitro preparations more amenable to intracellular recording than the in vivo preparations used so far to investigate the physiology of vasopressin-releasing pathways.

1. In Vitro PREPARATIONS

The in vitro slice preparation has been applied successfully to the study of the hypothalamic magnocellular nuclei by numerous authors (Hatton et al., 1978; Haller et al., 1978; Mason, 1983; Andrew and Dudek, 1983;

Yamashita *et al.*, 1983). Another method consists of perfused (Bourque and Renaud, 1985a) or superfused (Armstrong and Sladek, 1982) hypothalamic explants, which respect the integrity of the nucleus and of its connections with the neurohypophysis. All of these methods permit the study of adult neurosecretory cells under direct visual control of the nuclei (although not of the individual neurons), and respect to some extent the connectivity of the neurons.

Organotypic (Gahwiler and Dreifuss, 1979) or monolayer (Legendre *et al.*, 1982) cultures derived from neonates or fetuses have also been used. They allow direct visualization of the neurons themselves and direct pharmacological manipulation of individual cells. However, they cannot compare exactly with adult cells since they grow and differentiate in an *in vitro* environment.

2. GENERAL ELECTROPHYSIOLOGICAL CHARACTERISTICS

The basic resting membrane properties of magnocellular neurons do not differ greatly from those of other neurons in the CNS. In slices and explants from mature animals, the resting membrane potential is close to -70 mV, and the mean slope resistance in the depolarizing direction is about 150 MΩ. The current voltage curve is linear (Mason, 1983) in the hyperpolarizing direction, and up to the level of spike threshold in the depolarizing direction. In the presence of tetrodotoxin (TTX), which raises the threshold for spike generation, the slope resistance is reduced, revealing rectifying properties, which are at least partly Ca^{2+} dependent. In primary cultures, recordings of vasopressin neurons revealed identical basic membrane properties. However, vasopressin neurons that exhibit plateau potentials (see Section II,B,4) have a nonlinear (intensity voltage) IV curve in the hyperpolarizing direction (Legendre et al., 1982).

Synaptic potentials, inhibitory or excitatory, occur spontaneously in explants, slices, and cultures, which reveal the functional value of the synaptic connectivity in these preparations. As can be expected, synaptic events vary both in amplitude and rise time.

3. ACTION POTENTIALS

Spontaneous action potentials can be observed in all preparations. They are often preceded by a slow prepotential before the fast rising phase (Mason, 1983). Action potentials can also be evoked by injecting depolarizing currents at a threshold close to the resting potential (within 5–10 mV). Their duration is variable from spike to spike, ranging from 1.2 to 3.9 msec at one-third of peak amplitude (Bourque and Renaud, 1985a,b). Pharmacological analysis reveals two components: One component is Na^+

dependent and can be blocked with TTX; the other is Ca^{2+} dependent, can be blocked by calcium blockers, but is TTX resistant. This component appears as a shoulder on the repolarizing phase of the spike. A noteworthy feature of action potentials in magnocellular neurons is that they enlarge progressively during fast firing periods. Thus, during a burst of action potentials, they increase rapidly at the onset of the burst and remain enlarged for the duration of the burst. This enlargement concerns exclusively the Ca^{2+}-dependent component (Bourque and Renaud, 1985b). This mechanism may account for the facilitation of hormone release at high frequency firing, as we saw in Section IIA (see also in the crab, Stuenkel, 1985).

Each action potential is followed by a prolonged (>100 msec) hyperpolarization (Andrew and Dudek, 1984b). After a burst or train of potentials, the afterhyperpolarization sums up and may last several hundreds of milliseconds. This afterhyperpolarization is dependent on a Ca^{2+}-activated K^+ conductance and can be blocked by chelation of intracellular Ca^{2+} with EGTA injected into the cell.

Following the afterhyperpolarization, a delayed afterdepolarization is often observed (Andrew and Dudek, 1983, 1984a). Its long duration (1–3 seconds) and its amplitude (<5 mV) vary little in each cell. Summation of delayed after-depolarization may lead to the formation of a slow potential (or plateau potential), a depolarizing drive for bursting.

4. BURSTING ACTIVITY *in Vitro*

Probably the most significant findings provided by *in vitro* recordings concern the evidence that bursting activity is controlled to a great extent by endogenous mechanisms. Indirect evidence for the existence of endogenous mechanisms was provided by extracellular recordings showing that neurons *in vitro* retained their ability to evolve a phasic activity when all long afferent pathways were severed (Haller *et al.*, 1978). This did not exclude the possibility, however, that a patterned input could be provided by local circuitry within, or close to, the magnocellular nuclei. Attempts to uncouple the cells from synaptic input included using Ca^{2+} blockers in the medium (Hatton, 1982). Whether uncoupling was thoroughly achieved in such preparations is questionable, because bursting activity was still present and yet, as we will see, the endogenous mechanisms involved are Ca^{2+} dependent.

More direct evidence was provided by the description in primary cultures of slow regenerative potentials, referred to as *plateau potentials* (Fig. 2). In vasopressin-containing neurons growing in primary culture, these potentials consist of slow depolarizations reaching within 50–100 msec an absolute amplitude of -20 mV, and lasting 5–90 seconds. Presumably, the main depolarizing and repolarizing ions are Ca^{2+} and K^+, respectively, because plateau generation is inhibited by Co^{2+} and Cd^{2+}, but not by TTX. The plateau itself is augmented and prolonged by tetraethylammonium chloride (TEA). Each plateau is followed by a refractory period, which depends on a Ca^{2+}-activated K^+ conductance. No action potentials occur during the depolarizing phase, except at the onset of the plateau. This is probably due to the high level of depolarization in these immature neurons, which would inactivate the mechanisms necessary for spike generations (Legendre et al., 1982; Theodosis et al., 1983).

In slices, it is clear that the phasic pattern is not the consequence of a patterned synaptic input (Andrew and Dudek, 1983). Mature vasopressin neurons, as neurons in primary cultures, exhibit plateau potentials, but of a much smaller amplitude ($\sim$10mV) than that seen in cultures (Andrew and Dudek, 1983, 1984a; Yamashita et al., 1983). These plateaus are not blocked by TTX. Plateau potentials seem to arise from summated delayed afterdepolarizations, which follow action potentials (Andrew and Dudek, 1983). On the other hand, termination of the burst is achieved through other mechanisms. Earlier in vivo electrophysiological studies had postulated the existence of a recurrent inhibitory pathway on the basis of the inhibition following each antidromic spike: This recurrent pathway could have been responsible for the refractory period following each burst (for discussion see Poulain and Wakerley, 1982). However, the inhibition of firing is due to a prolonged afterhyperpolarization following each spike, and not to an inhibitory postsynaptic potential (Andrew and Dudek, 1984b). It is, therefore, probably through summation of afterhyperpolarizations that inhibitory mechanisms terminate a burst.

III. ELECTRICAL ACTIVITY OF OXYTOCINERGIC NEURONS

Under physiological conditions, oxytocinergic neurons display a highly characteristic electrical activity during the milk-ejection reflex (Wakerley and Lincoln, 1973; Lincoln and Wakerley, 1974). In lactating rats, pups attach themselves to the nipples of the doe for long periods of time, from 30 minutes if the unanesthetized mother is free to move, to several hours if the mother is anesthetized. With the teat in their mouths, the pups doze for most of the time, and give a brief suck from time to time. However, even though the teat is thus continuously stimulated, milk ejections, detected by an increase in the intramammary pressure recorded from one cannulated gland, occur only intermittently, every 3 to 15 minutes. The intramammary pressure reverts back to baseline level in 20 to 40 seconds, where it remains until the next milk ejection a few minutes later. Each milk ejection represents the contraction of the mammary gland in response to a pulse of oxytocin released from the neurohypophysis (Lincoln et al., 1973).

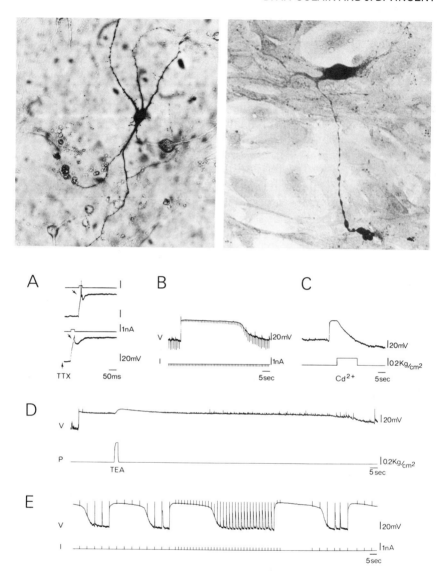

FIG. 2.   Vasopressin neurons in monolayer cultures. (Above left) Example of a dissociated hypothalamic neuron that displayed plateau potentials and was then injected with horseradish peroxidase. Note that numerous spines cover the cell body and dendritic processes but not the thin axonal process that emerges from one of the dendrites (arrow) (×168). (Above right) Hypothalamic culture treated with serum against vasopressin. The immunoprecipitate, made evident by peroxidase, was localized in cell bodies and axonal-like processes and their swellings. The darkly stained neuron had displayed electrical responses similar to those described below. Cells of the basal layer and adjacent neurons showed no or very faint immunoreactivity (×168) (adapted from Theodosis *et al.*, 1983). (A) Onset of plateau potentials triggered by a

In the absence of suckling, oxytocinergic neurons both in the paraventricular and the supraoptic nuclei display, as vasopressinergic neurons do, a a slow, irregular pattern of action potentials, with a mean firing rate most often below 2 Hz. When the young start to suckle, the neurons show no striking reaction for many minutes. Then, in relation to each milk ejection, each neuron exhibits a sudden burst of action potentials, or *high-frequency discharge*, reaching a peak of 30–100 Hz, and lasting 3–4 seconds (Fig. 1). Between two milk ejections, the neurons resume their slow irregular or fast continuous pattern of electrical activity, while the pups continue to suckle.

The electrical behavior of oxytocinergic neurons has two striking features: First, the response (high-frequency discharge) is highly synchronized in the whole neuronal population; second, the neuronal excitation is intermittent, whereas the stimulus is continuous. Although it is somewhat artificial to separate these different aspects, in the following sections we will consider in turn the evidence relative to the high-frequency discharge itself, then its synchronization, and finally its intermittent pattern.

A. The High-Frequency Discharge

The high-frequency discharge does not constitute an all-or-none phenomenon identical in all cells. In fact, from one cell to another, there is a whole range of bursts, some of them very brief with a fairly low peak frequency (10–15 Hz), others lasting 4–5 seconds, with peak rate over 100 Hz. In this case, at the maximum rate of discharge, spikes recorded extracellularly display a dramatic decrease in amplitude, probably linked to sodium inactivation (Lincoln and Wakerley, 1974). Statistically, oxytocinergic neurons in the paraventricular nucleus tend to have bigger bursts than cells of the supraoptic nucleus (Poulain *et al.*, 1977). For one given cell, its high-frequency discharges are very similar from one milk ejection to another. However, they tend to increase progressively in duration and rate of discharge through the first four to five milk ejections in a suckling

---

depolarizing pulse of current, in the absence (top) or presence (bottom) of TTX ($10^{-6}\,M$). The plateau was not blocked by TTX, but the initial spike arising from the depolarizing phase (arrow) was blocked by TTX. (B) Variations of input resistance during a plateau potential depolarization. Input resistance was visualized by passing hyperpolarizing pulses of current (I, 50 msec, 0.2 nA, 1 Hz). Note the important decrease in input resistance during the depolarizing phase. (C) Application of a calcium-conductance blocker (Cd^{2+}, 1 mM), which interrupts the plateau potential, whereas, (D) application of tetraethylammonium chloride (TEA), a K^{+}-conductance blocker, prolongs the duration of the plateau. (E) Plateau potentials can be evoked by depolarizing pulses of current (I, 50 msec, 0.4 nA). Each plateau arises directly from one depolarizing pulse. Between two plateaus, pulses can only trigger action potentials or short slow potentials. The refractory period can be increased by increasing the frequency of stimulation.

session. The duration and rate of discharge within the bursts increase when the number of pups suckling increases (Lincoln and Wakerley, 1975), and can be enhanced by intracerebroventricular administration of oxytocin (Freund-Mercier and Richard, 1984).

The mechanisms leading to a high-frequency discharge are unexplained. One possibility is that the bursts are entirely dependent on an external pacemaker. The spike activity in oxytocin cells would mirror the afferent volley in a one-to-one fashion, slightly modified by the excitability of the cell at that time. One has to assume, then, that it is at some level on the afferent pathway that the continuous stimulus of suckling is transformed in a burst of excitation, which in turn activates the oxytocinergic cells. Alternatively, an endogenous mechanism may underlie the burst of action potentials. In Section II,8,4, we saw that in cultures from fetal mouse hypothalamuses, neurosecretory cells display long-duration plateau potentials. They also display short calcium-dependent slow potentials, lasting 2 to 3 seconds (Legendre et al., 1982). Their duration and exponential decay correspond well to the pattern of firing rates within a high-frequency discharge. Although these slow potentials were described in immature vasopressinergic neurons, they could represent some property common to neurosecretory cells. This remains a matter of speculation, however, because no intracellular recording has yet been made during suckling. If the high-frequency discharge is superimposed on a slow regenerative potential, its occurrence would depend on (1) the capability and readiness of the membrane to display the slow potential and (2) the afferent input serving as a final trigger to the slow potential. For example, a noteworthy feature of the bursts is that they often start with one to three spikes at a low frequency before the discharge rate increases suddenly. One could postulate that depolarizing afterpotentials that follow each action potential (Andrew and Dudek, 1984b) serve to trigger the slow potentials as they trigger plateau potentials in vasopressin neurons.

B. Synchronization

At each milk ejection, a striking feature of the activity of oxytocin cells is the synchronization of the high-frequency discharges from one neuron to another. This is already apparent from the rather constant delay (12–15 seconds) between the bursts displayed by any one cell and the following milk ejection. Occasional double recordings at the tip of the same microelectrode also give evidence of a strong synchronization between adjacent cells (Lincoln and Wakerley, 1974). In a systematic study in which simultaneous extracellular recordings were made from a pair of electrodes, it was found that the delay between the onsets of the high-frequency discharges in two different cells ranges between 0 and 400 msec, whether the two cells were located in the same nucleus or in two different nuclei (Belin et al., 1984). A more precise evaluation of the degree of synchronization is not possible because we have no evidence of the intracellular events leading to the burst of activity. However, it is noteworthy that there is no spike-to-spike synchrony.

What then are the mechanisms that permit such a synchronization? First, it must be emphasized that to fire in synchrony depends as much on the functional organization of the milk-ejection reflex as on possible basic properties of oxytocin neurons. For instance, an increase in osmotic pressure or hemorrhage (usually known as vasopressin-releasing stimuli) are powerful stimuli for oxytocin release. However, they induce no high-frequency discharge, nor any synchrony in discharge (see discussion in Poulain and Wakerley, 1982): The neurons display a sustained increase in their electrical activity with a fast continuous pattern of action potentials reaching a mean firing rate of 4–10 Hz. In other words, excitation of the oxytocinergic neurons does not necessarily lead to their synchronization.

One possibility is that the synchronized bursting activity of oxytocinergic neurons is entirely coordinated by an external pacemaker (yet unidentified) connected to each individual oxytocinergic cell. Another possiblity is that synchronizing mechanisms exist within the oxytocinergic system itself. Among them, studies have described a number of morphological aspects that could at least facilitate synchronization at the level of the magnocellular nuclei themselves. These include gap junctions, neuronal membrane appositions, shared afferent synaptic connectivity, and synaptic connections between oxytocinergic neurons.

The existence of gap junctions was suggested by the demonstration of dye coupling between magnocellular neurons in hypothalamic slices in vitro (Andrew et al., 1981; Cobbet and Hatton, 1984). However, direct electrophysiological evidence for electrotonic coupling is still lacking and there is no morphological evidence for gap junctions between identified neurosecretory elements.

Another aspect concerns neuronal–glial relationships (Fig. 3). In nonlactating animals, supraoptic neurons are usually separated from one another by the neuropil and glia. In lactating animals, glial coverage of oxytocin cells diminishes, so that the surface membrane of numerous contiguous neurons become juxtaposed (Theodosis et al., 1981; Theodosis and Poulain, 1984a; Hatton and Tweedle, 1982). Thus, the proportion of oxytocin neurons directly apposed increases from 20% in virgin to 75% in lactating rats (Theodosis and Poulain, 1984b; Theodosis et al., 1986a). In normal animals, the extent of neuronal membrane in apposition is small; during lactation, it increases so that about 10% of the total surface membrane in the nucleus is in apposition. These numerous and extensive

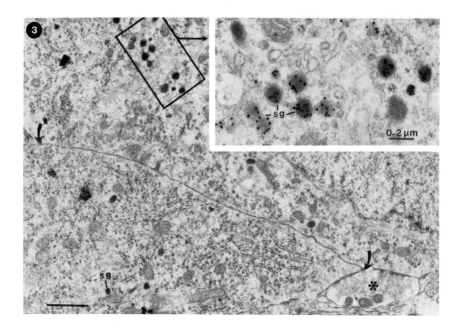

FIG. 3.   Electron micrograph of two oxytocin cell bodies in the supraoptic nucleus of a lactating rat, whose surface membranes are in extensive apposition (between arrows). The two neurons also share a common terminal (asterisk). The ultrathin sections had been incubated in serum raised against oxytocin I; immunoreactivity was revealed with immunoglobulin-coupled colloidal gold (postembedding method). Note that the gold particles cover only secretory granules (sg, and inset) containing the neuropeptide. Calibration bar, 1 μm (from Theodosis *et al.*, 1986a).

neuronal appositions could facilitate synchronization of adjacent elements by permitting ephaptic interaction or field effects between the contiguous cells, or extracellular accumulation of potassium (no longer buffered by the withdrawn glia). Nevertheless, they may also simply be the structural prerequisite for another modification that occurs at lactation, namely, the proliferation of "double" synapses between oxytocin cells (Fig. 3).

Double synapses are presynaptic terminals contacting two postsynaptic elements simultaneously (in the same plane of section) (Theodosis *et al.*, 1981; Theodosis and Poulain, 1984b). In virgin animals, these are extremely rare. At lactation, their incidence increases significantly: Less than 5% of neurons share the same synapse in virgin animals whereas in lactating animals over 20% of all oxytocin cells are thus contacted (Theodosis *et al.*, 1986a). Double synapses certainly constitute a powerful means to synchronize the activity of the neurons on which they impinge by distributing simultaneously afferent input to the postsynaptic elements. Whether this input is excitatory or inhibitory has still to be determined, but its proliferation at a time when the oxytocinergic system is hyperactive suggests that it may be excitatory. However, many of these double synapses contain γ-aminobutyric acid (GABA) (Theodosis *et al.*, 1986b), a neurotransmitter which inhibits the activity of magnocellular neurons (Arnauld *et al.*, 1983).

Finally, synaptic connections between oxytocinergic neurons have been demonstrated immunocytochemically: Within the supraoptic nucleus, oxytocin-containing synaptic terminals actually impinge onto oxytocinergic somata (Theodosis, 1985). This supports strongly the hypothesis that oxytocin itself may act upon oxytocinergic neurons to facilitate their activation. *In vitro*, oxytocin is released locally within the magnocellular nuclei, and addition of oxytocin or its analogs further enhances this local release (Moos *et al.*, 1984). *In vivo*, microiontophoretic application of oxytocin excites neurosecretory cells (Moss *et al.*, 1972), whereas intracerebroventricular injection of oxytocin increases the background electrical activity of oxytocinergic neurons and also the rate of action potential discharge within the high-frequency discharge (Freund-Mercier and Richard, 1984).

C. Intermittence

In the rat, the intermittent nature of the high-frequency discharges while the pups suckle continuously suggests that the oxytocin system possesses an inherent periodicity. However, the very notion of periodicity is questionable. In the rat, the stimulus–response relationship is blurred by the particular behavior of the young; in other species, the patterns of nursing, suckling, and milk ejections are quite different in that suckling is intermittent and milk ejections are closely time-locked to the periods of suckling. Nevertheless, in the pig or rabbit, for example, as in the rat, it is clearly apparent that the delay from the onset of suckling to the central response is much longer (of the order of one to several minutes) than would be expected were neural transmission dependent only on the conduction velocity of the afferent arc (Ellendorff *et al.*, 1982; Paisley and Summerlee, 1984). One has to assume, then, that the neural input originating from mammary receptors has to be integrated in a complex manner before oxytocin cells can react. Similarly, each milk ejection is obviously followed by a refractory period, during which the continuing suckling stimulus has no effect on the neurons.

The afferent pathway of the milk-ejection reflex is not yet fully traced, and its functional organization is still obscure (see Lincoln and Russell, 1985). The suckling stimulus seems to be transmitted ungated from the peripheral receptors to the first relays in the spinal cord: The electrical activity in the mammary nerve or in second-order spinal neurons excited by suckling follows in parallel the stimulation of the teats (Findlay, 1966; Poulain and Wakerley, 1986). At the spinal cord level, however, some degree of integration is already taking place, since the input from adjacent nipples summates at the level of second-order dorsal horn neurons (Poulain and Wakerley, 1986). This may explain, for instance, that suckling is more efficient when all the nipples are simultaneously sucked (Lincoln and Wakerley, 1975). The afferent pathway then follows either the contralateral anterolateral funiculi and/or the ipsilateral dorsolateral funiculi. The dorsal column system does not appear to be essential to the reflex (Eayrs and Baddeley, 1956; Fukuoka et al., 1984; Dubois-Dauphin et al., 1985), although it may play some role (Richard et al., 1970).

Is there any inhibitory mechanism in the spinal cord that would be responsible for the delayed response in oxytocinergic neurons? There is evidence that opiates may inhibit the reflex at the spinal level (Wright, 1985), but this inhibition does not seem to be tonic. Electrical stimulation of the lateral funiculi evokes an immediate activation of the oxytocin system, as if local inhibitory mechanisms were bypassed, but it may be simply that such a stimulus overcomes inhibitory mechanisms at higher levels (Poulain and Dyer, 1984). That dorsal horn neurons projecting to higher brain structures react immediately to nipple stimulation suggests, on the contrary, that no inhibition is exerted at this level during suckling.

In higher brain structures, the situation is more confused. It still has not been ascertained whether the pathway to the hypothalamus is discrete or not. Moreover, many structures, though not essential to the reflex, can markedly alter its pattern. The afferent pathway passes through the lateral tegmentum, the lesion of which blocks the reflex: Lesions of the central gray in the tegmentum do not block the reflex, but alter its pattern, permitting milk ejections to recur at very short intervals (Juss and Wakerley, 1981). Between the tegmentum and the hypothalamus, the pathway meanders through the data of conflicting reports. An interesting notion is that many structures that are not essential to the reflex (thalamus, cortex, septum) exert a powerful inhibitory role. For instance, the nursing mother has to be in a state of quiet arousal (in pigs and rabbits: Poulain et al., 1981; Neve et al., 1982) or slow wave sleep (in rats: Voloschin and Tramezzani, 1979; Lincoln et al., 1980), for alert arousal blocks the reflex; likewise, limbic structures involved in the control of vigilance (hippocampus, septum), when activated, may block the reflex (Lebrun et al., 1983; Tindal and Blake, 1980, 1984). However, any of these inhibitions cannot fully account for the intermittence of the reflex, because it remains intermittent after lesions of these structures have been made.

When the pathway finally reaches the hypothalamus, we find no magic center between the nipples and the magnocellular nuclei that would account entirely for the intermittent synchronized activity of oxytocin cells. Should the integration of the suckling stimulus then be coordinated within the pool of oxytocin neurons? The evidence for such a suggestion deserves some consideration. We have seen that oxytocin appears to be released locally within the magnocellular nuclei, perhaps through synaptic connections between oxytocin neurons (see Section III,B). We have also seen that intracerebroventricular administration of oxytocin enhances the background activity of oxytocin neurons and the intensity of the high-frequency discharges (see Section III,B); more importantly, it also accelerates the recurrence of these high-frequency discharges (Freund-Mercier and Richard, 1984). All these data suggest the existence of a positive feedback loop in the population of oxytocin neurons.

A possible model for synchronized activation of the whole population of neurons could therefore be as follows. Suckling would slightly change the electrical activity of the cells, thus permitting local release of oxytocin, which, in turn, would permit further activation until the cells reached the threshold necessary to fire in high-frequency discharges. These neurons would then serve as trigger cells to the rest of the population through synaptic connections between oxytocinergic cells within the same nucleus, or in different nuclei. Within each nucleus, double synapses, membrane appositions, and gap junctions would further facilitate the synchronization of excitation. The trigger neurons do not need to be a fixed subset of oxytocinergic cells. In simultaneous recordings from two different cells, there is no evidence that one cell precedes the other one in a systematic fashion at the time of a high-frequency discharge (Belin et al., 1984). Rather, the neurons acting as triggers would be the most excitable cells at the time of any particular milk ejection. In addition to this excitatory configuration there would be a powerful inhibitory circuitry, which would prevent the oxytocinergic system from reacting to unspecific stimuli with high-frequency discharges. Such inhibition would be removed by suckling and reactivated after each milk ejection.

REFERENCES

Andrew, R. D., and Dudek, F. E. (1983). Burst discharge in mammalian neuroendocrine cells involves an intrinsic regenerative mechanism. *Science* **221**, 1050–1052.

Andrew, R. D., and Dudek, F. E. (1984a). Analysis of intracellularly recorded phasic bursting by mammalian neuroendocrine cells. *J. Neurophysiol.* **51**, 552–566.

Andrew, R. D., and Dudek, F. E. (1984b) Intrinsic inhibition in magnocellular neuroendocrine cells of rat hypothalamus. *J. Physiol. (London)* **353**, 171–185.

Andrew, R. D., MacVicar, B. A., Dudek, F. E., and Hatton, G. I. (1981). Dye transfer through gap junctions between neuroendocrine cells of rat hypothalamus. *Science* **211**, 1187–1189.

Armstrong, W. E., and Sladek, C. D. (1982). Spontaneous "phasic-firing" in supraoptic neurons recorded from hypothalamo-neurohypophysial explants *in vitro*. *Neuroendocrinology* **34**, 405–409.

Arnauld, E., and Du Pont, J. (1982). Vasopressin release and firing of supraoptic neuro-secretory neurones during drinking in the dehydrated monkey. *Pflügers Arch.* **394**, 195–201.

Arnauld, E., Dufy, B., and Vincent, J. D. (1975). Hypothalamic supraoptic neurones: Rates and patterns of action potential firing during water deprivation in the unanaesthetized monkey. *Brain Res.* **100**, 315–325.

Arnauld, E., Cirino, M., Layton, B. S., and Renaud, L. P. (1983). Contrasting actions of amino-acids, acetylcholine, noradrenaline and leucine enkephalin on the excitability of supraoptic vasopressin-secreting neurons. A microiontophoretic study in the rat. *Neuroendocrinology* **36**, 187–196.

Belin, V., Moos, F., and Richard, P. (1984). Synchronization of oxytocin cells in the hypo-thalamic paraventricular and supraoptic nuclei in suckled rats: Direct proof with paired extracellular recordings. *Exp. Brain Res.* **57**, 201–203.

Bie, P. (1980). Osmoreceptors, vasopressin and control of renal water excretion. *Physiol. Rev.* **60**, 961–1048.

Bourque, C. W., and Renaud, L. P. (1985a). Calcium-dependent action potentials in rat supraoptic neurosecretory neurones recorded *in vitro*. *J. Physiol. (London)* **363**, 419–428.

Bourque, C. W., and Renaud, L. P. (1985b). Activity dependence of action potential dura-tion in rat supraoptic neurosecretory neurones recorded *in vitro*. *J. Physiol. (London)* **363**, 429–439.

Brimble, M. J., and Dyball, R. E. J. (1977). Characterization of the responses of oxytocin- and vasopressin-secreting neurones in the supraoptic nucleus to osmotic stimulation. *J. Physiol. (London)* **271**, 253–271.

Brimble, M. J., Dyball, R. E. J., and Forsling, M. L. (1978). Oxytocin release following osmotic activation of oxytocin neurones in the paraventricular and supraoptic nuclei. *J. Physiol. (London)* **278**, 69–78.

Cazalis, M., Dayanithi, D., and Nordmann, J. J. (1985). The role of patterned burst and interburst interval on the excitation–coupling mechanism in the isolated rat neural lobe. *J. Physiol. (London)* **369**, 45–60.

Cobbett, P., and Hatton, G. I. (1984). Dye coupling in hypothalamic slices: Dependence on *in vivo* hydration state and osmolality of incubation medium. *J. Neurosci.* **4**, 3034–3038.

Dubois-Dauphin, M., Armstrong, W. E., Tribollet, E., and Dreifuss, J. J. (1985). Somato-sensory systems and the milk ejection reflex in the rat. I. Lesions of the mesencephalic lateral tegmentum disrupt the reflex and damage mesencephalic somatosensory connec-tions. *Neuroscience* **15**, 1111–1129.

Dutton, A., and Dyball, R. E. J. (1979). Phasic firing enhances vasopressin release from the rat neurohypophysis. *J. Physiol. (London)* **290**, 433–440.

Eayrs, J. T., and Baddeley, R. M. (1956). Neural pathways in lactation. *J. Anat.* **90**, 161–171.

Ellendorff, F., Forsling, M. L., and Poulain, D. A. (1982). The milk ejection reflex in the pig. *J. Physiol. (London)* **333**, 577–594.

Findlay, A. L. R. (1966). Sensory discharges from lactating mammary glands. *Nature (London)* **211**, 1183–1184.

Freund-Mercier, M. J., and Richard, P. (1984). Electrophysiological evidence for facilitatory control of oxytocin neurones by oxytocin during suckling in the rat. *J. Physiol. (London)* **352**, 447–466.

Fukuoka, T., Negoro, H., Honda, K., Higuchi, T., and Nishida, E. (1984). Spinal pathway to the milk ejection reflex in the rat. *Biol. Reprod.* **30**, 74–81.

Gahwiler, B. H., and Dreifuss, J. J. (1979). Phasically firing neurons in long-term cultures of the rat hypothalamic supraoptic area: Pacemaker and follower cells. *Brain. Res.* **177**, 95–103.

Haller, E. W., Brimble, M. J, and Wakerley, J. B. (1978). Phasic discharge in supraoptic neurones recorded from hypothalamic slices. *Exp. Brain Res.* **33**, 131–134.

Hatton, G. I. (1982). Phasic bursting activity of rat paraventricular neurones in the absence of synaptic transmission. *J. Physiol. (London)* **327**, 273–284.

Hatton, G. I., and Tweedle, C. D. (1982). Magnocellular neuropeptidergic neurones in hypothalamus: Increases in membrane appositions and number of specialized synapses from pregnancy to lactation. *Brain Res. Bull.* **8**, 197–204.

Hatton, G. I., Armstrong, W. E., and Gregory, W. A. (1978). Spontaneous and osmotically-stimulated activity in slices of rat hypothalamus. *Brain Res. Bull.* **3**, 497–508.

Hayward, J. N., and Jennings, D. P. (1973). Activity of magnocellular neuroendocrine cells in the hypothalamus of unanaesthetized monkeys. I. Functional cell types and their anatomical distribution in the supraoptic nucleus and the internuclear zone. *J. Physiol (London)* **232**, 515–543.

Higuchi, T., Honda, K., Fukuoka, T., Negoro, H., Hosono, Y., and Nishida, E. (1983). Pulsatile secretion of prolactin and oxytocin during nursing in the lactating rat. *Endocrinol. Jpn.* **30**, 353–359.

Jennings, D. P., Haskins, J. T., and Rogers, J. M. (1978). Comparison of firing patterns and sensory responsiveness between supraoptic and other hypothalamic neurons in the unanesthetized sheep. *Brain. Res.* **149**, 347–364.

Juss, T. S., and Wakerley, J. B. (1981). Mesencephalic areas controlling pulsatile oxytocin release in the suckled rat. *J. Endocrinol.* **91**, 233–244.

Lebrun, C. J., Poulain, D. A., and Theodosis, D. T. (1983). The role of the septum in the control of the milk ejection reflex in the rat. Effects of lesions and electrical stimulation. *J. Physiol. (London)* **339**, 17–31.

Legendre, P., Cooke, I. M., and Vincent, J. D. (1982). Regenerative responses of long duration recorded intracellularly from dispersed cell cultures of fetal mouse hypothalamus. *J. Neurophysiol.* **48**, 1121–1141.

Leng, G., Mason, W. T., and Dyer, R. G. (1982). The supraoptic nucleus as an osmoreceptor. *Neuroendocrinology* **34**, 75–82.

Lincoln, D. W., and Russell, J. A. (1985). The electrophysiology of magnocellular oxytocin neurons. *In* "Oxytocin: Clinical and Laboratory Studies" (J. A. Amico and A. G. Robinson, eds.), pp. 53–76. Elsevier, Amsterdam.

Lincoln, D. W., and Wakerley, J. B. (1974). Electrophysiological evidence for the activation of supraoptic neurones during the release of oxytocin. *J. Physiol. (London)* **242**, 533–554.

Lincoln, D. W., and Wakerley, J. B. (1975). Factors governing the periodic activation of supraoptic and paraventricular neurosecretory cells during suckling in the rat. *J. Physiol. (London)* **250**, 443–461.

Lincoln, D. W., Hill, A., and Wakerley, J. B. (1973). The milk ejection reflex of the rat: An intermittent function not abolished by surgical levels of anaesthesia. *J. Endocrinol.* **57**, 459–476.

Lincoln, D. W., Hentzen, K., Hin, T., Van der Schoot, P., Clarke, G., and Summerlee, A. J. S. (1980). Sleep: A prerequisite for reflex milk ejection in the rat. *Exp. Brain Res.* **38**, 151–162.

Mason, W. T. (1983). Electrical properties of neurons recorded from the rat supraoptic nucleus *in vitro*. *Proc. R. Soc. London Ser. B* **217**, 141–161.

Moos, F., Freund-Mercier, M. J., Guerne, Y., Guerne, J. M., Stoeckel, M. E., and Richard, P. (1984). Release of oxytocin and vasopressin by magnocellular nuclei *in vitro*: Specific facilitatory effect of oxytocin on its own release. *J. Endocrinol.* **102**, 63–72.

Morris, J. F. (1983). Organization of neural inputs to the supraoptic and paraventricular nuclei: Anatomical aspects. *Prog. Brain Res.* **60**, 3–18.

Moss, R. L., Dyball, R. E. J., and Cross, B. A. (1972). Excitation of antidromically identified neurosecretory cells of the paraventricular nucleus by oxytocin applied iontophoretically. ' *xp. Neurol.* **34**, 95–102.

Negoro, H., Uchide, K., Honda, K., and Higuchi, T. (1985). Facilitatory effect of antidromic stimulation on milk ejection-related activation of oxytocin neurons during suckling in the rat. *Neurosci. Lett.* **59**, 21–25.

Neve, H. A., Paisley, A. C., and Summerlee, A. J. S. (1982). Arousal: A Prerequisite for suckling in the rabbit. *Physiol. Behav.* **28**, 213–217.

Nordmann, J. J. (1983). Stimulus–secretion coupling. *Prog. Brain Res.* **60**, 281–304.

Paisley, A. C., and Summerlee, A. J. S. (1984). Activity of putative oxytocin neurones during reflex milk ejection in conscious rabbits. *J. Physiol. (London)* **347**, 465–478.

Poulain, D. A. (1983). Electrophysiology of the afferent input to oxytocin- and vasopressin-secreting neurones. Facts and problems. *Prog. Brain. Res.* **60**, 39–52.

Poulain, D. A., and Dyer, R. G. (1984). Reproducible increase in intramammary pressure after spinal cord stimulation in lactating rats. *Exp. Brain Res.* **55**, 313–316.

Poulain, D. A., and Wakerley, J. B. (1982). Electrophysiology of hypothalamic magnocellular neurones secreting oxytocin and vasopressin. *Neuroscience* **7**, 773–808.

Poulain, D. A., and Wakerley, J. B. (1986). Sensory projections from the mammary gland to the spinal cord in lactating rat. II. Electrophysiological responses of dorsal horn neurones during stimulation of the nipples, including suckling. *Neuroscience* **19**, 511–521.

Poulain, D. A., Wakerley, J. B., and Dyball, R. E. J. (1977). Electrophysiological differentiation of oxytocin- and vasopressin-secreting neurones. *Proc. R. Soc. London Ser. B* **196**, 367–384.

Poulain, D. A., Rodriguez, F., and Ellendorff, F. (1981). Sleep is not a prerequisite for the milk ejection reflex in the pig. *Exp. Brain Res.* **43**, 107–110.

Richard, P., Urban, I., and Denamur, R. (1970). The role of the dorsal tracts of the spinal cord and of the mesencephalic and thalamic lemniscal system in the milk ejection reflex during milking in the ewe. *J. Endocrinol.* **47**, 45–53.

Shaw, F. D., Dyball, R. E. J., and Nordmann, J. J. (1983). Mechanisms of inactivation of neurohypophysical hormone release. *Prog. Brain. Res.* **60**, 305–317.

Shaw, F. D., Bicknell, R. J., and Dyball, R. E. J. (1984). Facilitation of vasopressin release from the neurohypophysis by application of electrical stimuli in bursts. Relevant stimulation parameters. *Neuroendocrinology* **39**, 371–376.

Sladek, C. D., and Armstrong, W. E. (1985). Osmotic control of vasopressin release. *Trends Neurosci.* **8**, 166–168.

Stuenkel, E. L. (1985). Simultaneous monitoring of electrical and secretory activity in peptidergic neurosecretory terminals of the crab. *J. Physiol. (London)* **359**, 163–187.

Theodosis, D. T. (1985). Oxytocin-immunoreactive terminals synapse on oxytocin neurones in the supraoptic nucleus. *Nature (London)* **313**, 682–684.

Theodosis, D. T., and Poulain, D. A. (1984a). Evidence for structural plasticity in the supraoptic nucleus of the rat hypothalamus in relation to gestation and lactation. *Neuroscience* **11**, 183–193.

Theodosis, D. T., and Poulain, D. A. (1984b). Evidence that oxytocin-secreting neurones are involved in the ultrastructural reorganisation of the rat supraoptic nucleus apparent at lactation. *Cell Tissue Res.* **235**, 217–219.

Theodosis, D. T., Poulain, D. A., and Vincent, J. D. (1981). Possible morphological bases for synchronisation of neuronal firing in the rat supraoptic nucleus during lactation. *Neuroscience* **6**, 919–929.

Theodosis, D. T., Legendre, P., Vincent, J. D., and Cooke, I. (1983). Immunocytochemically identified vasopressin neurons in culture show slow, calcium-dependent electrical responses. *Science* **221**, 1052–1054.

Theodosis, D. T., Chapman, D. B., Montagnese, C., Poulain, D. A., and Morris, J. F. (1986). Structural plasticity in the hypothalamic supraoptic nucleus at lactation affects oxytocin- but not vasopressin-secreting neurones. *Neuroscience* **17**, 661–678.

Theodosis, D. T., Paut, L., and Tappaz, M. L. (1986b). Immunocytochemical analysis of the GABAergic innervation of oxytocin- and vasopressin-secreting neurones in the rat supraoptic nucleus. *Neuroscience* **19**, 207–222.

Tindal, J. S., and Blake, L. A. (1980). A neural basis for central inhibition of milk ejection in the rabbit. *J. Endocrinol.* **86**, 525–531.

Tindal, J. S., and Blake, L. A. (1984). Central inhibition of milk ejection in the rabbit: Involvement of hippocampus and subiculum. *J. Endocrinol.* **100**, 125–129.

Voloschin, L. M., and Tramezzani, J. H. (1979). Milk ejection reflex linked to slow wave sleep in nursing rats. *Endocrinology* **105**, 1202–1207.

Wakerley, J. B., and Lincoln, D. W. (1971). Phasic discharge of antidromically identified units in the paraventricular nucleus of the hypothalamus. *Brain Res.* **25**, 192–194.

Wakerley, J. B., and Lincoln, D. W. (1973). The milk ejection reflex of the rat: A 20- to 40-fold acceleration in the firing of paraventricular neurones during oxytocin release. *J. Endocrinol.* **57**, 477–493.

Wakerley, J. B., Poulain, D. A., and Brown, D. (1978). Comparison of firing patterns in oxytocin- and vasopressin-releasing neurones during progressive dehydration. *Brain Res.* **148**, 425–440.

Wright, D. M. (1985). Evidence for a spinal site at which opioids may act to inhibit the milk-ejection reflex. *J. Endocrinol.* **106**, 401–407.

Yagi, K., Azuma, T., and Matsuda, K. (1966). Neurosecretory cell: Capable of conducting impulse in rats. *Science* **154**, 778–779.

Yamashita, H., Inenaga, K., Kawata, M., and Sano, Y. (1983). Phasically firing neurones in the supraoptic nucleus of the rat hypothalamus: *Neurosci. Lett.* **37**, 87–92.

Index

Contents of Recent Volumes